Ideology and Mass Killing

Ideology and Mass Killing

The Radicalized Security Politics of Genocides and Deadly Atrocities

JONATHAN LEADER MAYNARD

Great Clarendon Street, Oxford, OX2 6DP,
United Kingdom

Oxford University Press is a department of the University of Oxford.
It furthers the University's objective of excellence in research, scholarship,
and education by publishing worldwide. Oxford is a registered trade mark of
Oxford University Press in the UK and in certain other countries

© Jonathan Leader Maynard 2022

The moral rights of the author have been asserted

Impression: 2

All rights reserved. No part of this publication may be reproduced, stored in
a retrieval system, or transmitted, in any form or by any means, without the
prior permission in writing of Oxford University Press, or as expressly permitted
by law, by licence or under terms agreed with the appropriate reprographics
rights organization. Enquiries concerning reproduction outside the scope of the
above should be sent to the Rights Department, Oxford University Press, at the
address above

You must not circulate this work in any other form
and you must impose this same condition on any acquirer

Published in the United States of America by Oxford University Press
198 Madison Avenue, New York, NY 10016, United States of America

British Library Cataloguing in Publication Data

Data available

Library of Congress Control Number: 2021952418

ISBN 978–0–19–877679–6

DOI: 10.1093/oso/9780198776796.001.0001

Printed and bound by
CPI Group (UK) Ltd, Croydon, CR0 4YY

Links to third party websites are provided by Oxford in good faith and
for information only. Oxford disclaims any responsibility for the materials
contained in any third party website referenced in this work.

Acknowledgements

This book would not exist, let alone have taken anything like its eventual form, were it not for the support and assistance of dozens of friends and colleagues. In the first years of working on this project, two of the leading figures in the Department of Politics and International Relations at the University of Oxford inspired and guided my research. The first was Elizabeth Frazer, who served as both my master's and doctoral supervisor. It's very hard to put into words the depth and breadth of both Liz's intellect and her contribution to my life as an academic. She has exerted a major influence over my understandings of politics, violence, and ideology, and offered me great feedback on several sections of the book. But more than that, being supervised by Liz hugely improved my ability to think and the care and depth with which I thought about theory, argument, and method as a politics researcher. Like so many at the University of Oxford, I cannot adequately repay her. Second, I owe a huge debt to Jennifer Welsh, who as Professor of International Relations at Oxford initiated my scholarly engagement with international relations research on atrocities, norms, and ideas. In her subsequent roles as United Nations Special Advisor for the Responsibility to Protect, Professor of International Relations at the European University Institute, and Chair in Global Governance and Security at McGill University, Jennifer offered me much feedback and encouraged me to develop the normative and preventive implications of my research. My work owes a huge debt to her insight, advice, and scholarship, and I feel immensely privileged to know her.

Beyond Liz and Jennifer, I owe so much to Thomas (Tad) Homer-Dixon, who has been a source of inspiration, advice, and intellectual engagement since I first met him in Oxford in April 2012. Tad took a major leap of faith in inviting me to join a network of researchers interested in ideology based at the University of Waterloo, Ontario, and he and that network have tremendously enriched my thinking. Beyond that, Tad has become a great friend, and was immensely generous in reading half the book manuscript and giving me invaluable feedback. I cannot thank him enough. I extend that thanks, moreover, to all those scholars Tad brought together to create such exciting conversations as part of the Ideological Conflict Project at the University of Waterloo, in particular Marisa Beck, Esra Çudahar, Scott Janzwood, David Last, Matto Mildenberger, Manjana Milkoreit, Steven Mock, Jinelle Piereder, Steve Quilley, Tobias Schröder, and Paul Thagard. My appreciation also goes to the Balsillie School of International Affairs and the

Waterloo Institute for Complexity and Innovation for hosting and supporting our discussions.

Many scholars at the University of Oxford have offered sustained input and encouragement for my research. I am profoundly indebted to Michael Freeden—both for offering me so much advice and support over the years, and more obviously for his preeminent scholarship on ideology, which transformed my understanding of the topic. I'm similarly grateful to Martin Ceadel, who gave me great support in my first years at New College, Oxford; to Andrea Ruggeri, with whom I had much intellectual engagement surrounding our overlapping research agendas on ideology and armed conflict; and to Stathis Kalyvas, who gave me much encouragement for the research project and tremendously invigorated the conflict studies community at Oxford. I have also gained a lot from broader engagement with Janina Dill, Todd Hall, Dominic Johnson, Juan Masullo, and Matt Zelina, and owe thanks to the organizers and members of the 'Global Ideas' workshop at Oxford: David Priestland, Faridah Zaman, Paul Betts, Dace Dzenovska, Faisal Devji, and Yaacov Yadgar. I owe both professional and intellectual debts to Hugo Slim and Martin Shaw, who acted as my examiners for my doctorate and gave me much input and encouragement for the whole project. The influence of their work is frequently visible in the pages that follow. The first stage of research for this book was conducted during my doctorate at Oxford funded by a research grant from the Arts and Humanities Research Council, and extended into a Junior Research Fellowship at New College, and a Departmental Lectureship at Oxford's Department of Politics and International Relations. All three institutions also have my deep gratitude. I'm incredibly appreciative to Oxford University Press and my editor Dominic Byatt for publishing the book, and for showing immense patience when I decided to heavily rework some of the research between 2019 and 2021. My thanks also go to three anonymous referees, who offered supportive and detailed feedback on the original book proposal.

Beyond Oxford, numerous leading scholars have been incredibly generous in giving various forms of support to my work. My own understanding of mass killing, and the role of ideology in it, is tremendously influenced by the pioneering work of Scott Straus, who has given me invaluable advice and feedback on the project and the draft manuscript. I am also grateful to the support of Lee Ann Fujii, whose work reshaped my thinking about political violence, who offered me encouragement and input into my work in the short time I knew her, and whose unexpected death in March 2018 devastated me, as it did the discipline of political science. I am profoundly indebted to both the research and assistance of a broader group of scholars—many of whom have become great friends—who offered feedback on chapters of the manuscript or in other ways gave key input and support to my research. I therefore offer huge thanks to Gary Ackerman, Alex Bellamy, Susan Benesch, Donald Bloxham, Rachel Brown, Zeynep Bulutgil, Sabina Čehajić-Clancy, Coline Covington, Alan Cromartie, Lesley-Ann Daniels,

Lorne Dawson, Mark Drumbl, Jonathan Glover, Barbora Hola, Erin Jessee, Siniša Malešević, Dominik Markl, Omar McDoom, Theo McLauchlin, Elisabeth Hope Murray, Hollie Nyseth Brehm, Emily Paddon, Vladimir Petrovic, Juliane Prade-Weiss, Scott Sagan, Lee Seymour, David Simon, Savina Sirik, Timea Spitka, Abbey Steele, Sidney Tarrow, Kai Thaler, Benjamin Valentino, Manuel Vogt, Sri Wahyuningroem, and Elisabeth Jean Wood. My thanks also to the numerous members of the International Association of Genocide Scholars, International Network of Genocide Scholars, Conflict Research Society, International Studies Association, and American Political Science Association, as well as attendees at a wide range of other workshops, for their comments on presentations of my work. Thanks also to Sadia Fatima Hameed and all members of the Dangerous Speech Network convened by Wellspring Advisors for providing an invaluable forum for discussing the dynamics of extreme violence and atrocity prevention.

It's been a delight over the last year of the book project to have joined King's College London, and I'm extremely grateful to several new colleagues for reading and/or discussing parts of the book—Samueldi Canio, Robin Douglass, Shaun Hargreaves-Heap, Colin Jennings, Steven Klein, and John Meadowcroft—as well as to Adrian Blau for his constant support. I'm also privileged over the last five years to have played a role in supervising three terrific doctoral students—Leah Owen, Paola Solimena, and Angharad Jones-Buxton—from whom I have learnt so much, and who have all made me a better scholar. I also thank all the friends who have read parts of the book and supported me on the long journey of writing it, particularly Ezgi Aydin, Noémi Blome, Caroline Crepin, Blake Ewing, Ryan Hanley, Irina Herb, Jeff Howard, Helena Ivanov, Katie Mann, Rhiannon Neilsen, and Willa Rae Witherow-Culpepper. I'm especially grateful to Lise Butler, for holding me together through the tough times and bringing endless fun and friendship to the good times; to Will Jones, for always making the world seem that much more interesting; to Raphael Lefevre, who offered feedback on several sections of the book, and spent many hours with me discussing ideology and helping me develop my ideas; to Matt Longo, for pushing me to be my best and making Amsterdam almost a second home; to Timothy Williams, for his faith in me and for his generosity in going through almost half the book with detailed comments; and to Alex Worsnip, for making me a better thinker and bringing so many years of adventure into my life. Finally, there are no adequate words to express my thanks to my parents, Carol Leader and Michael Maynard: for reading and offering feedback on entire drafts of the book, for giving me constant support on the journey of writing it, and above all, for raising me in an environment where it was always good to be curious, and always safe to be wrong.

Contents

List of Figures xi

1. Introduction 1
   1.1 The Puzzle of Mass Killing 1
   1.2 Ideology and Its Critics 4
       A Neo-Ideological Perspective 6
       Advancing the Debate 12
   1.3 Elaborating the Argument 15
       Conceptualizing 'Mass Killing' and 'Ideology' 15
       Ideology, Perpetrator Coalitions, and Political Crisis 18
       On Method 24
   1.4 Plan of the Book 26

2. Clarifying Ideology 28
   2.1 What Are Ideologies? 30
       Narrow Conceptualizations 30
       Broad Conceptualizations 32
   2.2 How Do Ideologies Influence Behaviour? 37
       Internalized Ideology: Commitment and Adoption 42
       Ideological Structures: Conformity and Instrumentalization 45
       Infrastructural Diversity 49
   2.3 Conclusion 51

3. How Does Ideology Explain Mass Killing? 53
   3.1 Explaining Mass Killing Without Ideology? 54
       Rationalist Explanations 54
       Situationist Explanations 59
   3.2 The Basic Argument: Ideologies and Political Crises 63
       The Emergence of Justificatory Narratives for Mass Killing 63
       Ideological Security Doctrines and Political Crises 65
   3.3 Ideology and Perpetrator Coalitions 70
       Building a Perpetrator Coalition 70
       Political Elites 75
       Rank-and-File Agents 79
       Public Constituencies 84
   3.4 The Roots of Hardline Ideology 87
   3.5 Conclusion 92

4. The Hardline Justification of Mass Killing 94
   4.1 What Kinds of Ideologies Matter? 95

The Traditional Focus: Revolutionary Transformations and Moral Disengagement	95
The Neo-Ideological Focus: Radicalized Security Politics	100
4.2 The Power of Justificatory Narratives	103
Portraying the Victims: Threat, Guilt, and Exclusion	105
Portraying the Violence: Virtue, Future Goods, and Inevitability	115
The Limitationist Alternative	127
4.3 Conclusion	131
5. Stalinist Repression	134
5.1 Overview	134
5.2 Explaining the Rise of Stalinism	138
Marxism, Leninism, and the Early Soviet Union	138
Stalinist Radicalization	143
5.3 Stalinism and Repressive Mass Killing	145
The Stalinist Elite	145
The Rank-and-File of Stalinist Violence	160
Publicly Justifying Stalinist Repression	167
5.4 Conclusion	176
6. Allied Area Bombing in World War II	179
6.1 Overview	179
6.2 The Roots of Allied Bombing Doctrines	181
Early Ideas of Air Power	182
The Development of Air Doctrine in Interwar Britain	185
The Development of Air Doctrine in the Interwar United States	188
6.3 Ideology and Area Bombing in World War II	191
Policymakers and the Ideology of Area Bombing	191
The Rank-and-File Bombers	205
The Public Legitimation of Area Bombing	210
Ideological Restraints	217
6.4 Conclusion	218
7. Mass Killing in Guatemala's Civil War	221
7.1 Overview	221
7.2 The Evolution of Military Ideology	228
7.3 Ideology and Counter-Insurgent Mass Killing	235
The Military's Political Elite	236
Rank-and-File Agents in Guatemala's Mass Killings	242
Public Justification of the Killings	251
The American Connection	256
7.4 Conclusion	259
8. The Rwandan Genocide	262
8.1 Overview	262

	8.2 The Evolution of Hardline Hutu Ethnonationalism	266
	Hutu and Tutsi Identity in Rwanda	266
	The Second Republic Under Habyarimana	271
	Hardline Radicalization in the 1990s	274
	8.3 Ideology in the Genocide	280
	The Hutu Power Elite	280
	Rank-and-File Agents in the Genocide	287
	Disseminating the Hardline Narrative	297
	8.4 Conclusion	305
9.	Conclusion	307
	9.1 The Role of Ideology in Mass Killing	307
	9.2 Moving Forward	312
	9.3 Prediction and Prevention	315
	9.4 The Problem of Extremism and Atrocity	318
Methodological Appendix		321
Comparative Historical Analysis		321
Studying Ideas and Violence		324
Conclusion		329
Bibliography		331
Index		372

List of Figures

2.1.	An infrastructural model of ideology's influence.	41
3.1.	The basic argument in brief.	65
3.2.	The spectrum of ideological security doctrines.	66
5.1.	Official Soviet executions and forced labour camp sentences, 1930–53.	135

1
Introduction

1.1 The Puzzle of Mass Killing

Since the start of the twentieth century, between eighty million and two hundred million people have died in *mass killings*: large-scale coordinated campaigns of lethal violence which systematically target civilians.[1] These mass killings have taken many forms: from genocides to major terrorist campaigns and from aerial bombardments to massacres by paramilitary organizations. They have occurred on every continent bar Antarctica: from the Holocaust in Europe to Mao's Cultural Revolution in China, and from mass violence against the indigenous peoples of the Americas and Australasia to the Rwandan Genocide in central Africa. Such organized killing of civilians represents one of the deadliest categories of political violence, its victims heavily outnumbering the thirty-four million soldiers who died in twentieth-century battlefield warfare.[2] While there has been some recent decline in mass killings, they continue to recur.[3] In 2003, the Darfur region of Sudan was subjected to the twenty-first century's first major genocide, with 300,000 killed, while mass killings have also scarred Iraq, Syria, Myanmar, the Democratic Republic of the Congo, and numerous other states since the turn of the millennium.[4] These campaigns involve the most absolute violations of victims' human rights, and constitute the severest 'atrocity crimes' in international law.[5]

Why do mass killings occur? How do human beings come to *initiate*, *participate in*, and *support* such atrocities against unarmed men, women, and children? In popular commentary, films, and media coverage, three rough-and-ready answers to these questions are common. First, the perpetrators are often presented as either individually insane—as psychopaths and sadists—or as whipped up into a kind of

---

[1] Anderton and Brauer 2016b, 4. The variation in estimates reflects data problems and controversies over how to classify such violence—higher estimates are open to criticism; see Gerlach 2010, 256–8 & 468, fn.6.
[2] Valentino 2004, 1.
[3] Bellamy 2012b, 4–9. Overall, states have targeted civilians in one fifth to one third of all wars; see Downes and McNabb Cochran 2010, 23.
[4] Butcher et al. 2020.
[5] Scheffer 2006; United Nations 2014; Sharma and Welsh 2015; Dieng and Welsh 2016; Gordon 2017. I will therefore often refer to mass killings as atrocities, but not all atrocities are mass killings.

social madness of collective rage and hatred.[6] Second, mass killings are sometimes thought to expose humanity's innately aggressive and destructive nature. When the restraints of law and order are peeled away, it is suggested, this innate propensity towards violence is unleashed.[7] Finally, it is sometimes suggested that perpetrators of mass killing are simply acting under coercion.[8] As members of totalitarian societies or harsh military or paramilitary organizations, they kill because they themselves have the threat of death hanging over them should they disobey.

These explanations might seem superficially plausible, but five decades of scholarship on mass killings has shown all three to be inaccurate. Mental illness or mindless rage amongst perpetrators of organized violence is rare. In fact, as the psychologist James Waller puts it: 'it is ordinary people, like you and me, who commit genocide and mass killing'.[9] Although disturbing, this should not really be surprising. Mass killing generally requires the support or acquiescence of substantial sections of societies over periods of months or years.[10] It is unlikely that this number of people could be psychologically abnormal in any meaningful sense, or successfully participate in sustained coordinated violence while consumed by blind rage. Indeed, the organizations that recruit perpetrators of mass killing, such as secret police departments, state militaries, and insurgent groups, sometimes go to great lengths to *weed out* psychopaths, sadists, and uncontrollably hate-fuelled individuals from their ranks.[11]

Modern research also refutes claims that human beings are innately predisposed to violence.[12] If anything, as psychologists Rebecca Littman and Elizabeth Levy Paluck summarize, 'military history and scientific evidence show that most people *avoid* physically harming others, even at personal cost'.[13] Even in war, when there are overwhelming reasons to kill in order to stay alive, soldiers often struggle to do so. This is not a matter of cowardice: such soldiers often run immense personal risks, even throwing themselves on grenades, to aid comrades.[14] But they struggle to fire their weapons at the enemy, and often suffer serious trauma for doing so. If mass killings were really produced by innate human destructiveness,

---

[6] Aronson 1984; Kressel 2002; Wilshire 2006; Orange 2011; Bradshaw 2014; https://www.bbc.co.uk/news/world-europe-34840699. For further examples and critique see Kalyvas 2006, 32–4; Valentino 2014, 92.

[7] E.g. Ghiglieri 1999.

[8] Most commonly, perpetrators themselves make such claims; see, for example, Anderson 2017b, 51–2 & 56–7; Jessee 2017, 168–73. For scholarly accounts that emphasize coercive state power see Brzezinski 1956; Rummel 1994; Rummel 1995, 4–5.

[9] Waller 2007, 20. See also Staub 1989, 67; Browning 1992/2001; Smeulers 2008, 234; Alvarez 2008, 217–18; McDoom 2013, 455–6; Littman and Paluck 2015. Atrocities may still be 'evil'; see Card 2002; Vetlesen 2005; Russell 2014.

[10] How much support is needed is, however, debatable; see Mueller 2000; Valentino 2004, 2–3; Kalyvas 2006, 102–3.

[11] Schirmer 1998, 165; Valentino 2004, 42–4 & 57–8; Dutton 2007, 136; Waller 2007, 71; Baum 2008, 77.

[12] For summaries see Collins 2008; Grossman 2009.

[13] Littman and Paluck 2015, 84.

[14] Grossman 2009, 4.

moreover, they should occur in almost all instances of war and social breakdown. Yet, while mass killings are tragically recurrent across world history, most periods of war and upheaval pass by without them.[15] Rather than an uncontrolled consequence of human nature, then, mass killings are what the historian Christopher Browning terms 'atrocity by policy': organized collective campaigns deliberately implemented by *certain* people, at *certain* times.[16]

The third popular explanation, that killers are simply coerced, is not quite so misguided. Organizers of mass killing do deploy forceful coercion to suppress opposition, and sometimes to compel people to participate in violence. Nevertheless, in research on over a hundred years of modern mass killings, only a small minority of perpetrators seem to have reluctantly obeyed orders to kill issued on pain of death.[17] Even the most powerful totalitarian regimes in history have generally been unable to micromanage violence through coercion alone, relying instead on considerable support and willing compliance from their subordinates and broader populations.[18] Where perpetrators are coerced, moreover, this remains only a partial explanation, because campaigns of mass killing are not coercive 'all the way up'. Someone (and usually not just one person) has to decide that violence is the right course to take, and many others have to decide to support them. Coercion does not explain such decisions.

The inadequacy of these rough-and-ready explanations generates the central puzzle of mass killing. Mass killings are widely thought to be morally abhorrent, and typically involve acts (such as the killing of children) that run against established cultural norms across the world. The violence is typically psychologically arduous, at least initially, for those who carry it out. Perhaps most puzzlingly of all, mass killings often seem irrational for the very regimes and groups that perpetrate them. In the Soviet Union in the 1930s, Joseph Stalin's Great Terror included a large-scale purge of the Red Army which left it desperately weakened in the face of Nazi invasion four years later. In 1970s Guatemala, the military regime responded to a left-wing guerrilla insurgency with brutal massacres of the country's indigenous Maya communities, prompting many Maya to join the guerrillas and thereby strengthening the insurgency. Sometimes mass killings prove disastrous for perpetrators by antagonizing other states and encouraging outside intervention, as in Khmer Rouge Cambodia in 1979, or in recent ISIS atrocities in Iraq and Syria. Even when not clearly self-defeating, mass killings are risky courses of action, almost always wildly disproportionate to any actual challenges their perpetrators face, and target individuals who present no obvious threat. So why do they occur? Why do certain political leaders initiate these policies of extreme violence? Why

---

[15] Straus 2012b; Straus 2015a, ch.2.
[16] Browning 1992/2001, 161.
[17] Ibid. 170; Valentino 2004, 48; Szejnmann 2008, 31; Goldhagen 2010, 148–50.
[18] Overy 2004, chs.5 & 8.

do their subordinates willingly implement them? Why do broader sectors of society support or acquiesce to the violence? These are the questions I seek to address in this book.

1.2 Ideology and Its Critics

I argue that effective answers to such questions must analyse the role of ideologies—broadly defined as the *distinctive political worldviews of individuals, groups, and organizations, that provide sets of interpretive and evaluative ideas for guiding political thought and action*. Ideologies are not the only key cause of mass killing. Indeed, scholars have identified many others, including circumstances of war, political instability, and crisis;[19] discriminatory processes of nation-building;[20] psychological tendencies to follow authorities, conform to peer-pressure, or denigrate minorities;[21] and various self-interested motives for violence.[22] All of these factors matter. But they matter in interaction with ideology, because ideologies play a central role in determining both how people privately think about mass killings and how such violence can be publicly legitimated and organized. In cases such as those mentioned above, mass killings may look, from an outside perspective, like strategic and moral catastrophes. But they *appeared to perpetrators* as strategically advantageous and morally defensible. That impression was not a 'natural' consequence of the circumstances in which perpetrators found themselves, but it was a likely consequence given their prevailing ideological frameworks. Ideologies are therefore crucial in explaining two key things: first, whether mass killings occur in the first place, and second, the character of mass killings when they do occur—i.e. who they target, what logic of violence is employed, and how the killing unfolds within different areas and organizations.

This argument divides expert opinion. Indeed, the role of ideology is one of the most disputed issues in current scholarship on mass killing. That dispute rests, I will suggest, on rather murky theoretical foundations. But most existing research can be roughly characterized as adopting one of two perspectives.

In what I will call *traditional-ideological perspectives*, ideologies are seen as a crucial driver of mass killings, because they provide the extremist goals and mentalities that motivate ideologically committed individuals to perpetrate the

---

[19] Kalyvas 1999; Harff 2003; Valentino 2004; Valentino, Huth, and Balch-Lindsay 2004; Downes 2008; Fjelde and Hultman 2014; Maat 2020.
[20] Mann 2005; Levene 2008; Segal 2018.
[21] Milgram 1974/2010; Kelman and Hamilton 1989; Bandura 1999; Waller 2007; Zimbardo 2007; Neilsen 2015; Williams 2021.
[22] Aly 2008; Gerlach 2010; Esteban, Morelli, and Rohner 2015; Williams 2021.

violence.[23] Emphasis is most commonly placed on revolutionary ideological goals to remake society, ideological hatreds towards certain victim groups, and the ideological reversal of traditional moral norms. In early post-Holocaust scholarship (including famous works by the likes of Karl Popper, Isaiah Berlin, and Hannah Arendt), such dangers were centrally associated with totalitarian ideologies such as Nazism, Stalinism, and Maoism, which guided arguably the three most destructive regimes in human history.[24] More recent work has broadened the focus beyond totalitarianism, but still emphasizes *extraordinary* 'utopian' or 'revolutionary' ideological projects that upend conventional morality and abandon pragmatic political considerations. Such claims take their most emphatic form in Daniel Goldhagen's contention that mass support for 'eliminationist anti-Semitism' amongst ordinary Germans provided the necessary and sufficient motivational cause for the Holocaust.[25] But many other scholars, while not going as far as Goldhagen, also focus on the role of unconventional ideological goals, mindsets, and hatreds that motivate ideologically committed perpetrators of mass killing.[26]

In opposition to such arguments, many scholars adopt what I term a *sceptical perspective* on ideology's role in mass killing. Without necessarily declaring it completely irrelevant, such sceptics downplay ideology's significance and largely exclude it from their explanations of such violence. Two main arguments have been offered here. First, sceptics contend that few perpetrators actually seem motivated by deep ideological commitments in the way traditional-ideological perspectives suggest. Second, sceptics suggest that even if radical ideologies do influence perpetrators, such ideologies are themselves largely a symptom of more fundamental social or political causes, such as societal upheaval, authoritarian governing institutions, or war. With either argument, 'non-ideological' motives or forces appear to be the key drivers of mass killing, and ideology is largely reduced to a pretext or 'post-hoc rationalization' for violence.

Such ideology-sceptics therefore offer alternative 'non-ideological' explanations of mass killing.[27] The most influential of these, on which I focus most attention in this book, come from *rationalist* theories. Rationalists argue that mass killings

---

[23] I include perspectives primarily orientated around concepts distinct from but closely related to ideology (such as culture, identity, hate propaganda, and so forth) that explain mass killing in essentially the same fashion.

[24] Popper 1945/2003; Arendt 1951/1976; Berlin 1954/2002; Brzezinski 1956; Arendt 1963/2006; Popper 1963/2002, ch.18; Linz 1975/2000, ch.2; Kirkpatrick 1979; Kuper 1981, ch.5; Shorten 2012; Berlin 2013; Richter, Markus, and Tait 2018.

[25] Goldhagen 1996. For similar perspectives applied to mass killings more broadly see Kressel 2002; Goldhagen 2010.

[26] See, for example, Melson 1992; Weiss 1997; Kiernan 2003; Weitz 2003; Hagan and Rymond-Richmond 2008; Midlarsky 2011.

[27] These sceptical accounts are compatible, since they typically address different 'levels of analysis'—with rationalists often focusing on why political decision-makers *initiate* policies of mass killings, while situationists focus on why followers *participate in* such policies.

occur because they can be a useful, albeit brutal, strategy for achieving important goals common to all regimes and groups, whatever their idiosyncratic ideologies—such as holding onto power, winning wars, or gaining material wealth. Not all rationalists side-line ideology, and some (correctly, I will argue) see rationality and ideology as importantly intertwined.[28] But most rationalists suggest that mass killings do not depend on any particular kind of ideological worldview.[29] Instead, they are explained by particular *strategic circumstances*, such as certain kinds of political crisis or armed conflict, which create incentives for governments or groups to target civilian populations with violence.

An alternative source of ideology-scepticism comes from what I term *situationist* theories.[30] For situationists, mass killing is best explained by various kinds of *situational social pressure* on individuals, such as bureaucratic routines, orders from authorities, peer-pressure, or group emotions. Again, not all situationists deny that ideology plays an important role.[31] But for sceptical situationists, these social pressures are so powerful that they can induce violence amongst different individuals and groups irrespective of their ideologies. Like rationalists, situationists tend to emphasize how certain contexts of crisis and war create or intensify such situational pressures for violence. They often also stress the way such pressures encourage the relatively unplanned escalation of violent policies or practices within bureaucracies and local communities.

In this book, I challenge both the traditional-ideological perspective and the ideology-sceptics. Against the sceptics, I argue that ideology is essential in explaining mass killings. I do not reject rationalist or situationist theories per se. They are quite correct to emphasize the role of strategic circumstances and situational social pressures. But whether such circumstances and pressures lead to mass killing or not *depends on ideology*. This is not just true of a subset of especially 'ideological' cases, moreover—*all* mass killings have an important ideological dimension. Yet, I simultaneously argue that traditional-ideological perspectives mischaracterize that dimension, wrongly rooting mass killings in 'extraordinary' ideological goals and values. This book therefore advances a different account of ideology's role in mass killings—one that stresses the interdependence of ideology, strategic circumstances, and situational pressures. I term this a *neo-ideological perspective*.

A Neo-Ideological Perspective

This neo-ideological perspective revises the more traditional portrayal of how ideology might feed into mass killing in two key ways.

---

[28] See, for example, Valentino 2004; Maat 2020.
[29] E.g. Downes 2008, 11.
[30] See also Fujii 2009.
[31] See, for example, Zimbardo 2007.

First, I argue that the crucial ideological foundations for mass killing are not utopian ambitions, revolutionary values, or extraordinary hatreds that *contrast with* conventional strategic and moral concerns. Instead, the primary ideological foundations of mass killing *exploit conventional strategic and moral ideas*—specifically ideas associated with security, war, and political order. It is not the abandonment of strategic pragmatism and traditional morality in favour of extraordinary ideological goals that matters, in other words, but the radical reinterpretation of such conventional ideas within extreme ideological narratives of threat, criminal conspiracy, patriotic valour, and military necessity. These *justificatory narratives* for mass killing, as I shall call them, are thus largely security-orientated and yet vitally ideological—embedded in broader political worldviews and making little sense if stripped from their particular ideological context. Such narratives are critical, both in guiding the formulation of policies of mass killing by political elites, and in mobilizing, legitimating, and organizing the violence amongst broader sections of society.

Mass killing is not best understood, therefore, as a revolutionary project to transform society (as many traditional-ideological approaches suggest), an instrumental strategy largely dictated by circumstantial incentives (as rationalist-sceptics portray it), or an escalatory campaign primarily driven by social pressure (as situationist-sceptics often imply). There is truth in each of these portrayals, but none accurately characterizes ideology's role in mass killing. Instead, mass killing is best understood as a form of *ideologically radicalized security politics*. It is rooted in the ideological and institutional architecture of war-waging, policing, and national security found in all complex human societies. It is driven by the familiar strategic and moral concerns of such activities: the perception of threats and criminality and an assessment of violence as a necessary way of defending the political order against them. It is typically carried out by state security apparatuses (or their non-state analogues) with all the organizational norms, capacities, and tendencies typical of such institutions. But in mass killings, all these familiar features of security politics have become radicalized by extreme 'hardline' ideological worldviews, which make *civilians* appear justified targets of *mass* violence.

I therefore characterize the central agents of mass killing—both among political leaders and in wider society—as security hardliners, promoting massive violence against civilian populations to advance the safety and interests of the society, regime, or group they identify with. But I argue that it is imperative to *understand such hardliners as an ideological category*. Hardliners are not merely responding rationally to the objective situations they find themselves in, nor are they 'unthinkingly' following orders or bureaucratic procedures. They are guided by distinctive sets of ideas about security and politics which ideologically distinguish them from less hardline groups. This reflects the fact, ignored by too many scholars, that the politics of war and national security are just as 'ideological' as any other branch of politics, with different factions of society guided by different sets of ideas about security and how to achieve it.³² I identify six clusters of hardline ideas as most crucial here: (i) the portrayal of civilian targets of violence as *threats*; (ii) the assertion that such civilian targets are *guilty of serious crimes*; (iii) the *denial of common links of identity* between civilian targets and the primary political community; (iv) the *valorization of violence* against civilians as dutiful, tough, and soldierly; (v) the assertion that such violence will generate tremendous *future strategic benefits*; and (vi) the *destruction of meaningful alternatives* to mass killing, so that it is portrayed as essentially unavoidable. When strongly hardline factions guided by such ideas are able to achieve political dominance, mass killing becomes a serious possibility in times of crisis. But when dominant political factions have little sympathy for hardline ideas, they will almost always opt for less extreme (though not necessarily benign) courses of action.

This book therefore challenges a dominant assumption of existing scholarship on political violence: namely, that a fundamental contrast separates 'ideological' motives for violence (associated with the pursuit of ultimate political ideals) from more 'strategic' or 'pragmatic' concerns (associated with the pursuit of security, power, or military victory).³³ Though common, this assumption is a profound conceptual handicap that distorts prevailing understandings of mass killing and political violence more generally. In reality, few if any real-world ideologies simply ignore strategic concerns in favour of dogmatic implementation of their ultimate ideals. Nor is the pursuit of power and security ever governed by a self-evident pragmatism free from the influence of ideology.³⁴ Instead, ideologies critically influence decision-makers' strategic thinking—shaping their perceptions of threats, their assessments of the appropriate and effective policies for neutralizing those threats, and their assumptions about the moral basis for and limits to those policies.³⁵

I therefore agree with the rationalist claim that mass killing is a *strategic* form of violence.³⁶ It is, as Ben Valentino puts it, 'an instrumental policy ... designed to accomplish leaders' most important ideological or political objectives and counter

---

[32] This point has been emphasized by constructivist, critical, and feminist security scholars; see Katzenstein 1996a; Owen 1997; Campbell 1998; Smith 2004; Haas 2005; Sjoberg and Via 2010; Haas 2012.

[33] For similar critiques of this dichotomy see Straus 2012a, 549; Verdeja 2012, 315–16; Staniland 2015, 771–2; Straus 2015a, 11–12. So entrenched is the dichotomy that some scholars write as though the mere fact that mass killing is instrumental suggests that ideology's role must be minimal—as if ideology *requires* a kind of killing for killing's sake campaign; see Roemer 1985; du Preez 1994, 69–70; Mitchell 2004, 38 & 41. But this is a caricature. Even exterminatory mass killings rooted in the most egregious ideological fantasies—such as the Nazi belief in a Jewish world conspiracy—are still committed in pursuit of *ends*, such as the protection of the state and the community against (imagined) enemies.

[34] Jabri 1996; Campbell 1998.

[35] Straus 2012a; Straus 2015a.

[36] I.e. it is a means for achieving certain goals, chosen because it is deemed a consequentially (and perhaps also deontologically) superior means for doing so compared to perceived alternatives in the relevant material and social environment. I do not assume that strategic violence must be driven purely by cost–benefit calculations or serve the objective interests of unitary actors—such claims should be

what they see as their most dangerous threats'.[37] Mass killings are not 'wanton and senseless', but deliberate policies engaged in by certain individuals and groups and guided by comprehensible logics of violence.[38] Yet it is rarely plausible to portray mass killings as simply the most ruthlessly efficient way for perpetrators to achieve purportedly 'non-ideological' goals—as though anyone in their position would have favoured such an extreme course. What is crucial is the way mass killings can come to *appear* strategically rational (as well as normatively legitimate), within a certain set of hardline ideological assumptions, narratives, and institutions.

Take, for example, the Ottoman Empire's genocidal attacks on its Armenian population in 1915–17. These were motivated, in part, by Ottoman fears that Armenian nationalist groups might side with neighbouring Russia, the Ottomans' enemy, in World War I. Given such fears, genocide might appear like a rationally comprehensible strategy for a ruthless Ottoman state to eliminate a threat to its security. Yet the vast bulk of the Armenian population engaged in no such collaboration with Russia. Various less extreme options existed through which Ottoman leaders could have secured themselves against any anticipated Armenian rebellion.[39] Other European empires, with similar concerns about their ethnic minorities, did not employ such policies of annihilation.[40] So what mattered was not simply the strategic circumstances of war and crisis that the Ottoman Empire confronted. Pursuing genocide as a response to such circumstances only made sense in light of certain ideological narratives about the Armenian population and Ottoman security, which were adhered to by key political elites, were institutionalized within the organizations they commanded, and promoted the broader social mobilization and escalation of violence.[41] Understanding this interdependence of ideology and security politics in contexts of crisis is, I argue, essential for effective explanation of all mass killings.

The second way this book revises traditional portrayals of ideology's role in mass killing is by retheorizing the basic psychological and social processes that link ideologies to violent action. Such processes rarely receive explicit dissection in research on mass killing or, indeed, political violence more broadly. But scholars often work with a tacit picture of ideology that I will refer to as the 'true believer model'. The true believer model depicts ideologies as rigid belief-systems which primarily shape political behaviour through strong ideological commitments to a

---

understood as possibly true explanatory arguments, to be assessed against alternatives, not a 'default' image written into the very notion of strategic choice. See also Sjoberg 2013, 187–91.

[37] Valentino 2004, 3. This in no way denies that violence *also* has self-perpetuating qualities—contra Wolfgang Sofsky, who leaps from the correct claim that violence often exceeds its instrumental intentions to the erroneous conclusion that instrumental motives therefore play no causal role; see Sofsky 2002, 18–19. On self-perpetuation see Arendt 1970; Sjoberg and Via 2010; Littman and Paluck 2015; Eastwood 2018.

[38] Kalyvas 1999; Straus 2012a.
[39] Bloxham 2005, 86.
[40] Bulutgil 2017.
[41] Bloxham 2005.

certain 'ideal vision of society'. Consequently, the question of ideology's importance in mass killing is principally about the intensity of ideological belief in such visions among perpetrators.[42] Both traditional-ideological and sceptical scholars of mass killing tend to implicitly adopt this model of ideology—with traditional-ideological theorists emphasizing perpetrators' strong ideological beliefs, while sceptics contend that ideology is unimportant *precisely because* they find few 'true believers' amongst the perpetrators they study.

By contrast, I argue that this focus on true believers is a mistake, for two reasons. First, even sincerely held ideas do not need to be endorsed with particularly deep conviction in order to shape political behaviour. 'True believers' do play a role in mass killing, especially amongst political elites, but most perpetrators lie between the extremes of intense ideological devotion on the one hand and complete ideological disinterest on the other. They are frequently conflicted and participate in mass killing in part for non-ideological motives. They have often only come to accept justifications of extreme violence relatively recently, and as tacitly endorsed notions or taken-for-granted assumptions rather than substantive, self-conscious 'beliefs'. But this does not make such perpetrators 'unideological'. Most still internalize key hardline justifications of mass killing—even if rather selectively or half-heartedly—which can be *vital* in explaining their participation in violence.

But this is still only half the story. Some people may not internalize ideological justifications of mass killing to *any* serious extent. Yet they often find themselves perpetrating or supporting the violence, and espousing the ideologies that justify it, nevertheless. It is typical to think that this proves that ideologies don't really matter. If killers don't really endorse an ideology, but kill anyway, it is easy to conclude that other factors or motives must have been the 'true' cause of violence. This is a fundamental error. We need to examine ideologies precisely because they can exert powerful influence *even over those who do not believe in them*. When an ideology becomes embedded in the institutions, norms, and discourses of a group, organization, or society, even non-believers are subject to considerable social or 'structural' pressure to comply with that ideology. Such individuals might consciously disbelieve in the ideology, or they might hold more ambivalent and/or ambiguous views, but they act *as if* they believe.[43] Social pressure may be the immediate driver of action, but the *direction* of that pressure—the specific behaviour it encourages—cannot be explained if the ideology in question is removed from the picture.

By unpacking these two central ways—*internalized* and *structural*—in which ideologies may matter, I seek to move the locus of debates over ideology away from disputes over the number of true believers. Ideologies are crucial, not because

---

[42] An alternative tendency is essentially the opposite—to treat ideologies as purely instrumental tools largely unrelated to actual belief. I address this in Chapter 2.
[43] See also Wedeen 2019, 4–8.

they provide a single 'ideological motive' for violence found only amongst their most devoted followers, but because they bind diverse coalitions of perpetrators into collective violence through multiple causal mechanisms.[44] Ideologies are deeply believed in by some, but more shallowly internalized, complied with under pressure, or cynically manipulated by others. They typically operate alongside, and sometimes through, careerist, conformist, and self-interested concerns. For some perpetrators, ideological justifications for mass killing may directly *motivate* violence—whether by internalized personal convictions that draw individuals into violence, or by generating structural pressure within ideological groups, organizations, and institutions that induces individuals to perpetrate. For other perpetrators, ideological justifications may primarily *legitimate* violence—whether by allowing individuals to sincerely see their own participation in violence as legitimate, or by creating a structural context in which violence is publicly legitimizable.[45] For some, ideological justifications may even amount to little more than a kind of 'negative legitimation'—sowing confusion about the violence and producing an 'atmosphere of epistemic and affective murk', as Lisa Wedeen puts it, that obstructs effective opposition.[46] But in *all* these circumstances, the specific ideas that make up an ideology can crucially shape whether and how mass killings occur.

So, mass killings do not require mass ideological enthusiasm. But they do depend, as I shall put it, on a kind of *ideological infrastructure*: a mutually reinforcing mix of both sincerely accepted hardline ideas *and* hardline norms and institutions that, together, sustain and guide extreme collective violence. Compared to the association of ideology with true believers, this multifaceted characterization of ideology's influence is more consistent with modern social, psychological, and economic theories of the impact of ideas on human behaviour. It rejects several weary scholarly dichotomies: ideology is not associated solely with 'structure' *or* 'agency', 'intentions' *or* 'functions', 'micro' *or* 'macro', but is powerful precisely because it operates across such factors and levels of analysis. This also fits much better with leading empirical research on mass killings, which strongly emphasizes the diversity of perpetrators. I thus agree with several claims made by ideology-sceptics: that perpetrators of mass killing act from various motives and are not typically guided by longstanding devotion to a single monolithic ideology. But it is a mistake to think that these findings imply that ideology's role in violence is peripheral.

---

[44] For similar emphases of ideology's multifaceted role in violent coalitions, see Gutiérrez Sanín and Wood 2014; Anderson 2017a, chs.4–5; Williams 2021.
[45] Skinner 1974, 292–300; Jost and Major 2001.
[46] Wedeen 2019, 4.

Advancing the Debate

This neo-ideological perspective is not wholly unprecedented. A few scholars—principally Alex Bellamy, Donald Bloxham, Zeynep Bulutgil, Omar McDoom, Elisabeth Hope Murray, Jacques Sémelin, Scott Straus, and Ben Valentino—have argued that ideology matters in mass killings, but in ways that do not fully match the traditional-ideological perspective.[47] In presenting mass killing as a form of ideologically radicalized security politics, my perspective also has strong affinities with Martin Shaw's conception of genocide as a form of 'degenerate war',[48] and with critical and feminist scholars' emphasis on the expansive justificatory potential of modern discourses of security and violence.[49] As noted, some rationalists and situationists also embrace ideology, and even the sceptics are often only really disputing the traditional-ideological portrayal of mass killing and may be receptive to the kind of neo-ideological perspective developed here.[50]

While indebted to these precedents, this book attempts to go further than existing studies in advancing the debate over ideology's role in mass killing. That role remains so disputed, I suggest, because there is huge uncertainty amongst scholars over the very nature of ideologies and their influence. Even amongst those who agree that ideology matters, *how* they think it matters varies considerably. Some scholars identify pernicious ideological themes, which vary from theorist to theorist. Thus, Eric Weitz focuses on 'utopias of race and nation';[51] Ben Kiernan on 'racism', 'territorial expansionism', 'cults of cultivation', and 'purity';[52] Alex Alvarez on 'nationalism', 'past victimization', 'dehumanization', 'scapegoating', 'absolutist worldview', and 'utopianism';[53] Hugo Slim on a set of twelve 'anti-civilian ideologies';[54] and Gerard Saucier and Laura Akers on twenty major elements of the 'democidal thinking' behind mass killing.[55] Other scholars focus on specific linkages between certain ideologies and violence. Valentino emphasizes the way 'radical communization' and 'racist or nationalist beliefs' generate ideological goals which may be furthered by mass killing.[56] Straus suggests that ideological 'founding narratives' of the political community are key—with exclusivist

---

[47] How these scholars theorize ideology's role varies considerably, however; see Valentino 2004; Bloxham 2005; Semelin 2007; Bloxham 2008; Bellamy 2012b; Murray 2015; Straus 2015a; Bulutgil 2017; McDoom 2021. More sophisticated understandings of ideology can also be found in specialist literature on Nazism and Communism; see, for example, Schull 1992; Kershaw 1993; Kotkin 1995; Priestland 2007; Roseman 2007.
[48] Shaw 2003. See also Moses 2021.
[49] Jabri 1996; Sjoberg and Via 2010; Sjoberg 2013.
[50] See, for example, Newman and Erber 2002; Valentino 2004; Zimbardo 2007; Balcells 2017.
[51] Weitz 2003.
[52] Kiernan 2003.
[53] Alvarez 2008, 220–7.
[54] Slim 2007.
[55] Saucier and Akers 2018.
[56] Valentino 2004, 4–5.

narratives portraying outgroups as pathological dangers in times of crisis.[57] Bulutgil contends that political parties ideologically orientated around ethnic cleavages, due to a lack of class-based or other 'cross-cutting' concerns, are most likely to engage in (ethnic) mass killings.[58] Though there are common ideas here, it is hard to know how these various accounts might be reconciled.

Such uncertainty in scholarship is mirrored in political practice. The prevention of mass killing has become a central concern of international governmental and non-governmental organizations.[59] Yet ideology remains an area of perennial weakness in understanding. The United Nations' *Framework of Analysis for Atrocity Crimes*, for example, affirms the relevance of ideology under its fourth risk factor: 'Motives or Incentives'. Yet its comments are characteristically vague:

> [I]t is extremely important to be able to identify motivations, aims or drivers that could influence certain individuals or groups to resort to massive violence as a way to achieve goals, feed an ideology or respond to real or perceived threats ... [such as] those that are based on exclusionary ideology, which is revealed in the construction of identities in terms of 'us' and 'them' to accentuate differences. The historical, political, economic or even cultural environment in which such ideologies develop can also be relevant.[60]

This does not take us very far in assessing ideological risks of mass killing. 'Us' and 'them' differences are ubiquitous in global politics—how could we tell when they represent dangerous exclusionary ideologies?[61] Is this construction of us–them differences the only significant hallmark of such ideologies? How is the 'political, economic, or even cultural environment' relevant, and how do we assess when it, in combination with ideology, truly promotes atrocities? This paragraph, from the only page of the *Framework* that mentions ideology, provides no way of even beginning to answer such questions.[62]

Ideological dynamics have also become important for legal practitioners, because efforts to address ideological propaganda, hate speech, and extremism are increasingly central concerns of domestic and international law.[63] Yet legal analyses often rest on vaguely substantiated claims about how ideology and speech relate to violence. In the foremost study of 'atrocity speech law', for example, Gregory S. Gordon affirms a 'compelling connection between hate speech and mass atrocity' and contends that '[p]erpetrator conditioning through speech is a *sine qua non* for

---

[57] Straus 2015a.
[58] Bulutgil 2016; Bulutgil 2017.
[59] Welsh 2010; Welsh 2016.
[60] United Nations 2014, 13.
[61] Valentino 2004, 17–18.
[62] Such problems also characterize the literature on counter-extremism and radicalization. For discussion see Neumann 2013; Schuurman and Taylor 2018.
[63] Gordon 2017; Wilson 2017.

mass atrocity'.[64] But Gordon generally supports such claims with purely descriptive observations of the *extent* of hate speech surrounding atrocities, not causal analysis of the *difference* such speech actually makes. Conversely, other legal scholars argue that speech has little to no impact on political violence but base this conclusion on highly selective reference to relevant empirical research.[65] Similarly, as Susan Benesch and Richard Ashby Wilson observe, international tribunals have often made bold assertions that certain instances of ideological speech caused, or did not cause, violence, but on the basis of minimal evidence and little apparent understanding of what role such speech might play.[66]

These problems, both in academic research and in political and legal practice, are rooted in a common cause: namely, that ideology's role in mass killing remains *under-theorized*. Scholars of mass killing often mention 'ideology', but they rarely engage in detailed analysis of what ideologies actually are, nor systematically consider the range of ways in which ideological factors *could* encourage violence. They also tend to ignore specialist work on ideology from other fields, such as political psychology, political communications research, intellectual history, social movement studies, and political theory.[67] Consequently, ideology's importance is often dismissed for very muddled reasons. Some suggest, for example, that because a single ideology (such as Nazism) is an overarching feature of a particular mass killing (such as the Holocaust), that ideology cannot explain the variation of violence over time, or in different areas, or against different victim groups.[68] This would only be true, however, if ideologies never changed over time, never varied in strength or form across different locations, and never contained different ideas about different categories of victim. Clearly this is not right. Other scholars downplay ideology simply because they find that other motives or considerations appear to have played a role—as though ideology must either exclusively and deterministically guide violence or else be deemed essentially irrelevant. When rendered explicit, such assumptions seem obviously mistaken, but they persist because of the lack of theoretical clarity over what ideologies are and how they operate.

This book is an attempt to tackle these problems. Rather than addressing ideology only as part of a general discussion of mass killings or only in one particular case, I make the role of ideology in mass killing the central focus of a comparative study. I seek to demonstrate that ideologies are indeed critical, but as dynamic sets of ideas about security politics, operating through multiple forms of influence, which interact with other important causes on the complex path of radicalization to mass killing. I make reference to a wide range of empirical research but

---

[64] Gordon 2017, 6 & 24.
[65] E.g. Danning 2019.
[66] Benesch 2012a; Wilson 2016; Wilson 2017. See also Straus 2007b.
[67] There are partial exceptions to such neglect; see Malešević 2006; Priestland 2007, 16–21; Alvarez 2008, 216–17; Malešević 2010; Cohrs 2012; Ryan 2012, 10–15.
[68] Hiebert 2008, 8; King 2012, 331; Maat 2020, 777–8.

develop detailed evidential support for my account by examining four quite different campaigns of mass killing: Stalinist repression in the Soviet Union, the Allied area bombing of Germany and Japan in World War II, mass atrocities in the Guatemalan civil war, and the Rwandan Genocide. Several other cases—such as the Holocaust, ethnic cleansing in the former Yugoslavia, and Cambodia under the Khmer Rouge—also figure repeatedly. These cases involve contrasting perpetrators, from different parts of the globe, influenced by various ideologies. If we find ideological patterns across such diverse contexts, there is a good chance that those patterns apply to mass killings more generally.

1.3 Elaborating the Argument

Conceptualizing 'Mass Killing' and 'Ideology'

Several parts of this book are relevant for thinking about ideology's role in political violence in general—there is also, after all, much dispute about its place in war, civil war, terrorism, and revolution. But I am focused on 'mass killings': *large-scale coordinated campaigns of lethal violence which systematically target civilians*.[69] Numerical criteria for 'large-scale' violence vary: Valentino suggests 50,000 civilian deaths over the space of five years,[70] Bellamy opts for 5,000 deaths in a particular campaign,[71] while Valentino and Jay Ulfelder's statistical analysis of mass killings and Charles Anderton's dataset of 'mass atrocities' examine cases involving over 1,000 deaths a year.[72] The cases focused on in this book all meet Valentino's higher threshold, but where one draws the line is rather arbitrary.[73] Where relevant, I assume a simpler threshold of 10,000 civilian deaths a year. What matters is that mass killing involves deadly violence against civilians which is systematic and widespread rather than sporadic or uncoordinated.[74] I focus on modern (twentieth- and twenty-first-century) mass killings, although much of the analysis

---

[69] In defining civilians, I use Valentino, Huth, and Balch-Lindsay's description of 'non-combatants' as 'any unarmed person who is not a member of a professional or guerrilla military group and who does not actively participate in hostilities by intending to cause physical harm to enemy personnel or property'; see Valentino, Huth, and Balch-Lindsay 2004, 378–9. Outside armed conflict, talking of 'civilians' or 'non-combatants' is something of a misapplication of International Humanitarian Law. But in lieu of any other obvious term, I follow other scholarship on mass killing in using 'civilians' to refer to unarmed, non-military populations even in peacetime contexts. My thanks to Jennifer Welsh for alerting me to this point. For similar conceptualizations of mass killing see Valentino 2004, 10–15; Owens, Su, and Snow 2013, 71–2.
[70] Valentino 2004, 10–13.
[71] Bellamy 2011, 2.
[72] Ulfelder and Valentino 2008, 2; Anderton 2016.
[73] Amongst other problems, numerical thresholds should ideally scale with relevant population size.
[74] Straus 2015a, 22–4. Like most scholars, I assume that such violence differs from large quantities of uncoordinated and privately motivated abuses; see Humphreys and Weinstein 2006, 433 & 445. See also, however, Barnes 2017. I use the term 'massacres' to refer to smaller-scale killings of ten or more civilians within the space of 24 hours, whether as part of a campaign of mass killings or not.

could be extended, with some modification, to earlier cases. Unlike many studies, I do not restrict my focus solely to genocidal mass killings. Genocides have distinctive features, which I discuss, vis-à-vis ideology, in Chapter 4. But studying them in isolation from other forms of violence against civilians often yields misleading conclusions, exaggerating the centrality of such distinctive features and obscuring crucial links to non-genocidal forms of violence.[75]

Mass killings include a range of phenomena—genocide, total war, state repression, ethnic cleansing, and the deadliest campaigns of terrorism and civilian victimization in civil wars—which are distinct, and unlikely to be fully explained by a single overarching theory. My contention, though, is that the key questions scholars ask about *ideology* are broadly consistent across those phenomena. My focus is squarely on such questions, not the full gamut of all relevant causes of mass killing.[76] Chapter 3 and the book's Conclusion clarify how my account interacts with other explanations of mass killing—in particular those focused on the origins, incentives, and dynamics of various kinds of political crisis.

The cases I study in this book all involve mass killings perpetrated by *states*—domestically and internationally recognized governments and their agencies—albeit often with significant collaboration by non-state actors. While my argument is not limited to state violence, this focus contrasts with much recent research centred on rebel insurgencies, civil war factions (whether rebel or government), or terrorist organizations. That recent trend is understandable since insurgencies and terrorist organizations have become increasingly prevalent in global conflict. But there is still a compelling reason to focus on states: they are, by far, the worst perpetrators of violence against civilians.[77] In focusing on *mass* killings, I also exclude lower-level forms of 'one-sided violence'.[78] This is consequential for theory-building. Lower-level violence against civilians requires fewer perpetrators, carries lower political costs and risks, and might suggest greater efforts at discrimination. Theories that sideline ideology may therefore look more plausible here. Nevertheless, all forms of political violence have a relevant ideological dimension.[79] Hopefully, readers primarily interested in other forms will still find this book relevant.

Of course, whether you think ideology matters depends on what you mean by 'ideology'. This is a problem, because few words have been so varyingly defined by

---

[75] Powell 2011, 90–5; King 2012, 324–5 & 330–1; Straus 2012a; Verdeja 2012, 311–12. Exclusively focusing on genocide can also have problematic *political* consequences; see Straus 2019; Moses 2021.

[76] In this sense I seek to provide a 'focused theory' that specifically explicates ideology's role in such violence; see George and Bennett 2005, 67 & 70.

[77] Davenport 2007, 1 & 12.

[78] Eck and Hultman 2007, 235.

[79] On ideological dimensions of terrorism see, for example, Drake 1998; Asal and Rethemeyer 2008; Stepanova 2008; Chenoweth and Moore 2018, ch.5; Ackerman and Burnham 2019; Holbrook and Horgan 2019. On the ideological dynamics of armed groups see, for example, Bosi and Della Porta 2012; Gutiérrez Sanín and Wood 2014; Costalli and Ruggeri 2017; Schubiger and Zelina 2017; Lefèvre 2021; Parkinson 2021.

scholars.[80] As already suggested, my conception of ideology is a broad one: ideologies are distinctive political worldviews, and therefore ubiquitous and ordinary features of political life. Individuals, groups, and organizations generally require ideologies, both to make sense of their political worlds and to mobilize, coordinate, and sustain collective action. In this usage, to say that violence is 'ideological' is not to necessarily impute especially dogmatic or idealistic motives or justifications to it. It is instead to emphasize that the motives and justifications, *whatever they are*, are vitally embedded in broader distinctive sets of ideas about politics, without which the violence cannot be properly understood or causally explained. This sort of broad conception of ideology is increasingly popular, but some readers may be more familiar with a narrower conception, where ideologies denote tightly consistent belief-systems that provide detailed visions of ideal social order. I explain my rejection of this narrower conception in Chapter 2. But in brief, it is inconsistent with what we now know about those familiar real-world phenomena—like conservatism, environmentalism, neoliberalism, and feminism—that almost everyone agrees are ideologies. These rarely take the form of tightly consistent belief-systems providing detailed blueprints for society, but are looser sets of ideas, values, and narratives, that nevertheless constitute profoundly distinct orientations to politics and society, and often generate distinct political norms and institutions.

This point is about more than mere semantics: it is part of my plea for comparative scholars of political violence to adopt a more sophisticated view of what ideologies are and how they shape politics—one closer to that of many historians, ethnographers, political theorists, and scholars of social movements.[81] Notably, almost identical arguments have also been made in both terrorism studies and research on civil wars, where scholars similarly warn that a preconception of ideology as a rigidly consistent and idealistic belief-system obscures more than it reveals.[82] This argument also highlights the tight interrelationship between ideology and many other important focal points of 'ideational' research on political violence, such as propaganda,[83] discourse/speech,[84] norms,[85] identity,[86] and organizational culture.[87] Ideology does not offer some sort of competing 'alternative' to these concepts in explaining mass killing, but operates through and alongside them. All these concepts are therefore tied together in my analysis of mass killing.

---

[80] McLellan 1995, 1; Gerring 1997.
[81] For leading examples of such scholarship see Skinner 1965; Skinner 1974; Snow and Benford 1988; Boudon 1989; Eagleton 1991; Freeden 1996; Wedeen 1999; Snow 2004; Priestland 2007; Wedeen 2019.
[82] Gutiérrez Sanín and Wood 2014; Holbrook and Horgan 2019.
[83] McKinney 2002; O'Shaughnessy 2004; Timmermann 2005; Yanagizawa-Drott 2014; Stanley 2015.
[84] Jabri 1996; Scutari 2009; Benesch 2012a; Benesch 2012b; Waldron 2012.
[85] Fujii 2004; Morrow 2015; Morrow 2020.
[86] Fearon and Laitin 2000; Suny 2004; Gartzke and Gleditsch 2006; Volkan 2006.
[87] Johnston 1995; Katzenstein 1996a; Hull 2003; Long 2016.

Ideology, Perpetrator Coalitions, and Political Crisis

Mass killings are complex campaigns of collective action that cannot be reduced to one set of characters.[88] They are generally initiated by *political elites*, implemented by various kinds of *rank-and-file subordinates*, and tacitly or actively supported by *broader segments of the societies or communities* in which mass killing occurs.[89] Individuals in all three of these groups, moreover, are guided by a diverse range of motives and considerations, not a single shared 'perpetrator mindset'. As Thomas Kühne writes of the Holocaust and Stalin's Great Terror:

> Not all ... embraced mass murder unanimously. Carrying out mass murder meant integrating different individuals and social entities, varying degrees of willingness to participate, different perpetrators, collaborators and accomplices, sadists, fanatics, cold-blooded killers, occasional doubters, more serious dissenters, and unwilling yet submissive collaborators.[90]

Mass killings are best understood, in other words, as a product of what Kjell Anderson terms 'perpetrator coalitions'.[91] Explaining mass killing is consequently not a matter of identifying 'the reason' for civilians being killed, but of identifying why such internally diverse perpetrator coalitions come into being, and how they are held together and organized so as to carry out systematic violence against civilians.

Here, I share the view of most contemporary scholars that perpetrator coalitions generally emerge in response to certain kinds of political crisis. But crises will not produce mass killing in the absence of some kind of hardline ideology. This is because of two key properties of mass killing: first, their highly destructive, uncertain, and risky consequences (what I refer to as their *strategic indeterminacy*), and second, their exceptionally brutal and troubling moral character (what I call their *normative extremity*). Since mass killings are strategically indeterminate and normatively extreme, it is never obvious that regimes or groups will resort to them, even in dire emergencies. For sure, crises may open the opportunity for mass killing, and could make it look potentially useful. We can often, as such, tell a plausible story as to why perpetrators might 'rationally' target civilians in such a crisis. But there is almost always *at least as plausible a story* as to why perpetrators should have rationally avoided such violence. It is the particular way crises are ideologically interpreted and mobilized by hardliners—to generate a justificatory narrative for mass killing—that is crucial.[92]

---

[88] Harff 2003; Owens, Su, and Snow 2013; Williams 2021.
[89] Mann 2005, 8–9.
[90] Kühne 2012, 141.
[91] Anderson 2017a, 99–101.
[92] See also Mann 2005, 7–8. I therefore present ideology in a way consistent with 'INUS' ('Insufficient but Necessary causes in Unnecessary but Sufficient causal sets') or 'NESS' ('Necessary Element of a

I therefore show that hardline ideologies and justificatory narratives are not mere 'symptoms' of certain 'deeper causes' of mass killing, easily generated and manipulated to rationalize whatever course of action political elites prefer. Nor, however, do ideologies generally provide longstanding preformed plans for extermination for which political crises are no more than a pretext. Instead, ideologies radicalize (or deradicalize) over time according to *both* broader material and social conditions and the existing character of an ideology itself. Extremist ideologies often, for example, flourish in times of economic depression or social conflict, and the onset of violence itself typically radicalizes social norms and intergroup attitudes. But the existing strength of sympathy for, or opposition to, hardline ideas is also a critical catalyst for or constraint on radicalization. Frequently, indeed, ideologies shape the onsets of crises just as much as crises shape prevailing ideologies. In my account of mass killing, this escalatory interaction between crisis and ideology therefore takes centre stage. Ideologies matter because they determine whether a justificatory narrative for mass killing, capable of binding together and sustaining a perpetrator coalition, emerges in serious strength in times of crisis.

Within perpetrator coalitions, both 'elites'—the political leaders and high-ranking officials who generally initiate and organize mass killing—and 'masses'—the rank-and-file subordinates and broader communities who implement or support the violence, matter. This book is about both. I argue that ideologies, by motivating and legitimating decisions by individuals across these groups, are crucial to the creation, maintenance, and activities of the perpetrator coalitions needed for mass killing to occur. But there are debates in contemporary scholarship between comparatively 'top-down' or *elitist* theories of mass killing, which focus on decisions by senior political decision-makers,[93] and more 'bottom-up' or *societal* theories, which emphasize either strong public support for mass killing[94] or diffuse patterns of local violence which become interconnected into joint campaigns.[95]

In truth, mass killings vary in this respect,[96] but I generally lean towards elitist accounts.[97] More purely bottom-up forms of violence against civilians, such as lynchings and ethnic riots, are possible. But they rarely escalate to mass killing

---

Sufficient Set') causation. I.e., mass killings may occur via multiple different sets of causal factors, with ideological justifications of the violence a necessary but insufficient component of each of those causal sets. See Mackie 1965; Wright 2013.

[93] Valentino 2004; Straus 2015a; Bulutgil 2017; Maat 2020.

[94] Goldhagen 1996; Su 2011. Goldhagen places somewhat more emphasis on elite leadership in Goldhagen 2010. Su's landmark study of China's Cultural Revolution strongly stresses the role of 'willing communities', but also emphasizes how 'the perpetrators invariably were organised by local authorities' (Su 2011, 65), so his account mixes elitist and societal elements.

[95] Gerlach 2010; Karstedt 2012.

[96] Anderson thus distinguishes, for example, between 'specialized' and 'participatory' genocides; see Anderson 2017a, 46.

[97] For similar perspectives see Valentino 2004; Straus 2015a; McLoughlin 2020.

without elite organization.[98] Even when there is public pressure for discriminatory policies against certain groups, elites generally possess significant latitude in deciding how to satisfy such popular pressures, and capacities to dampen, mobilize, or funnel them using state authority and propaganda.[99] Consequently, many mass killings occur despite little initial public pressure, but few cases appear to involve highly reluctant leaders pushed into mass killings by a clamouring public. I therefore place primary emphasis on ideology's role in shaping elite perceptions and decision-making.

Yet there are dangers in overly elitist accounts.[100] Waller contends, for example, that 'political, social or religious groups wanting to commit mass murder do. Though there may be other obstacles, they are never hindered by a lack of willing executioners.'[101] This is somewhat misleading. While states are always able to mobilize *some* willing executioners, how effectively and extensively they can do so varies. Security or military forces may refuse to enact violence against civilians, or even turn against elites: as in Iran in 1979 or Egypt in 2011.[102] When mass killing is perpetrated, moreover, bottom-up escalatory dynamics often shape violence independent of elite intentions,[103] and in a few cases, elites largely just authorize or tolerate mass killing, with enthusiastic rank-and-file groups—such as the mercenary gangs who massacred indigenous populations in nineteenth-century California—taking centre stage.[104] Even in more centralized mass killings, elites can rarely *coercively micromanage* rank-and-file perpetrators or broader mass publics, so non-elite individuals retain significant agency.[105] Consequently, hardline justificatory narratives are important, not just in shaping elite decision-making, but also in mobilizing and organizing rank-and-file perpetrators of the violence, and in sustaining vital support and legitimacy for mass killing amongst broader publics.[106]

In emphasizing that these key hardline justificatory narratives revolve around familiar but radicalized strategic and moral ideas about security politics, I am revising rather than entirely rejecting traditional-ideological perspectives. Most

---

[98] Even in these forms of violence, however, elites tend to play critical roles. See, for example, Brass 2005; Dumitru and Johnson 2011, 9–11.
[99] Gagnon 2004; Valentino 2004, ch.2. See also Zaller 1992.
[100] See also Mann 2005, 8–9.
[101] Waller 2007, 15.
[102] Weiss 2014, 2–3.
[103] Gerlach 2010; Karstedt, Nyseth Brehm, and Frizzell 2021, 10.6–8.
[104] Madley 2004.
[105] Aerial bombing campaigns are something of an exception, as discussed in Chapter 6.
[106] In many cases, the ideological orientation of *external actors* may also matter. The United States heavily facilitated mass killing in Guatemala and Indonesia, for example, as did the French in the Rwandan Genocide, due in part to ideology; see Wallis 2006/2014; Grandin 2011; Robinson 2018. But this is not consistently significant across mass killings. External actors generally matter more when perpetrators depend on external patron–client support relationships, are vulnerable to intervention, or when international norms prohibiting mass killing are relatively strong. See also Welsh 2010; Bellamy 2012b; Dill 2014; Salehyan, Siroky, and Wood 2014; Stanton 2016.

traditional-ideological perspectives also emphasize the framing of victims as threatening, guilty, and 'other'. Ideological conceptions of security politics are always linked, moreover, to broader political goals and visions in some way, because *what one seeks to secure* in security politics depends on one's account of the political community and its ultimate purposes. Nevertheless, I contend that traditional-ideological perspectives have wrongly emphasized the radicality of perpetrators' ultimate goals, when what really matters is the radicality of how perpetrators understand the pursuit of that most conventional political goal: the securing of a given political order.

I argue, indeed, that existing scholarship often overfocuses on the more 'emotional' aspects of ideologies: utopian dreams, dogmatic absolutism, intense hatreds, and so on. Emotional dynamics are certainly important. But this book places equal emphasis on ideologies' 'epistemic' aspects: the way they shape purportedly factual *narratives about the world*.[107] Such ideological narratives are politically crucial in complex societies, because our ability to ground our political beliefs in direct experience is highly constrained. Given limited time and expertise, we rely on both our ideological preconceptions and the claims of prominent ideological producers such as state authorities, political parties, and the media. This is especially true in security politics, where most citizens can make little use of their personal experiences to assess the truth or falsity of claims about 'national security'. But more broadly, conservatives, liberals, communists, and fascists are divided, not simply by different aims and values, but by contrasting narratives about their political worlds. In this sense, ideologies provide *imagined realities*—visions of politics rooted in existing beliefs, indirect testimony, and story-telling, more than direct experience or hard evidence.[108] In this respect, my analysis dovetails with the recent growth of research on 'fake news', conspiracy theories, and political misinformation.[109] Since false beliefs are pervasive, persistent, and powerful even in the freest and most prosperous societies, it is hardly surprising that they can prove crucial in the crisis-ridden contexts of mass killing.

Such ideological narratives remain, of course, emotionally and morally charged. This is crucial, since the psychological sciences have now generated a wide degree of consensus that emotions are *essential* foundations for collective political action,[110] including violence.[111] But our emotions and moral judgements are deeply intertwined with our narratives about reality—as Jennifer Hochschild puts it:

---

[107] See also Holbrook and Horgan 2019. For broader work on political narratives see Patterson and Monroe 1998; Hammack 2008; Haidt, Graham, and Joseph 2009; Krebs 2015.
[108] I intentionally allude to Anderson 1983/2006, 6.
[109] It also aligns with those who associate ideology with the employment of power/knowledge; see, variously, Simonds 1989; Howarth, Norval, and Stavrakakis 2000; Fricker 2009.
[110] McDermott 2004; Mercer 2010; McDoom 2012; Ross 2014; Hall and Ross 2015.
[111] Chirot and McCauley 2006; Dutton 2007; Collins 2008; Grossman 2009; Klusemann 2010; Costalli and Ruggeri 2017.

'Where you stand depends on what you see.'[112] Moreover, since people's underlying values change slowly, yet their perceptions of the world can change quickly, rapid ideological radicalization towards violence is more likely to be a product of changing narratives than a wholesale moral reorientation. Those who think mass killings are justified and those who think them unconscionable are divided, I argue, as much by fundamentally different imagined realities as by contrasting values.[113] 'When viewed from divergent perspectives,' eminent psychologist Albert Bandura reminds us, 'the same violent acts are different things to different people.'[114]

Take, for example, recent mass violence against the Rohingya Muslims of Myanmar's Rakhine region, which peaked in large-scale expulsions and numerous killings in 2017.[115] Outside of Myanmar, this was widely condemned as the indiscriminate ethnic cleansing of an unarmed civilian population long victimized by Myanmar's government. But supporters of the violence from within Myanmar—including Buddhist religious leaders—operate inside an entirely different ideological narrative of reality.[116] The Rohingya—whom are typically referred to simply as 'Muslims' or 'Bengalis'—'stole our land, our food and our water', stated one Buddhist abbot; a member of Myanmar's Parliament asserted that 'all the Bengalis learn in their religious schools is to brutally kill and attack'; while a local administrator of a 'Muslim-free' village explained that Muslims 'are not welcome here because they are violent and they multiply like crazy'.[117] A mother working for the *Patriotic Association of Myanmar* likewise argued that:

> [Muslims] are swallowing our religion ... Their religion is terrorism ... They have been taught this since they were children, so it's very terrifying. We say, 'don't kill' ... They say, 'kill, if you kill you will be blessed' ... Now, in the news, we see about their Jihad in other countries, cutting off people's heads ... I don't want to see our Buddhists suffer like that. That's why I want to show people the horror of their religion. I want everyone to know.[118]

Most individuals expressing such sentiments had little familiarity with actual Rohingya. But *the ideological portrayal of Rohingya*, rooted in years of rumour, story-telling, and, increasingly, fake news on social media, drove support for their violent persecution and forced expulsion.[119] The aforementioned administrator

---

[112] Hochschild 2001. See also Zaller 1992, 24.
[113] See also Crelinsten 2003.
[114] Bandura 1999, 195.
[115] On this case in general see Lee 2021.
[116] Schissler, Walton, and Thi 2015, 10.
[117] Beech 2017.
[118] Schissler, Walton, and Thi 2015, 9–10.
[119] See also Mozur 2018.

acknowledged that he had never met a Muslim, but observed that 'I have to thank Facebook because it is giving me the true information in Myanmar'.[120] Another interviewee commented that: 'According to [what I hear from] other people, I am worried that ISIS will affect us, and in our country we have many Muslims.' Asked when she started feeling scared of Muslims, she answered: 'It happened after seeing that news and the Rakhine problem. Since then the news always pops up about it.'[121]

Again, some scholars tend towards the view that such ideological narratives are merely post-hoc rationalizations for self-interest, longstanding hatreds, or underlying value orientations. But this interpretation is generally implausible. For most of the interviewees quoted above, little personal stake in the Rakhine region existed, and threatening perceptions of Muslims appeared to emerge only in *response* to propaganda and events.[122] In most cases of mass killing, indeed, evidence of longstanding hatreds or purely self-interested motives for violence is surprisingly scarce. The post-hoc rationalization interpretation fails to appreciate how deeply individuals rely on socially disseminated narratives to interpret the world around them—and how easily baseless claims about matters of fact can therefore come to look plausible within the right ideological context.

In focusing on ideologically radicalized security politics, I also oppose the tendency of many traditional-ideological perspectives to implicitly 'other' mass killing by presenting it as essentially a pathology of manifestly 'totalitarian', 'authoritarian', or 'evil' ideologies.[123] A broad range of regimes and groups, including liberal ones, have engaged in mass killing. The ideological detail of different cases varies in important ways—and a central argument of this book is that different mass killings take radically different forms due to the different ideological contexts in which they occur. Yet the most basic ideological processes through which mass killings are justified are largely consistent across cases. As Neil Mitchell observes: 'Human beings are uninventive when it comes to reasons for atrocity.'[124] This is meant *neither* to suggest that all mass killings are fundamentally the same, nor to imply that they are morally equivalent. While my research has convinced me, for example, that the Allied area bombing of civilians in World War II was brutal, ineffective, and unjustified, it was obviously not morally akin to the Holocaust. But there are a range of reasons for agreeing with Alex Bellamy's contention that 'whilst the precise contours of justification shift from case to case, it is important to recognise the family resemblances between them'.[125]

---

[120] Beech 2017.
[121] Schissler, Walton, and Thi 2015, 12.
[122] Ibid. 11, 15–17, & 21–2.
[123] Somewhat contra Kuper 1981, ch.5; Fein 1990, ch.4; Rummel 1994; Kressel 2002; Midlarsky 2011; Richter, Markus, and Tait 2018. See also Powell 2011, 95–7.
[124] Mitchell 2004, 53.
[125] Bellamy 2012a, 180.

On Method

A detailed discussion of my methodological approach can be found in the Methodological Appendix at the end of the book, but a few points should be clarified from the outset. In the chapters that follow, I address a mixture of 'what?', 'why?', and 'how?' questions,[126] and seek to make both broad generalizations about ideology's role across mass killings and context-specific claims about the ideological dynamics of individual cases.[127] My principal aim is to advance a causal argument. I claim that ideologies are crucial to explaining *why* mass killings occur, and I show *how* particular hardline ideas shape the initiation and implementation of violence. In contrast to some books in political science, I do not present my inquiry as a kind of 'experiment' for testing preformed hypotheses. Such an approach is not the only valid method of causal inquiry in social science and is often a rather inaccurate presentation of how research actually proceeds.[128] Much social science is more analogous to a detective unravelling a crime than a natural scientist working in a laboratory—in that it uses established foundational knowledge to interrogate available evidence concerning particular events and reach the most plausible causal conclusions. I embrace this approach, drawing on empirical research from across the humanities and social sciences to examine specific mass killings and make the best causal inferences about ideology's role.[129]

I provide a broad range of evidence for my arguments across the first, theoretical half of the book, but then delve in much more detail into the ideological dynamics of mass killing through my four historical case studies. The available evidence, I argue, counts against traditional-ideological, rationalist-sceptical, and situationist-sceptical explanations of mass killing, and supports the neo-ideological synthesis I propose. Across cases of mass killing, key ideological justifications pre-date the violence, are closely linked to patterns of violence which cannot be explained if ideology is ignored, and involve ideas recognized in psychological science as capable of increasing support for violence. Even in the most obviously 'strategic' cases, such as the Allied area bombing of Germany and Japan or the Guatemalan civil war, killing civilians in their hundreds and thousands was hugely disproportionate to the actual benefits, if any, that such violence yielded, and was not an obvious logical response to the pressures of war. In both these cases, mass killing *was* a strategy for military victory, but one that vitally depended on distinctive

---

[126] In social scientific jargon, I am interested in both causal and constitutive, and explanatory and interpretive, forms of inquiry. But these are more closely connected than often assumed; see Ylikoski 2013; Elster 2015, ch.3; Jackson 2016; Norman 2021. I reject the view that such forms of inquiry are ultimately incompatible, but see Hollis and Smith 1990; Bevir and Blakely 2018.

[127] Tilly and Goodin 2006.

[128] King, Keohane, and Verba 1994, 7, fn.1; Yom 2015; Norman 2021.

[129] As discussed in the Methodological Appendix, this involves a form of 'iterative induction' revolving around comparative-historical analysis; see Skocpol and Somers 1980; Mahoney and Terrie 2008; Mahoney 2015; Yom 2015.

ideological understandings of warfare and crisis which significantly preceded the violence. Yet the primary ideological justifications for mass killings consistently revolve around conventional arguments about security, punishment, necessity, and valour. Even the most 'revolutionary' violence of the Stalinist terror was not part of a longstanding 'utopian' programme to transform society, but fundamentally an effort to secure the Soviet state in response to perceived threats and crisis.

My methodological approach does not generate some sort of knock-down 'proof' of ideology's impact in the way that one might prove the role of haemoglobin in blood or the relationship between a planet's mass and its gravitational pull. But such strong proofs are relatively rare in social science. My argument is that the available empirical evidence renders a neo-ideological perspective the *most plausible* characterization of ideology's role in mass killing. It is better supported than the sceptic's dismissal of ideology as playing a marginal role, or the traditional-ideological focus on deep ideological commitments to extraordinary goals or values.[130]

But this book is not solely concerned with causal claims. The very mindset of perpetrators of mass killing, and the meaning of the violence they implement, is mysterious. *How* could these people perpetrate? *What* were they thinking? *To what extent* did they support the violence? To demystify perpetrators, and make their violence intelligible, we must trace their narratives, assumptions, and claims—identifying the reasons and sentiments through which they appear to have understood their own actions and showing how such ideas can gain currency in particular ideological contexts. For this purpose, I draw on interpretive techniques of intellectual history, discourse analysis, and political theory. Such inquiry also exposes the relative ease with which commonplace justifications of violence can be twisted to support horrific atrocities. This imparts an ethical dimension to my contribution, since this justificatory capacity carries implications for debates in political theory over the moral regulation of political violence and war. I return to these ethical implications in the book's Conclusion.

My entire argument depends on examining *ideas*—the building blocks of ideologies. Some scholars worry about this, objecting that we cannot rigorously study ideas because they operate in the human mind and are therefore not directly observable. As explained in my Methodological Appendix, this objection is misplaced. Problems of direct observation are common in science, and a matter of degree. Like many scholars, I believe that analysis of discourse, the use of psychological science, the examination of observable behaviour, and close attention to historical context can collectively allow us to make inferences—albeit somewhat tentative ones—about the role ideas play in human action.[131] Moreover, we

---

[130] See also Lipton 2004; Douven 2011.
[131] Others worry that ideational explanation is 'tautological'—but this critique only holds when the evidence used to determine ideas is the very behaviour those ideas are then used to explain. Sophisticated ideational research avoids this error.

generally *have* to make such inferences.[132] Refusing to study ideas rarely results in scholars neutrally reserving judgement about their impact. Instead, scholars either implicitly treat ideas as unimportant, or make tacit assumptions about the ideas that guide human action without grounding such assumptions in evidence. Neither approach is justifiable. Instead, we should use actual empirical research to make the best inferences we can about the ideas and ideologies that appear to influence those we study.

1.4 Plan of the Book

The rest of the book is organized into two halves. The first develops my *retheorization* of ideology's role in mass killing. Chapter 2 (Clarifying Ideology) begins this task by defending a broad conceptualization of ideology and theorizing the multiple ways in which ideologies can influence political behaviour. Chapter 3 (How Does Ideology Explain Mass Killing?) presents the core argument of the book in more detail, showing why accounts that ignore ideology fail to explain mass killings, and detailing how 'hardline ideologies' generate and hold together perpetrating coalitions in times of political crisis. Chapter 4 (The Hardline Justification of Mass Killing) then delves deeper into the actual character of those hardline ideologies, the key justificatory narratives through which they promote mass killing in times of crisis, and the basis for thinking that such narratives have genuine 'causal power' to encourage violence.

The second half of the book provides deeper *empirical* support for my account through my four case studies. I intentionally focus on four quite different cases of mass killing, which should prove collectively difficult for an ideology-centred account to coherently explain. Chapter 5 examines Stalinist Repression, which represents what might be thought of as a classic 'ideological' mass killing, although one that has received less attention in comparative research than more canonical cases like the Holocaust or Armenian Genocide. While this represents a relatively easy case for my argument that ideology is crucial, it is a tougher case for my claims that ideology's most important role in mass killing revolves around security politics rather than revolutionary goals. By contrast, I examine the next two cases, Allied Area Bombing in World War II in Chapter 6 and the Guatemalan Civil War in Chapter 7, because they should be much harder cases for arguments asserting ideology's importance. Again, these are classic 'strategic' mass killings, of the kind that many scholars suggest can be explained without reference to ideology, but I show that ideology remained essential in both. Chapter 8 examines the Rwandan Genocide, which lies somewhere between the other cases. Most scholars recognize the strength of racist ideology in Rwanda, but several downplay the importance of

---

[132] See also Mercer 2005.

that ideology in explaining why and how the genocide unfolded. I again show that ideology's role in Rwanda was nuanced, but crucial.

There are important limits to what I offer over these chapters. This book is not an exercise in new primary research on particular cases—involving fieldwork and new data collection—but an attempt to use the best existing scholarship on the cases I examine to advance debates over the role of ideology in mass killing. Since I am focused squarely on *ideology's role*, a range of further important dynamics necessarily get limited attention, although I try to highlight them where appropriate. Several issues—such as the role of ideology in cases where mass killing does *not* occur, or the deeper societal roots of hardline ideological radicalization—I do address, but will need to return to in future research to elaborate in full.[133] The book also sides with scholars who favour complex, context-sensitive, historically detailed, and interpretively rich theories of political violence, as opposed to those more focused on building relatively simple and general law-like predictive models.[134] Ultimately, different studies contribute to our ability to make sense of political violence in different ways. But a focused and comparative study aimed at advancing our understanding of ideology's role in mass killing is, I believe, overdue.

---

[133] I discuss 'negative cases' where mass killings do not occur in Chapters 3 and 4. Some readers may worry that without a dedicated case study of such a 'negative case', I problematically 'select on the dependent variable', but this complaint misunderstands the way I am making causal inferences in this book, as explained in the Methodological Appendix. For four excellent comparative studies incorporating negative cases see Kaufman 2015; Straus 2015a; Bulutgil 2017; Hiebert 2017.

[134] Pierson 2004; Tilly and Goodin 2006; Owens, Su, and Snow 2013; Kaufman 2015, 6–11; Williams and Pfeiffer 2017; Williams 2021.

2
Clarifying Ideology

What exactly does it mean to argue that ideology is, or is not, an important driver of mass killing? For all the debate over ideology's role, most existing research does not answer this question explicitly. Yet there is a way of thinking and talking about ideology that, at least implicitly, many scholars employ. Although more of a rough set of analytical tendencies than a firmly specified account of ideology, for simplicity of discussion I'm going to label this way of thinking and talking *the true believer model* of ideology. It revolves around three broad assumptions:

1. that ideologies are *rigidly specified belief-systems* that provide detailed prescriptions for political action;
2. that those belief-systems revolve around *special ideological goals*—typically, to implement a particular 'vision of the ideal society'—that contrast with more 'pragmatic' concerns with, for example, security, power, and material self-interest;
3. that such ideologies influence behaviour to the degree that individuals are *motivated by strong convictions* about the need to achieve these special ideological goals.

These assumptions imply that ideology's role in mass killing centres around the activity of those who possess such deep convictions in the special goals of elaborated belief-systems—i.e. an ideology's true believers. Such 'ideological' perpetrators are presumed to stand in stark contrast to 'unideological' perpetrators motivated by self-interest, opportunism, or social pressure. The question of ideology's importance thus largely becomes a question of how numerous and influential the true believers are.

Traditional-ideological perspectives on mass killing often tacitly employ the true believer model: suggesting that ideology is important because many perpetrators—whether elite ideologues or their indoctrinated followers—do indeed fit this picture. In recent scholarship, however, the true believer model has more often encouraged scepticism about ideology's importance. Many scholars downplay ideology precisely because few perpetrators appear to match stereotypes of the 'raving ideologue'[1] motivated by 'deeply held ideological beliefs',[2] or 'fanaticism'.[3] Such sceptics suggest that *by contrast with ideology*, 'pragmatic', 'strategic', 'economic', 'material', 'bureaucratic', 'local', 'communal', or a host of other kinds of concern are the real motor of violent action, and that ideological justifications are likely to offer little more than a 'pretext'[4] or 'rationalization'[5] for violence.[6]

Against both perspectives, I argue that all three assumptions of the true believer model are faulty. It is true that only a minority of perpetrators of mass killing match the stereotype of ideological zealots employing violence to bring about a political utopia. But this is now an established finding, and if the debate over ideology is reduced to it, then the debate is dead and theoretically uninteresting. More substantively, the fact that most perpetrators are not ideological true believers does not mean that the ideological content of, for example, 'communism' or 'Nazism' or 'ethnonationalism' cannot profoundly shape the violence they perpetrate. Indeed, *despite* the relative rarity of true believers driven by special ideological goals, mass killings appear to be influenced in important ways by particular ideological claims and processes. This should be unsurprising, since there is now a widespread recognition in social science that ideas do not only matter when people deeply believe in them. Such insights need to be applied to ideology.

This chapter proceeds in two parts. Section 2.1 challenges the true believer model's narrow *conceptualization* of ideologies as rigid belief-systems revolving around special goals. Instead, ideologies are better understood more broadly, as distinctive interpretive and evaluative frameworks about politics which vary—in patterned ways—across individuals and groups. Section 2.2 then turns to a more extensive retheorization of ideology's *influence over behaviour*. Rather than operating only through strong convictions, ideologies influence behaviour both through sincere acceptance of their ideas *and* through the social pressures and instrumental incentives created when ideologies become embedded in norms, discourses, and institutions. Taken together, these forms of influence allow ideologies to act *infrastructurally*, binding diverse individuals into collective political action, including campaigns of collective violence.

This understanding challenges the dominant way of thinking about ideology in *some* branches of political science. But it is closer to the prevalent way of thinking about ideology in most other research fields, including history, political psychology, political theory, social movement research, and discourse analysis,[7] as

---

[1] Waller 2007, 102.
[2] Straus 2006, 96.
[3] Kalyvas 2006, 64–6.
[4] Fujii 2008, 570.
[5] Waller 2007, 49. See also ibid. 53.
[6] See also Kalyvas 1999, 247–51; Mueller 2000; Kalyvas 2006, 44–8, 64–6, & 130; Gerlach 2010, 5–6 & 273.
[7] Lane 1962; Geertz 1964; Freeden 1996; van Dijk 1998; Oliver and Johnston 2000; Norval 2000; Zald 2000; Jost and Major 2001; Snow 2004; Freeden 2005; Jost 2006; Jost, Federico, and Napier 2009; Freeden, Tower Sargent, and Stears 2013; Knott and Lee 2020.

well as some specialist work on terrorism and armed conflict.[8] It also incorporates key insights from scholarship on both Nazi Germany and the Soviet Union, where historians have long recognized that the influence of Nazi and communist ideology was immense, yet not reducible to mass convictions in rigid belief-systems.[9] Even in these fields, however, the broader and more multifaceted understanding of ideologies is often left implicit or underspecified. By integrating and systematizing such a perspective, I diffuse several sources of scepticism over ideology's importance and provide key foundations for the neo-ideological account of ideology's role in mass killings provided in Chapters 3 and 4.

2.1 What Are Ideologies?

Narrow Conceptualizations

The 'true believer model' characterizes ideologies *narrowly*, as rigidly consistent and explicitly elaborated belief-systems which provide detailed prescriptions or 'blueprints' for action, typically rooted in an ideal vision of society. This conceptualization, rooted in early post-World War II political science, presents ideologies as a special type of systematic political worldview that only certain doctrinally self-conscious people—typically elites—possess.[10] Ideologies, in this narrow sense, are assumed to contrast with more 'pragmatic' or 'strategic' political stances, centred around concerns with power, security, or material self-interest rather than 'visions of the ideal society'.[11] As noted above, this dichotomy is often central to scepticism about ideology's importance. It is *because* many scholars see pragmatic/strategic/material factors as the 'real motive' for violence that they downplay ideology's role.

Though common, this narrow conceptualization of ideology suffers from major problems. Most fundamentally, it is irreconcilable with actual characteristics of those systems of ideas that everyone agrees are ideologies, such as liberalism, conservatism, socialism, and libertarianism. We rarely assume that such ideologies and those influenced by them must possess some well-defined blueprint of the 'ideal society', and most self-identified liberals, conservatives, socialists, or libertarians clearly do not. Even classically 'utopian' ideologies such as communism and fascism are infamously hazy over the specific details of their utopias or how to get there.[12] What binds followers of such ideologies together are not rigidly elaborated

---

[8] Gutiérrez Sanín and Wood 2014; Ackerman and Burnham 2019; Holbrook and Horgan 2019.
[9] See, for example, Schull 1992; Kershaw 1993; Kotkin 1995; Robinson 1995; Casier 1999; Hedin 2004; Priestland 2007, 16–20; Roseman 2007.
[10] Converse 1964; Glazer and Grofman 1989; Kinder and Kalmoe 2017.
[11] du Preez 1994, 65–78; Valentino 2004, 72.
[12] Kershaw 1993; Kotkin 1995, 225–30; Priestland 2007, ch.1; Roseman 2007, 95–6.

belief-systems, but looser bundles of values, narratives, goals, and frameworks that generate distinctive perceptions and evaluations of politics.[13] These looser bundles of ideas still generate profound and patterned differences in people's durable tendencies to respond to political events, policies, or ideas in different ways—what Stuart Kaufman calls their 'symbolic predispositions'.[14] So ideological contrasts between liberals, conservatives, socialists, and libertarians are real and consequential. They are just not sustained by highly systematic belief-systems or detailed blueprints for action.

In addition, while it is common for scholars to oppose 'ideological' aims to 'pragmatic' or 'strategic' concerns, this dichotomy rests on deeply implausible assumptions. Pragmatic or strategic action is only intrinsically 'unideological' if ideologies are presumed, *by definition*, to be rigidly dogmatic, idealistic, and wholly silent on strategy—as though 'ideological' decision-making involves the mechanical implementation of abstract ideals with a total disregard of the practical consequences. Few, *if any*, real-world ideologies look like this caricature. Even famously doctrinaire ideologues such as Vladimir Lenin, Mao Zedong, and Osama bin Laden devoted tremendous intellectual attention to practical strategy in their writings. Such strategic consideration was not a 'sacrifice' of their ideological precepts, but *a core part of their ideological thought*. Most ideologies, indeed, contain specific ideas about power, security, and material interests and the best strategies for achieving them. American liberals and conservatives differ in their views of national security and US interests, for example, and disagree over what policies are most appropriate for pursuing them.[15] Marxist-Leninism, Irish revolutionary republicanism, or militant jihadism all likewise advance particular ideological understandings of strategy, security, and self-interest rather than blindly disregarding such considerations in favour of the dogmatic imposition of their ultimate ideological goals.[16] Indeed, this interpenetration of ideology and pragmatic strategy is inevitable, since what *seems pragmatically sensible* depends on your assumptions about the relative importance of different political goals and the efficacy of different policies—assumptions that vary across ideological worldviews.[17] 'There are', as Neta Crawford reminds us, 'few "real" material interests that cannot be viewed in more than one way.'[18]

Such dubious dichotomies have troubled many influential typologies proposed by scholars for classifying different sorts of cases or perpetrators of mass killing. There may well, for example, be a distinction between what Michael Mann calls

---

[13] Jost 2006, 653–4; Snow and Byrd 2007, 132; Thorisdottir, Jost, and Kay 2009, 4–7; Holbrook and Horgan 2019.
[14] Kaufman 2015, 11–15.
[15] Gries 2014; Mirilovic and Kim 2017.
[16] See, for example, Gerges 2009; Whiting 2012; Balcells and Kalyvas 2015; Byman 2016.
[17] For studies emphasizing this interpenetration see Owen 1997; Gause 2003; Haas 2005; Haas 2014; Straus 2015s.
[18] Crawford 2002, 79.

'ideological', 'bigoted', 'disciplined', 'comradely', and 'bureaucratic' killers.[19] But can't bigotry and comradeship be ideologically rooted? Can't bureaucrats be influenced by ideologies? The same could be said of Barbara Harff's separation of 'ideological' and 'retributive' genocides or Frank Chalk and Kurt Jonassohn's distinction between genocides that 'implement ... an ideology' and those that 'eliminate a real or potential threat ... spread terror among real or potential enemies ... [or] acquire economic wealth'.[20] Retribution, elimination of threats and enemies, and the pursuit of economic wealth *can all be deeply ideological activities.* The United States' massacres of Native Americans, Belgium's murderous exploitation of the Congo, and Germany's annihilation of the Herero people of South West Africa were, for example, variously motivated by economic interests and the perceived need to punish dangerous 'rebels'. But they were also inextricably bound up with European colonialist ideology and its denigration of indigenous populations as inferior beings deprived of moral rights.[21] Since such ideological legitimations were inapplicable to white Christian Europeans, the same states with the same interests deployed quite different policies when pursuing material gains and security in Europe. By associating 'ideology' only with the imposition of 'special ideological goals', the true believer model obscures such links, painting an unhelpfully compartmentalized picture of ideology's potential relevance to violence.[22]

Broad Conceptualizations

In large part due to these problems, many scholars now embrace a broad conceptualization of ideology as denoting the distinctive political worldviews of particular individuals, groups, and organizations, *whatever form these take.*[23] The leading theorist of ideology Michael Freeden, for example, describes ideologies as:

> those systems of political thinking, loose or rigid, deliberate or unintended, through which individuals and groups construct an understanding of the political world they, or those who preoccupy their thoughts, inhabit, and then act on that understanding.[24]

---

[19] Mann 2005, 27–9.
[20] Chalk and Jonassohn 1990, 29; Harff 2003, 61. For further examples see Fein 1990; du Preez 1994, 66–78; Smeulers 2008.
[21] Bellamy 2012b, 81–95. See also Ndlovu-Gatsheni 2013, ch.5; Madley 2016.
[22] Alvarez 2008, 215 & 220. See also Monroe 2011, 12; Powell 2011, 92–3.
[23] There is a third way of understanding ideology, namely through pejorative conceptualizations in which ideology, by definition, legitimates dominant interests, distorts understanding of reality, or is rooted in non-epistemic motives; see Larrain 1979; Thompson 1984; McLellan 1995. This is less prominent in research on mass killing, but I eschew it because, contra the assertions of its advocates (e.g. Larrain 1979, 77; McLellan 1995, 23 & 82; Rosen 2000, 395), there is simply no need to *define* ideology this way in order to critically analyse how ideologies do in fact legitimate dominant interests, distort understanding, or reflect non-epistemic motives; see van Dijk 1998, 11.
[24] Freeden 1996, 3.

Rather than being limited to true believers crusading for their ideal society, ideologies in this broad sense are fundamentally *ordinary* and, indeed, *ubiquitous*.[25] As Francisco Gutierrez Sanín and Elisabeth Jean Wood argue: 'all armed groups engaged in *political* violence ... do so on the basis of an ideology, that is, a set of ideas that include preferences (possibly including means toward realizing those preferences) and beliefs'.[26] Alex Alvarez similarly affirms that 'all genocides have an ideological component that is integral to enabling and facilitating the perpetration of this particular form of group violence'[27] since 'all communal life is, to some extent, ideological'.[28] Individuals need such ideologies to provide frameworks for understanding a highly complex political world, and as a source of meaning, direction, self-esteem, and coordination with like-minded allies in operating within that world.[29]

This is a better way of thinking about ideology. It does not conceptualize ideology in a way that is flatly inconsistent with observable features of those worldviews that almost everyone would call ideologies. It does not encode the particular empirical assumptions and parochial concerns of Anglo-American political science after World War II into our very definition of what ideologies are. It does not assume that strategic calculation or material interests are somehow self-evident or deny that they are contested between different ideological orientations. Importantly, it also renders the concept of ideology much more consistent with contemporary psychological science. If you assume that ideologies only denote rigidly consistent and well-elaborated belief-systems, this effectively *guarantees* that few people will be deemed ideological, since human beings rarely think this way.[30] On the contrary, psychologists now heavily emphasize the intuitive, implicit, and inchoate foundations of most ideological views.[31] Of course, if individuals' political attitudes were utterly *chaotic*, it would make little sense to speak of them having a meaningful political worldview at all.[32] But this does not mean that ideologies must adhere to high standards of intellectual sophistication, logical consistency, or self-conscious belief to be identifiable or to influence political behaviour. Most of the time, human beings rely on a much messier mix of cultural scripts, mental models, frameworks of meaning, narratives about the

---

[25] Žižek 1994; van Dijk 1998; Norval 2000, 316; Jost 2006; Steger 2013, 216–17; Thagard 2014, 12.
[26] Gutiérrez Sanín and Wood 2014, 214.
[27] Alvarez 2008, 215.
[28] ibid. 217.
[29] See also Covington 2017, 168–71.
[30] Jost 2006, 653; Dukalskis and Gerschewski 2018, 4. Even Philip Converse—perhaps the most influential scholar within the political science paradigm focused on rigidly consistent belief-systems found primarily amongst political elites—recognized that more vernacular 'folk' ideologies can still be identified amongst mass publics; see Converse 1964, 255–6.
[31] Haidt, Graham, and Joseph 2009; Jost, Federico, and Napier 2009; Haidt 2012; Gries 2014, 43–8; Thagard 2014.
[32] Gerring 1997, 980.

world, and emotional values. But these are nevertheless patterned across individuals and groups, and are how most people internalize elements of familiar ideologies like liberalism, socialism, or conservatism. This psychologically realistic picture of ideology is central to my approach in this book.

I therefore define political ideologies broadly, as *the distinctive political worldviews of individuals, groups, and organizations, that provide sets of interpretive and evaluative ideas for guiding political thought and action.*[33] Numerous similar definitions can be found in leading work on ideology.[34] Ideologies, in this sense, come in many forms: dogmatic and pragmatic, systematic and inchoate, dominant and insurgent, accurate and distortionary, intellectual and vernacular, elaborate and simple, religious and secular, nationalist and internationalist. Rather than being limited to 'special' political goals, ideologies are broad interpretive and evaluative frameworks, which offer purportedly factual narratives and beliefs about the world as well as underlying preferences, values, and ideals.[35] 'To study ideology', in other words, 'is to focus on systems of ideas which couple understandings of how the world works with ethical, moral, and normative principles that guide personal and collective action.'[36]

This broad definition of ideology is neutral on the 'scale' at which particular ideologies are conceptualized. In research on mass killing, scholars have generally focused on familiar 'big isms' such as fascism, communism, and nationalism, which are typically portrayed as singular 'macro-level' variables—i.e. unified belief-systems attached to whole societies. But this monolithic image tends to overaggregate ideologies, and has led many scholars to erroneously assume that ideologies cannot explain variation in violence *within* a society.[37] Charles King, for example, suggests that 'the constancy of ideology cannot account for the variable timing, pace, and completeness of mass murder',[38] while Maureen Hiebert writes that an emphasis of ideology cannot 'explain why the same [Nazi] ideology singled out the Jews for complete extermination while the Poles were slated for perpetual servitude'.[39] Such claims rest on a confusion: thinking that because a single overarching ideological *label*—e.g. 'Nazism'—is constant in a given case, the *influence and contents* of that ideology must be singular, uniform, and unchanging.[40]

---

[33] In earlier work I used a somewhat more elaborate definition but with almost identical meaning; see Leader Maynard 2014, 825.
[34] Hamilton 1987, 38; Cassels 1996, 4–6; Hall 1996, 29; Walt 1996, 25; Drake 1998, 54–5; Kramer 1999, 540, fn.7; Oliver and Johnston 2000, 43; Ugarriza and Craig 2013, 450; Gutiérrez Sanín and Wood 2014, 214.
[35] Jost, Federico, and Napier 2009, 309; Cohrs 2012, 54; Saucier 2013, 922–3. See also Long 2016, 15–19.
[36] Oliver and Johnston 2000, 44.
[37] Anderson 2017a, 7; Swidler 2001, 97 & 101, fn.5. On similar problems in terrorism research see Ackerman and Burnham 2019.
[38] King 2012, 331. See also Maat 2020, 777–8.
[39] Hiebert 2008, 8.
[40] Marchak even more sweepingly argues that because Nazi Germany perpetrated genocide but Fascist Italy did not, ideology cannot explain such outcomes—under the seeming misconception

But broad ideological systems like Nazism are not monoliths. They are internalized in different ways amongst different sections of society, evolve and radicalize over time, and contain different ideas about different victim groups. Ideologies can, as such, be essential to explaining 'intracase' variation in violence.

Ultimately, no two people possess exactly the same set of ideas about politics, so we may sometimes want to examine variation in individuals' *personal ideologies*.[41] Most of the time, however, ideologies are of interest precisely because they are patterned at the social level. Personal ideologies tend to cluster—whether around two broad 'left' and 'right' poles or more numerous distinct orientations—because people who share common sociocultural influences, life experiences, and personalities are often drawn towards similar ideological outlooks.[42] In addition, as I discuss in section 2.2, ideologies are often socially institutionalized in ways that transcend individuals' personal outlooks. It is for this reason that scholars tend to be most interested in *shared ideologies*, such as liberalism, communism, or Nazism. I stick with this trend, but instead of portraying shared ideologies as monolithic macro variables slapped onto whole societies, I emphasize the need to unpack or 'disaggregate' them.[43] Ideological heterogeneity, contestation, and change are typical in all societies, and in all mass killings. Ideological 'hardliners', on whom I focus, tend to represent a coalitional faction within their societies, not the entire population. Even Nazism—a highly dogmatic ideology orientated towards Hitler as a focal point of tremendous authority—evolved over time and subsumed individuals with a range of personal ideological views.[44] As Johannes Steizinger observes, 'a political ideology works best as a controlled plurality'.[45]

Some scholars worry that a broad conceptualization trivially guarantees that ideology matters by making 'everything' ideology, draining the concept of any theoretical value.[46] This is incorrect. Many phenomena—including apolitical interests or ideas, interpersonal relationships, physiology, universal psychological processes, immediate situational perceptions, and basic material forces—are not parts of ideology, though they may become causally related to it. But the move from narrower to broader conceptualizations does change the theoretical role ideology plays in explaining behaviour. Ideology is no longer a placeholder for certain idealistic political motives, possessed only by certain ideologically committed actors. Instead, ideology becomes akin to concepts like 'personality', 'identity', and

---

that Nazism and Italian Fascism were 'the same' ideology; see Marchak 2003, 100. For a contrasting explanation of how different forms of fascism generated different forms of state violence see Kallis 2009.

[41] Cohrs 2012, 56. See also Schull 1992.
[42] Bar-Tal 2007, 1444; Jost, Federico, and Napier 2009; Jost 2017.
[43] See also Hinton 2005, 26–8.
[44] Kallis 2005, 11; Kallis 2009, Parts III–IV; Steizinger 2018, 4–6.
[45] Steizinger 2018, 143.
[46] See, for example, Mullins 1972, 498; Larrain 1979, 15 & 129; Rosen 2000, 395.

'culture'[47]—a property of *all* political actors, but the *specific contents of which* vary across actors.[48] Debates over ideology's significance should not, therefore, revolve around competing claims as to whether political actors 'are ideological or not'. This is like asking whether actors 'are cultural' or 'have identities' or not. The core question is instead: *are the particular contents of an individual's, group's, or organization's ideology relevant to their behaviour?* To what extent do differences (or similarities) between actors' ideologies explain differences (or similarities) in what they do?

A broad conceptualization does emphasize that ideologies are mutually constitutive with other important types of ideas, such as identities, norms, and frames, rather than standing in some sort of explanatory competition with them.[49] This is particularly worth emphasizing with respect to *identities*, since scholars often talk as though identities contrast with ideologies. A common suggestion, for example, is that conflicts in the post-Cold War era are generally about identity *rather than* ideology.[50] This is a mistake. Ideologies and identities are not the same, but the impact of identities on political violence depends on how they are constructed and mobilized within particular ideological movements and narratives.[51] Nazism invoked a very specific understanding of 'Germans' and 'Jews', just as Islamist fundamentalism relies on particular ideological constructions of 'Muslim' and 'infidel'. Ideologies, in turn, may be components of identity—individuals may see specific ideological affiliations ('communist') or broad partisan alignments ('left') as critical in defining the social categorization of themselves or others.

For this reason, statistical tests of the role of identity in violent conflict have tended to produce negative results. Needing to measure something quantifiable, such tests have typically employed relatively 'objective' demographic measures such as the ethnic or religious diversity of a population, and found that these fail to predict levels or forms of violence.[52] This is an important finding, since it disproves 'primordialist' theories of identity which see intergroup divisions as inherently conflict prone. But while the mere *existence* of intergroup distinctions is a poor predictor of conflict and mass killing, there is abundant evidence that certain *subjective ideological understandings* of ethnic, religious, or political identity

---

[47] Some understandings of *culture*—as sets of values and beliefs held by a group—are similar to this conception of ideology. But I follow Lisa Wedeen and William Sewell Jr. in conceptualizing culture more generally as all the aggregated 'semiotic practices' through which human collectives construct mutually intelligible meanings of the world. See Geertz 1964; Balkin 1998; Wedeen 2002; Sewell Jr. 2005.

[48] In more technical language, ideology denotes a *dimension* of 'actor-specificity' or 'actor-heterogeneity'; see Hudson 2005; Hafner-Burton et al. 2017, S4 & S13–18.

[49] Cohrs 2012, 54–6.

[50] Huntington 2002, 22; Kaldor 2012, 7; Blair 2014. See also Hiebert 2008, 8–9.

[51] Malešević 2006; Hammack 2008; Malešević 2010, 325; Gutiérrez Sanín and Wood 2014, 215; Vollhardt 2015, 99–101. Louis Althusser's conception of ideological 'interpellation' addresses similar dynamics; see Althusser 1971; Wedeen 2019.

[52] Krain 1997; Fearon and Laitin 2003; Harff 2003; Gartzke and Gleditsch 2006; Downes 2008, 48–56. See also Montalvo and Reynal-Querol 2008.

can promote violence.[53] What matters is not the mere existence of groups with different identities—a ubiquitous feature of human social life—but the way, as Randall Collins puts it, 'allegedly long-standing ethnic hostilities nevertheless are ideologically mobilized at particular points in time'.[54]

One final conceptual clarification: in this book I treat political ideologies as subsuming what could alternatively be termed 'military ideologies', i.e. distinctive sets of ideas about military action. This is important, since I present mass killings as profoundly rooted in the mixture of certain military-strategic ideas with more manifestly 'political' concepts and claims. I treat all such ideas as part of ideology in recognition of the fact that military action is political in character, and always embedded in broader political relations, projects, and values. Different ideologies promote different approaches to the use of military force just as much as they promote different approaches to education or welfare or economic management. Liberal ideology penetrates deeply into the military practice of Western democracies, for example, while virulent anticommunism guided the military doctrines of numerous dictatorships in Latin America and South East Asia during the Cold War.[55] In addition, militaries are rarely if ever truly apolitical, and in many countries they are deeply embroiled in broader politics.[56] Nevertheless, certain scholars may prefer to distinguish political and military ideologies, in which case my argument in this book should be seen as simultaneously addressing the role of both.

2.2 How Do Ideologies Influence Behaviour?

Not all research on ideologies and mass killing addresses the *causal links* between ideology and violence in detail. Many discussions simply focus on identifying the major recurring ideological themes in mass killings, and while this may carry causal implications, it does not directly assess the impact such themes have.[57] But to determine how ideology actually affects mass killing we need some underlying theoretical story of the causal power of ideology: how it *could* actually influence political behaviour. Although not always explicit, two main causal stories of this kind can be identified in existing scholarship.

The first is that of the true believer model, which assumes that ideologies shape behaviour through *strong ideological convictions that motivate action*.[58] Daniel

---

[53] Volkan 2006; Moshman 2007; Hammack 2008, 238–9. Tezcür 2016, 253–4.
[54] Collins 2009, 20. See also Ugarriza 2009, 84; Malešević 2010, 325–6; Gutiérrez Sanín and Wood 2014, 215.
[55] See, for example, Pion-Berlin 1988; Dillon and Reid 2009; Cromartie 2015; Robinson 2017; Scharpf 2018.
[56] Huntington 1957; Finer 1962/2017; Koonings and Kruijt 2002.
[57] See, for example, Kiernan 2003; Alvarez 2008; Saucier and Akers 2018.
[58] E.g. Marchak 2003, 87–94; Mann 2005, 27; Smeulers 2008, 241 & 246–7.

Goldhagen, for example, stresses the Holocaust's roots in 'the autonomous motivating force of Nazi ideology',[59] with violence explained by the fact that 'the perpetrators, "ordinary Germans," were animated by antisemitism, by a particular *type* of antisemitism that led them to conclude that the Jews *ought to die*'.[60] This is certainly one way in which ideology might matter, but there is no reason to assume that it is the sole or even primary one. Indeed, such conviction-centric accounts leave us with a major puzzle. Research on mass killings generally finds that perpetrators are diverse, acting from various motives and with only a minority displaying intense ideological convictions.[61] Nevertheless, strong ideological patterns in the overall violence are often visible, with mass killing implemented according to identifiable ideological rationales, with ideological training and propaganda shaping how it unfolds, and with important ideological differences separating regimes that readily employ mass killing from those that don't. In short, ideology's influence on collective behaviour often seems *disproportionate* to the underlying strength of ideological convictions amongst the individuals who actually produce that behaviour.[62]

The second story of ideology's causal power is, interestingly, almost the opposite of the conviction-centric account. Some rationalist theorists of armed conflict do acknowledge a role for ideologies, but present them as merely *instrumental tools*—deployed not because perpetrators believe in them, but because they are a useful means to an end. They might provide, for example, a legitimating cover-story for violence, a mechanism of mobilizing public support behind violence, or a recruitment tool for encouraging willing participants to step forward.[63] Mass atrocities in the Yugoslav Wars during the 1990s are, for example, often portrayed in such terms, with ethnonationalism serving as a way for opportunist political leaders like Slobodan Milošević to garner public support, or for mercenaries to legitimate self-interested looting and hooliganism.[64]

This narrow focus on ideology as an instrumental tool runs into as many problems as the focus on ideological convictions. By contrast with the conviction-centric account, it can readily acknowledge that ideologies have significant effects even when individuals do not deeply believe in the ideology. But it struggles to explain *how* ideology has such effects. Legitimating claims will only neutralize opposition to mass killing if they actually lead people who would otherwise have engaged in such opposition to now avoid it. Ideological mobilization of support for mass killing will only work if people actually respond to ideological appeals by engaging in supportive actions. But if ideology is merely a tool, and no one really

---

[59] Goldhagen 1996, 13.
[60] ibid. 14 (*emphasis in original*).
[61] Gerlach 2010, 1–5; Williams and Pfeiffer 2017; Williams 2021.
[62] See also Barnes 1966; Schull 1992.
[63] Gutiérrez Sanín and Wood 2014, 217–20. See also Skinner 1974.
[64] See, in general, Silber and Little 1997; Mueller 2000; Gagnon 2004.

cares about it, why would anyone change their behaviour in these ways? What makes instrumental deployments of ideology work? Most instrumental accounts of ideology fail to convincingly answer these questions.[65]

A few theorists of mass killing offer a third causal story, where ideologies provide *collective scripts or frames* for violence.[66] In his analysis of ethnic conflict in Yugoslavia, for example, Anthony Oberschall emphasizes that deep ethnonationalist commitments were lacking. But he proposes that an ethnonationalist 'crisis frame', resonant in light of historical legacies of conflict, could be activated by elites to transform interethnic relations and generate ethnic violence.[67] John Hagan and Wenona Rymond-Richmond similarly emphasize the role of racial dehumanization, theorized as a form of 'collective framing', in genocidal violence in Darfur.[68] Lee Ann Fujii's leading study of the Rwandan Genocide argues that state-sponsored ethnonationalist constructions of identity were not deeply believed by perpetrators, but nevertheless provided a 'script or dramaturgical blueprint for violence'.[69] In his analysis of genocidal dehumanization, Rowan Savage likewise emphasizes that 'the only necessity is a narrative that legitimizes the action in question. In order to fulfil a psychological need such a "script" does not actually require wholehearted or long-term belief.'[70]

This third story is more nuanced, and significantly influences my own analysis. Like the instrumentalist account, however, it still requires an explanation of *why individuals respond to certain scripts and frames*. Scripts and frames do not possess innate force, and many fail to exert serious political influence. In Oberschall's case, the explanation tacks back towards conviction: he presents the 'crisis frame' as possessing real resonance and persisting, albeit 'dormant', in Yugoslavia. Hagan and Rymond-Richmond more explicitly present frames as sources of motivational conviction, arguing that 'racial epithets are important ... because they capture attackers' motivation and intent'[71] and produce 'the heightened vulnerability and fanatical fury that lead to genocide'.[72] This resurrects some of the concerns with conviction-centric accounts. By contrast, Fujii and Savage reject the claim that frames and scripts are underpinned by strong beliefs, but this leaves their influence somewhat mysterious. What exactly does it mean to say, as Fujii does, that

---

[65] As observed by Fearon and Laitin 2000, 846 & 853–5; Kaufman 2001, 6; Bubandt 2008, 791 & 813; Gutiérrez Sanín and Wood 2014, 222. Instrumental accounts might assume that while *elites* purely use ideological claims instrumentally, this works because the claims really resonate with target *mass audiences*. But there's no good theoretical or empirical basis for this assumption. Indeed, it inverts the conventional wisdom within political science research (however problematic) that it is elites who tend to be ideological and masses who are not; see Converse 1964; Glazer and Grofman 1989; Kinder and Kalmoe 2017.
[66] Owens, Su, and Snow 2013, 78. See also Schull 1992; Fearon and Laitin 2000, 851–3.
[67] Oberschall 2000.
[68] Hagan and Rymond-Richmond 2008.
[69] Fujii 2009, 104. Fujii does not present such identity-constructs as ideological, however.
[70] Savage 2013, 149.
[71] Hagan and Rymond-Richmond 2008, 878.
[72] ibid. 886, fn.14. See also ibid. 895–6.

'[p]eople could kill with scripted ethnic claims, without believing those claims to be true or accurate, because as a script, ethnicity *required certain performances*'?[73] What makes individuals follow such 'requirements'? In a similar fashion, Savage argues that: 'a dehumanized discursive system regarding an out-group *constitutes the (moral) reality* within which members of the in-group act'.[74] But how does a certain 'constitution of moral reality' shape behaviour if individuals do not believe in it?[75]

To answer such questions, we need a more systematic account of ideology's influence on political behaviour. I offer such an account, built on claims that are familiar in social, economic, and psychological theory, but which have not been systematically applied to ideology.[76] This account starts, as explained in Chapter 1, by distinguishing two key forms of ideological influence.[77] First, ideologies may be sincerely *internalized* to some degree, genuinely guiding people's interpretations and evaluations of politics and political action.[78] Second, ideologies may become embedded in *structural* features of individuals' social environments—in ideological norms, discourses, and institutions which create pressures, incentives, or opportunities for individuals to act ideologically, irrespective of their sincere private views.[79] I further suggest that each of these forms of influence can shape behaviour in two main ways. Internalization may generate relatively strong ideological *commitments* to certain ideas for which individuals feel real intrinsic attachment, but it may also involve a much shallower and contingent form of sincere acceptance that I will term ideological *adoption*. Ideological structures, meanwhile, may encourage fairly unreflexive *conformity* to an ideology's ideas, or more calculated *instrumentalization* of those ideas for personal gain.[80] These four processes—discussed in more detail below—often work in tandem, and most individuals will be guided by several of them simultaneously.[81]

---

[73] Fujii 2009, 104 (*my emphasis*).

[74] Savage 2013, 149 (*emphasis in original*).

[75] The same issue problematizes Schull's theorization of ideology as a legitimating discourse; see Schull 1992, 736–8.

[76] A few scholars of mass killing do make arguments in a similar direction, however; see Browder 2003, 494–5; Adamson 2005; Hinton 2005, 25–31; Murray 2015, 14–16 & 185–6; Anderson 2017a, 7–8. To these might be added research that emphasizes the power of 'discourses' as constitutively enabling and structuring political violence; for example, Jabri 1996; Der Derian 2009; Peterson 2010; Sjoberg 2010.

[77] For related distinctions see Tetlock and Manstead 1985; Tannenwald 1999, 437–41; Tannenwald 2005; Wunderlich 2013; Checkel 2017, 597.

[78] I say 'sincerely internalized' to denote ideas that, however tacitly, tentatively, or unreflectively, individuals actually *accept or treat* as true, valid, legitimate, or appropriate. I recognize that there is a much more minimal sense in which even structural norms and institutions depend on the internalization of information about them, in order for individuals to understand their behavioural implications, but this does not require any degree of sincere acceptance.

[79] I do not assume that norms, discourses, and institutions are *only* structural—they may also promote ideological internalization. For the two best studies of norms and mass atrocities see Bellamy 2012b; Morrow 2020.

[80] Note that by ideological structures, I denote a kind of *social structure* that is ideological in nature—not the *internal conceptual or logical structure* of an ideology's content—by contrast with, e.g., John Wilson's meaning, as discussed in Holbrook and Horgan 2019, 6–7.

[81] See also Kelman and Hamilton 1989, 109–10.

Tying these claims together, I argue that ideologies should therefore be analysed as *infrastructures in collective action*. What I will therefore term 'ideological infrastructures' are particular configurations of sincerely internalized ideas and structural norms and institutions, which sustain and shape collective action through the mutually reinforcing interaction of ideological commitment, adoption, conformity, and instrumentalization.[82] It is this mutually reinforcing interaction that often gives ideological infrastructures a disproportionate 'greater-than-the-sum-of-their-parts' influence over political behaviour—although it can also sometimes explain why ideological infrastructures collapse or transform with remarkable speed. Ideological infrastructures are especially powerful within formal organizations, like political parties, armed groups, or state agencies, which more or less intensely institutionalize ideologies and socialize their members to accept and understand them.[83] But more diffuse ideological infrastructures also inhabit the social networks of societies at large, embedded in widespread political practices that serve to reproduce (and sometimes modify) the infrastructure. This way of understanding ideology is antimonolithic in a double sense. First, the people influenced by any ideological infrastructure are diverse—they do not all internalize the ideology in the same way or to the same extent, and they act out of multiple motives. Second, multiple ideological infrastructures generally coexist in a society's social networks, competing and/or overlapping in ongoing political practices.

This 'infrastructural' model of ideology is summarized in Figure 2.1. The central point is that when we seek to assess the impact of a given ideology—Nazism, say—we should not simply go about looking for intense commitments to Nazi ideas, *or* focus on the way those ideas could be manipulated for personal gains. We should

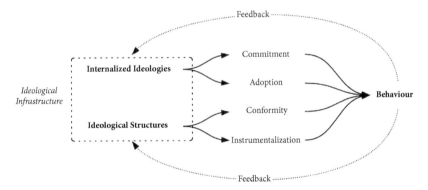

Fig. 2.1 An infrastructural model of ideology's influence.

---

[82] The language of ideological infrastructures is somewhat inspired by, but quite different in meaning to Mann 1984; Snow and Benford 1988, 205-7. For a technical definition of infrastructure, which can be largely analogized to my usage, see Star 1999, 381-2.

[83] Gutiérrez Sanín and Wood 2014; Checkel 2017; Schubiger and Zelina 2017; Revkin and Wood 2020; Lefèvre 2021; Parkinson 2021.

instead investigate how political behaviour was shaped by the *interaction* of (i) sincerely internalized Nazi ideas in the minds of various individuals and (ii) Nazi norms, discourses, and institutions and the conformity pressures and instrumental incentives they generated. It is this multifaceted Nazi ideological infrastructure that fundamentally constituted Nazism's political power.

To make this infrastructural account of ideology a little less abstract, consider the example of mass atrocities in the former Yugoslavia in the early 1990s. One of the great puzzles of the Yugoslav conflicts is a version of the disproportionality problem I noted above. Contrary to early assertions that 'ancient hatreds' motivated violence in Yugoslavia, most Yugoslav citizens attested to positive ethnic relations in the years preceding the conflicts, marriage and friendships across ethnic groups were widespread, and many rank-and-file perpetrators seem to have been guided more by loot and thuggishness than politics. Yet ethnonationalist logics remained central: violence was organized around ethnic groups seeking to maximize national territory and saw atrocities that targeted individuals based on their ethnicity. My contention is that ethnonationalism rapidly became, in the early 1990s, a dominant ideological infrastructure of the Yugoslav conflicts. It was powerful because (i) some people, although probably a minority, were committed to ethnonationalist claims and aims; (ii) many others selectively accepted certain ethnonationalist ideas and largely supported the more committed ethnonationalists; and (iii) others conformed to such ideas in a context where ethnonationalism suddenly seemed dominant, while (iv) others actively manipulated ethnonationalist claims for political and personal benefit. Individually, none of these processes may have been sufficient for ethnic conflict. But together they created an 'emergent' greater-than-the-sum-of-their-parts power to ethnonationalist narratives in the 1990s crisis.[84] In what follows, I delve deeper into the four distinct kinds of ideological influence than imbue ideological infrastructures with this power.

Internalized Ideology: Commitment and Adoption

Internalization is the most familiar way ideology matters. Many ideological claims, beliefs, and values are powerful because individuals do, to varying degrees, sincerely accept them.[85] Indeed, a point of consensus across modern psychological and social theory is that people necessarily rely on internalized systems of ideas to make sense of the world around them, to attach emotional meaning to events

---

[84] This paragraph draws broadly on: Hodson et al. 1994; Silber and Little 1997; Oberschall 2000; Semelin 2003; Malešević 2006.

[85] The specialized literature on ideologies' internalized effects is vast; for overviews see van Dijk 1998; Jost and Major 2001; Bar-Tal 2007, 1435–40; Jost, Federico, and Napier 2009; Cohrs 2012; Thagard 2014.

and outcomes in that world, and to decide what courses of action to take.[86] Consequently, people do not all see the world the same way, and do not all mean the same things by the same terms, concepts, or values.[87] Indeed, what seems self-evident and mere 'common sense' to those operating under one ideological worldview looks incoherent and delusional to those under others. Internalized ideologies provide, in short, the frameworks through which individuals engage in 'actual political thinking'.[88] Such internalized frameworks help explain how violence against civilians can come to elicit approval, satisfaction, or indifference from perpetrators, rather than disapproval, shame, or indignation.[89]

Yet such internalization need not denote especially deep or self-conscious conviction. As Paul Staniland observes, individuals 'need not be sophisticated ideologues, but ideology can nevertheless shape their general preferences and predispositions, often through unreflective or habitual practices'.[90] Individuals internalize ideas about politics to varying degrees, and in two principal ways.

First, internalized ideas may shape decision-making when individuals feel some degree of direct and relatively stable *commitment* to the ideas involved.[91] Such ideological commitments typically form the most longstanding core of individuals' personal ideological frameworks: their central values, convictions, and politically relevant identities. Ideological commitments need not be systematic or fanatical, and many operate in a relatively intuitive or taken-for-granted manner.[92] Indeed, while individuals with well-defined and systematized ideologies are relatively rare, political psychologists have nevertheless found that most ordinary citizens possess committed values and beliefs that are stably patterned along ideological lines.[93] But sincere ideological commitments carry intrinsic resonance for individuals, are resistant to change, generate strong emotional responses, and bear directly on processes of perception and decision-making.

Though not the limit of ideology's influence, varying degrees of real ideological commitment are crucial in mass killings. Many of the key orchestrators of the violence, and their most enthusiastic followers, display evident commitment to its central ideological justifications. No serious scholar denies, for example, that leading Nazis were intensely committed to the key tenets of Nazi ideology—displaying

---

[86] In contemporary psychological science, interpretation and evaluation depend on 'cognitive-affective' thought processes involving both reasoning *and* emotion—in contrast to the traditional reason–emotion dichotomy. See also McDermott 2004; Cohrs 2012; McDoom 2012; Ross 2014; Thagard 2014.
[87] Skinner 1974; Skinner 1978; Tully 1983; Freeden 1996.
[88] Freeden 2008, 197.
[89] This paragraph draws in general on Geertz 1964; Berger and Luckmann 1967; Blumer 1969; Edelman 1977; Kahneman and Tversky 1984; Boudon 1989; Freeden 1996; van Dijk 1998; Tomasello 1999; Jost and Major 2001; Skinner 2002.
[90] Staniland 2015, 778.
[91] Gutiérrez Sanín and Wood 2014, 220–2.
[92] See also Monroe 2011.
[93] See, in general, Zaller 1992; Jost and Major 2001; Jost, Federico, and Napier 2009; Haidt 2012; Gries 2014; Morgan and Wisneski 2017.

fervent belief, in both public and private settings, in Hitler's vision of a Judeo-Bolshevik conspiracy against the German people, and in the overriding moral obligation to serve the Führer. Although ideological commitments need not be irrational, they are often most visible when they elicit behaviour that seems hard to explain in instrumentally rational terms. Islamic State's brutal atrocities have often proved highly counterproductive, for example, alienating potential supporters and provoking opposition from almost all other Middle Eastern actors.[94] Many scholars suggest that the organization's persistence with such tactics reflects, in part, sincere ideological commitments amongst its membership.[95]

But there is a second broad way in which ideological elements may be internalized. Political psychologists, sociologists, and theorists of political communication have long recognized that individuals often sincerely accept ideological positions even though they do not feel any intrinsic commitment to the ideas involved.[96] Individuals nevertheless *adopt* those ideas, as I put it, because the ideas have become connected to deeper beliefs, values, and needs, are advanced by trusted authorities, or simply fill in gaps in an individual's political worldview left unaddressed by their intrinsic commitments. Peter Neumann notes, for example, how many recruits to jihadist terrorist organizations do not express longstanding religious or political commitments, but join a terrorist group out of a core political grievance or sense of Islamic identity, on the basis of which they then adopt the wider ideology of jihadism.[97] Adoption is often rooted in 'identification'—individuals adopt ideas that are associated with identities they feel genuinely committed to.[98] Democratic and Republican voters in the United States will typically, for example, sincerely support an idea or policy when it is advocated by a representative of 'their' party, yet reject and oppose the exact same idea or policy if it is proposed by their political opponents.[99] Similarly, perpetrators of mass killing may sincerely adopt certain justificatory narratives for violence, not because they are committed to such narratives themselves, but because those narratives are successfully linked to identities or ideas that the individual does feel real commitment to.[100]

Since individuals do not act out of the intrinsic appeal of adopted ideological ideas, those ideas may not provide their deepest motives, and the individual's broader personal history may show little longstanding sympathy for those ideas. Yet adopted ideological notions are still internalized—sincerely accepted by individuals and often offering 'frames', 'roadmaps', or 'organizing scripts' for action,

---

[94] Hafez 2017; Gade, Hafez, and Gabbay 2019.
[95] Byman 2016, 132–9 & 150–3. See also Revkin and Wood 2020.
[96] Somewhat contra Schull, who conflates accounts of ideology centred around strong commitments and beliefs with any emphasis on psychological internalization: Schull 1992, 729–33.
[97] Neumann 2013, 882. See also Kalyvas 2006, 45–6.
[98] Kelman and Hamilton 1989, ch.5. On identification and violence see Littman and Paluck 2015.
[99] Cohen 2003.
[100] Checkel 2017.

which may provide the crucial link between a more diverse array of private motives and actual participation in collective violence.[101] Ideological adoption may therefore be essential in explaining when mass killings occur and why individuals participate in them.

Ideological Structures: Conformity and Instrumentalization

But internalization only takes us so far. It is a long-established finding of both social theory and psychological science that people's actions are influenced not just by their genuine private preferences and beliefs, but also by their understandings of socially expected behaviour, and their anticipations about the likely beliefs and behaviour of other people. This is central to the power of ideology, with the ideological discourses, norms, and institutions of an individual's social environment creating pressures, incentives, opportunities, and constraints on that individual even if they do not internalize the ideologies in question.[102] Members of a political organization may all act to further their organization's ideological goals even if none of them feels personal attachment to those goals. They may recognize that a certain ideological 'norm' exists between them even if none of them privately approves of that norm.[103] They are likely to feel significant pressure to legitimate their actions in terms of the organization's prevailing ideological discourses even if they harbour personal doubts about the ideas involved.[104] In short, individuals' behaviour is influenced not only by their personal internalized ideologies, but also by the *ideological structures* of their social environment.[105]

Where do such structural pressures and incentives come from? Structures are often rooted in the genuinely internalized ideologies of powerful agents or groups.[106] The Cold War was, for example, characterized by a powerful capitalist-communist ideological structure, which induced many groups not deeply committed to those ideologies to nevertheless present themselves as bona fide capitalists or communists. The most important foundation of this ideological structure was the two superpowers' willingness to offer extensive material and political support to those who claimed to act in the name of their ideologies.[107] Most commonly of all, a society's dominant ideological structures are upheld by those in control

---

[101] Costalli and Ruggeri 2017, 924–5. See also Snow and Benford 1988; Oberschall 2000.
[102] The argument I make in this section parallels longstanding claims in various fields—methodological holism, interpretivism, hermeneutics, and so on—that action must be understood in terms of constitutive social rules, norms, reasons, or meanings; see Hollis and Smith 1990, ch.4; Bevir and Blakely 2018. My presentation of the argument in a more causal form simply reflects my rejection of firm explanatory–interpretive, or causal–constitutive, dichotomies.
[103] Morrow 2020.
[104] Skinner 1974.
[105] Kuran 1989; Anderson 2017a, 69; Dukalskis and Gerschewski 2018, 3.
[106] Owen 2010; Costalli and Ruggeri 2017.
[107] Gould-Davies 1999; Westad 2005; Kalyvas and Balcells 2010, 420–1; Owen 2010, ch.6.

of the state or other powerful organizations, through the distribution of resources and setting of rules and policies.[108] As committed Nazis consolidated control over the German state, for example, this relatively small vanguard was able to profoundly shape prevailing ideological norms, bringing the vast bulk of German society—and much of occupied Europe—into line with Nazi ideology.

Like most social structures, however, ideological structures are not reducible to the beliefs of powerful elites. Sometimes, indeed, we can observe powerful ideological structures in a society even when *its own leaders* appear to have little sincere belief in them. In the latter years of the Soviet Union, for example, it is widely accepted that hardly anyone—neither political elites nor the mass public—retained much belief in official Soviet ideology.[109] Yet that official ideology continued to exert significant influence over the behaviour of millions of individuals and hundreds of state agencies. This highlights how ideological structures are ultimately constituted by *convergent social expectations*.[110] If *most people expect* that *most people will comply* with a given ideology, this creates strong social pressures and incentives to do likewise, creating a self-reinforcing dynamic that reproduces such expectations and sustains the ideological structure.[111] Even though few people sincerely believed in orthodox Soviet ideology in the early 1980s, everyone did *expect* that this ideology would continue to structure the discourses and practices of state agencies and the broader population. The risk involved in any individual being a 'first-mover' to challenge the established structure was therefore extremely high, while habituation of the constituent ideological performances was strong. Unable to coordinate effective opposition, individuals (whether amongst the elite or the masses) experience such dominant ideological structures as a 'social fact', imposing real costs and benefits on their actions in light of the anticipated responses of others.[112] Ideological structures often, therefore, have a force that is disproportionate to the levels of sincere internalization that underlie them.[113] This is a familiar insight from research on social norms but is rarely linked to ideology, even though it is essential to explain many large-scale ideological patterns in politics.

Of course, the dramatic eventual collapse of communist ideology in the Soviet bloc illustrates how ideological structures that are totally disconnected from underlying internalized ideologies are often brittle. As the strongest dissenters start to deviate from ideological norms, this erodes the expectations that sustain the ideological structure, and a threshold is reached where such deviations can

---

[108] Mann 1984; Revkin and Wood 2020.
[109] Schull 1992.
[110] This is sometimes referred to as 'social knowledge'. I prefer the language of convergent expectations since it does not imply a singular set of social norms and standards to be 'known'.
[111] Waltz 1979, 91; Giddens 1984; Waltz 1990, 29 & 34; Jabri 1996, ch.3; Wendt 1999, chs.1, 3, & 4; Legro 2005; Owen 2010, 51–2.
[112] See also Wendt 1999, chs.1, 3, & 4; Legro 2005; Owen 2010, 51–2; Lickel 2012, 99–100.
[113] Noelle-Neumann 1974; Kuran 1989.

snowball into a mass defection from the ideology. Amongst the Soviet leadership, circles of moderates around Mikhail Gorbachev had been sincerely persuaded, by the mid-1980s, of the need for political change. Orthodox hardliners initially created fundamental obstacles to such change—vigorously maintaining the existing ideological structure by demanding that communist beliefs, policy stances, and rituals be reproduced, with the result that opposition appeared costly. But as Gorbachev's reforms weakened such expectations, the societies of the Soviet Union and its allies crossed a threshold or 'tipping point' (much more dramatically than Gorbachev intended), provoking mass defection from communist discourses and norms across the Eastern bloc and the resulting collapse of its transnational communist ideological structure.[114] This is why it is crucial to focus on the *infrastructural* interaction of internalization and structure. Usually, however, ideological structures are not so totally disconnected from internalized ideologies, but have many real supporters (though often still a minority) who will actively work to sustain the structure by noisily propagating the ideology in public discourse and creating costs for those who do not comply.

As with internalized ideologies, ideological structures operate through two principal pathways. First, structural influence is often rooted in *conformity* effects: the widely researched tendency of human beings to relatively unreflectively follow explicit or implicit expectations of behaviour generated by peer pressure, orders from authority, organizational roles, or similar social influences.[115] Much of the time, individuals will simply go along with such social pressures, since doing otherwise risks interpersonal tension, loss of face, or heightened personal accountability. This power of conformity is one of the most famous findings of social psychology, associated with Solomon Asch's experiments on individual submission to group judgements, Stanley Milgram's experiments on obedience to authority, and Philip Zimbardo's 'Stanford Prison Experiment'.[116] All these experiments found that under the pressure of a group consensus, orders from authority, or official roles in a simulated 'prison' institution, individuals would often abandon their personal convictions and simply *act as was expected of them*.

In research on mass killing, scholars have sometimes pointed to social conformity as a reason to downplay the importance of ideology.[117] But this is a mistake. Conformity effects do show how individuals can perpetrate mass killing in response to social pressure rather than from deep ideological convictions. But ideological structures profoundly affect the direction and force of such social pressure. The behaviour expected in a Khmer Rouge torture facility or Nazi concentration camp is very different from that expected in most contemporary

---

[114] Kuran 1991; Checkel 1997; Casier 1999; Kramer 1999, 563–76; English 2002; Haas 2005, ch.6.
[115] Newman 2002; Waller 2007, ch.8.
[116] Asch 1956; Milgram 1974/2010; Zimbardo 2007.
[117] E.g. Bauman 1989; Roth 2004; Roth 2005; Fujii 2009; Williams and Pfeiffer 2017, 80–1.

prisons in liberal democracies (abuses in the latter notwithstanding).[118] Similarly, the behaviour expected in internal organs of the Soviet Communist Party is very different than the behaviour expected in internal organs of the American Republican Party. The *varying ideological character of such organizations* creates different institutional norms, policies, and practices, resulting in varied pressures to conform to different ideological standards of behaviour.[119]

Second, ideological structures may encourage individuals to calculatedly *instrumentalize* ideology—i.e. to espouse and enact the ideology because they expect this to elicit beneficial responses from others. Like those who conform to ideological expectations, actors who are making instrumental use of an ideology need not have any sincere belief in it. But the apparent ideological content of prevailing norms, discourses, and institutions incentivizes them to instrumentally espouse or comply with the ideology in question. Such instrumentalization may not last long, however, unless some benefits really do follow—so what explains the *ideological responsiveness* of audiences to such instrumentalization? To return to the Yugoslav Wars, how can leaders such as Slobodan Milošević, who deploy a rhetoric of nationalism and hatred against other groups, use such rhetoric to achieve political control if most of their own group are not deeply motivated by nationalist sentiments or feelings of hate?

Again, an infrastructural account of ideology suggests two complementary answers to such questions. First, instrumental appeals may work because they actually do resonate—*to some degree*—with the audiences they target.[120] Milošević's nationalist claims seem to have appeared plausible for some ordinary Serbs, for example, and for key Bosnian-Serb nationalists like Radovan Karadžić and Ratko Mladić whose support Milošević sought.[121] The internalization involved here might be quite limited. Perhaps ethnonationalist slanders give *just enough* plausible justification for violence, or merely sow sufficient confusion and uncertainty, to stop large numbers of people from organizing protests and resistance against it. This would still be enough to make such appeals useful. Ideological justifications could motivate and mobilize a small but significant minority of true believers who drive forward the violence, and establish just enough sincere legitimation amongst the rest of the group to discourage opposition.

Second, however, instrumental appeals may be able to elicit useful behaviour from target audiences, because those audiences are themselves acting due to structural influence. Audience members assume that the ideological appeals are going to 'work', even if they do not personally believe in them, and therefore all go

---

[118] For discussion of abuses in America's prison facility at Abu Ghraib in Iraq, for example, see Zimbardo 2007.
[119] Revkin and Wood 2020.
[120] Hagan and Rymond-Richmond 2008, 892; Gutiérrez Sanín and Wood 2014, 222.
[121] For examples of local Serbian attitudes to nationalist claims see Dragojević 2019. See also Silber and Little 1997; Gagnon 2004.

along with them, creating a self-fulfilling prophecy.[122] Because it seemed as though a nationalist mood was becoming increasingly dominant in the late 1980s and early 1990s, most Serbs may have responded to nationalist rhetoric or policies out of conformity pressures or their own instrumental incentives. As Maria Petrova and David Yanagizawa-Drott observe: 'What everybody else is believed to think, and the likelihood that they might support the violence and actively participate in it ... is likely to shape any given [individual's] perceived costs and benefits and willingness to join (or not) in the attacks.'[123] Consequently, ideological appeals can prove instrumentally powerful even when they resonate with relatively few audience members.[124] In such circumstances the majority of disbelievers are, to paraphrase Michael Mann, *structurally outflanked*—they cannot overcome the collective action problems needed to successfully overturn the dominant expectations.[125] Indeed, as Russel Hardin explains: 'At the extreme, one might suppose that some group has no genuine believers in its fanatical views but that all members are coordinated on acting by the false sense that everyone else or most others do believe.'[126] More typically, however, the two explanations I have offered work together in ideological infrastructures: some respond to instrumental appeals out of sincere sympathies, others due to structural pressure, and many due to both. It is to this *interdependence* of different forms of ideological influence that I now turn.

Infrastructural Diversity

Ideologies, then, are infrastructures in collective politics, which shape outcomes through the interaction of both internalized beliefs and assumptions, and structural norms, discourses, and institutions. Specific ideological elements—such as Nazism's valorization of loyal service to the Führer, communist notions of the enemy of the people, and nationalist constructions of the political community—can therefore be highly consequential in shaping patterns of violence even when they are not sources of mass ideological conviction in a group or society. Individuals comply with such ideological elements for a range of reasons. But the overlap of the four forms of ideological influence analysed above—commitment, adoption, conformity, and instrumentalization—can collectively imbue ideologies with

---

[122] Wendt 1999, 42.
[123] Petrova and Yanagizawa-Drott 2016, 280.
[124] This modifies the contention of some scholars that instrumental use of ideology *requires* sincere commitments from the audiences it targets (e.g. Gutiérrez Sanín and Wood 2014, 222). Embedded expectations, unthinking conformity, and 'pluralistic ignorance' of genuine attitudes can induce positive responses to instrumental ideological appeals independent of sincere ideological commitments (see also Noelle-Neumann 1974; Granovetter 1978, 1438; Kuran 1989; Williams 2021, 74–5).
[125] Mann 1986, 7.
[126] Hardin 2002, 16. See also Elster 1982, 132.

tremendous political and social power. This allows us to recognize the deep influence of ideology on mass killings without effacing perpetrators' diverse motives and mindsets.

The interdependence of the four forms of ideological influence runs down to the individual level. In truth, almost no individuals act purely out of deep ideological commitment unaffected by structural pressures, or purely through instrumental calculations or unthinking conformity without any degree of sincere internalization. As Richard Overy observes regarding Nazi Germany and Stalin's Soviet Union: 'dissent, enthusiasm and compliance rubbed shoulders in Soviet and German society. They could inhabit the same individual as he faced the different things that society asked him to do, or as social and political obligations changed through time.'[127] Even those who largely conform to or instrumentalize ideology will typically sincerely adopt ideological justifications of their actions *to some degree*, since a range of psychological and social mechanisms make it difficult to continuously comply with ideologies publicly while entirely disbelieving them privately.[128]

Individuals' relationships to ideologies are also dynamic. Individuals who initially instrumentalize certain ideological claims may become socially pressured by their past public statements to maintain ideological positions they would now prefer to abandon.[129] Ideologically committed individuals can become disillusioned but have placed themselves in situations where they still face strong conformity pressures or instrumental incentives to comply with that ideology.[130] Many ordinary Germans may have espoused Nazi ideology out of opportunism in the early years of the regime, for example, with millions joining the party just after its takeover of power in 1933. But this tied many such individuals into institutions that promoted sincere adoption of or commitment to Nazi ideology—Michael Thad Allen notes how Nazi officials often 'identified their individual interest so strongly with those of "the German people" or other grand entities beyond themselves [that] they readily developed genuine attachments to the ideals of those organizations which promoted their careers'.[131] Others, meanwhile, may have abandoned or never developed such Nazi commitments, but become subject to immense conformity pressures to comply with the increasingly radical Nazi state even when this no longer served their self-interest or matched their private views.

We can rarely identify the relative balance between ideological commitment, adoption, conformity, and instrumentalization in large-scale collective action with any precision. The precise balance may not even matter that much in most circumstances: what is key is that a certain set of ideas comes, through the interaction

---

[127] Overy 2004, 308–9.
[128] See also Tetlock and Manstead 1985, 72; Kuran 1989.
[129] A form of what Frank Schimmelfennig terms 'rhetorical entrapment'; see Schimmelfennig 2001.
[130] See also Kelman 1958; Anderton 2010, 482–3; Bosi and Della Porta 2012, 362, fn.2; Checkel 2017, 596.
[131] Allen 2002, 114–15.

of such forms of influence, to shape collective action in observable ways. In the following chapters I argue that our available evidence from cases of mass killing suggests that, in general, most perpetrators do internalize ideological justifications of the violence to at least a significant degree, while also acting under the pressures, incentives, and opportunities of ideological structures. But commitment, adoption, conformity, and instrumentalization are all key parts of the story. To explain ideology's role in mass killings, we need to recognize all four of these forms of ideological influence, and the powerful interactions between them.

2.3 Conclusion

Scholars of political violence have raised various concerns with explanatory appeals to ideology. Christian Gerlach, for example, expresses scepticism about ideology because he fears that it implies a monocausal account of mass killing, where a single set of motives and ideas held by a uniform group produces violence.[132] David Snow and Scott Byrd articulate a similar concern: that 'the use of the concept of ideology is often encumbered by two misguided tendencies ... to view ideology in a homogenized, monochromatic manner [and] to conceptualize it as a tightly coupled inelastic set of values, beliefs, and ideas'.[133] Donald Bloxham and Dirk Moses, in their introduction to the *Oxford Handbook of Genocide Studies*, worry that overemphasizing 'the role of narrow political ideology in genocide' undermines comparative analysis, by placing the explanatory emphasis on case-specific ideologies rather than common causal dynamics.[134] We should be especially suspicious, Bloxham and Moses suggest, of 'a classically liberal understanding of genocide, where the crime results above all from aberrant political ideologies and oppressive political systems, and where the problem of genocide can be solved by the reassertion of the healthy norms of international democratic society'.[135] Waller raises the same concern. 'It is too easy', he argues, to focus on '*only* an extraordinary culture, like Germany, and *only* an allegedly extraordinary ideology, like eliminationist anti-Semitism.'[136] This sort of focus on extraordinary ideology detracts from the key finding that mass killings are performed by 'ordinary people', and are therefore a potential threat in societies across the globe.[137]

---

[132] Gerlach 2010, 3 & 5–6.
[133] Snow and Byrd 2007, 132.
[134] Bloxham and Moses 2010, 10. See also Valentino 2004, 16.
[135] Bloxham and Moses 2010, 9. See also Levene 2008, 9; Saucier and Akers 2018, 83.
[136] Waller 2007, 53 (*emphasis in original*).
[137] See also Browning 2007, xi.

A neo-ideological approach avoids these pitfalls. Whereas past studies often present ideologies as monolithic belief-systems attached to entire societies, I emphasize ideological heterogeneity within cases and ideological change over time. While ideologies are often associated with a singular kind of motivational state, I see their effects as rooted in ideological infrastructures that operate through multiple forms of internalized and structural influence. Although much scholarship has focused on 'extraordinary' ideologies that motivate revolutionary true believers, I emphasize the ordinary nature of ideological thinking, its broad role in shaping the way individuals understand and evaluate politics, and its interpenetration with strategic concerns surrounding security and self-interest. This all lays the groundwork for the rest of the book. Ideologies should not be reduced to 'special ideological motives' for violence rooted in unusually elaborate belief-systems, and they do not define a special 'category' of cases or perpetrators. There is an ideological dimension to all mass killings—as, indeed, to all political behaviour in general. The question is whether and how this ideological dimension explains the specific occurrence and character of the violence. The next chapter, Chapter 3, begins the task of answering that question.

3
How Does Ideology Explain Mass Killing?

Individuals and groups obviously differ in their political worldviews. But it is not obvious that such ideological differences significantly affect whether groups and individuals engage in mass killing. For all humanity's ideological diversity, many aspects of politics are remarkably consistent across different groups. Governments of all ideological stripes strive to hold onto power. Factions in all violent conflicts seek victory over their opponents. Indeed, civilians have been killed by individuals declaring allegiance to a wide range of ideologies: fascism, nationalism, liberalism, communism, and a host of less familiar doctrines. One could easily conclude that ideologies do not really affect individuals' or groups' engagement in violence.

Yet it is hardly surprising that we cannot just identify one or two ideologies which all perpetrators of mass killing adhere to or which always lead to mass killings irrespective of other social circumstances. This is not the right test for the claim that ideology matters. The key question is: would a given mass killing have occurred—and taken the same form—if the ideologies of the relevant individuals, groups, organizations, or societies had been substantially different? Are other causal factors, such as conditions of armed conflict, authoritarian political institutions, or longstanding intergroup divisions, so powerful that they will lead to mass killing irrespective of the ideological character of perpetrators and their social environment? Or are there essential ideological preconditions for mass killing?

In this chapter, I argue that the occurrence of mass killing does indeed depend on certain ideological preconditions: namely the dominance of certain *hardline ideas about security politics* that, under conditions of political crisis, generate a justificatory narrative for killing civilians. After the clarifications of ideology offered in the last chapter, this argument may seem intuitively plausible. Yet many scholars attempt to explain mass killing without significant reference to ideology. This chapter begins, in section 3.1, by explaining why this neglect of ideology by 'sceptical' perspectives is problematic. I critically examine rationalist and situationist explanations of mass killing, and show that while both have significantly enhanced our understanding of such violence, they must be combined with a focus on ideology to genuinely explain it.

I then develop my own neo-ideological account of ideology's role in mass killing in three steps. Section 3.2 sets out my basic argument—presenting the interaction of hardline ideologies and political crises as crucial in generating justificatory narratives for mass killing, and stressing the importance of those justificatory narratives in influencing both *whether mass killing occurs* and the *character of mass killings that do occur*. Section 3.3 then unpacks the way in which ideology shapes decisions to initiate, implement, and support mass killing, explaining how hardline ideas hold together a 'perpetrating coalition' of *political elites, rank-and-file agents*, and *public constituencies*, and explicating the distinct role ideology plays for each of these three categories. Section 3.4 completes the central explanatory argument by discussing the roots of hardline ideology, countering the suggestion that hardline ideas are merely 'symptoms' of deeper, non-ideological causes.

3.1. Explaining Mass Killing Without Ideology?

Many explanations of mass killing, while not necessarily declaring ideology irrelevant, largely ignore it. As noted in Chapter 1, two categories of such explanations have proved most influential in contemporary scholarship: rationalist explanations, which emphasize strategic circumstances that create cost–benefit incentives for mass killing; and situationist explanations, which emphasize tendencies to comply with social pressures, expectations, and roles as driving the initiation and implementation of violence. Again, I do not critique rationalist or situationist explanations of mass killing per se. Several rationalists and situationists—including some of those whose work I discuss below—do acknowledge an important role for ideology, and many others may only really intend to deny claims associated with traditional-ideological perspectives.[1] But most rationalists, and some situationists, either explicitly downplay ideology's significance or implicitly exclude it from their accounts of mass killing. This, I argue, is fatal for efforts to genuinely explain the occurrence and character of such violence. Strategic circumstances and situational pressures *are* crucial. But whether they encourage mass killing depends upon the internalized and structural ideologies of individuals, groups, and organizations.

Rationalist Explanations

Rationalist explanations contend that mass killings occur because violence against civilians can appear useful—a ruthless but rational strategy for the pursuit of self-interested goals.[2] Most rationalists therefore suggest that it is not variation in

---

[1] Most rationalists assume that people are 'boundedly' rather than 'perfectly' rational, making 'the best choice they can while subject to more or less severe limitations on their abilities' (Anderton and Brauer 2016b, 8). Ideologies could, therefore, be seen as 'building in the valuational and cognitive/belief components' that inform boundedly rational actors; see Zald 2000, 4. For rationalist or situationist approaches that acknowledge ideology's importance see Newman and Erber 2002; Valentino 2004; Zimbardo 2007; Balcells 2010; Maat 2020. See also Elster 1986.

[2] This section surveys all rationalist research on violence against civilians which does not *exclude* mass killing, but some of these scholars mainly focus on smaller-scale 'one-sided violence'.

ideologies which primarily explains why some groups and regimes engage in mass killings while others do not, but variation in the political and material circumstances those groups and regimes find themselves in, and the varying incentives for targeting civilians such circumstances generate.[3] Such rationalist arguments have taken three main forms.

Many rationalist theories emphasize *military* incentives for targeting civilians. Stathis Kalyvas has influentially argued that violence against civilians in civil wars is deployed by armed factions as a means of controlling civilian populations, encouraging civilians to support the perpetrators rather than their opponents.[4] Ben Valentino, Paul Huth, and Dylan Balch-Lindsay argue that the 'intentional killing of civilians during war is often a calculated military strategy designed to combat powerful guerrilla insurgencies' which draw support from those civilians.[5] Dongsuk Kim, Hanna Fjelde, and Lisa Hultman likewise suggest that violence against civilians is a way of attacking broad ethnic groups so as to weaken the support base for an enemy in conflict.[6] Focusing on interstate wars, Alexander Downes argues that 'civilian victimization is driven by perceived strategic necessity'[7] under two main conditions: first, in desperate wars of attrition, where battlefield stalemate encourages the targeting of civilians as an alternative path to victory, and second, when a belligerent seeks to annex territory inhabited by another ethnic group.[8]

Military incentives cannot be the only basis for killing civilians, since up to a third of mass killings occur outside of war.[9] But a second set of rationalist theories focuses on domestic *political* incentives for such violence. Rationalist theories of state repression often, for example, emphasize that violence against the civilian population can be an effective mechanism for maintaining a dominant elite's power and interests in the face of challenges from rival elites or social groups.[10] In a slight variation, 'elite manipulation' theories of mass violence focus not so much on the repressive effects of violence itself, but on the way elites may espouse extremist and sectarian ideologies to mobilize support and outflank challengers.[11] Violence may then be a corollary or escalatory outcome of such sectarian mobilization.

A third set of rationalist theories focuses more on *material* incentives and motives. Joan Esteban, Massimo Morelli, and Dominic Rohner suggest that mass killings result from perpetrators' desire for a larger share of future economic output in a society, and are an attempt to reduce the population of opposing groups so as

---

[3] Downes 2008, 10–11.
[4] Kalyvas 1999; Kalyvas 2006; Kalyvas 2012. There are also situationist and ideological elements in Kalyvas' work, however; see Kalyvas 2003; Kalyvas 2015; Kalyvas 2018.
[5] Valentino, Huth, and Balch-Lindsay 2004, 376.
[6] Kim 2010; Fjelde and Hultman 2014. See also Uzonyi and Demir 2020.
[7] Downes 2006, 170.
[8] Downes 2008, 3–5.
[9] Bellamy 2011, 2.
[10] Marchak 2003; Mitchell 2004, 30–4; Davenport 2007; Anderton 2010; Svolik 2012; Maat 2020. See also Valentino 2014, 96–8.
[11] Gagnon 1994/1995; de Figueiredo and Weingast 1999; Snyder 2000; Gagnon 2004; Brass 2005.

to weaken their capacity to demand available resources.[12] Chyanda Querido finds that easily lootable resources like oil and diamonds increase both the likelihood and intensity of mass killings in Africa, and suggests that material greed is therefore a principal driver of such violence.[13] Several scholars of genocides similarly contend that they are fuelled, in part, by the desire to appropriate the accumulated wealth of victim populations.[14]

This research has significantly enhanced scholarly understanding of mass killings. In particular, rationalists demonstrate how even horrific atrocities are often intelligible *instrumental strategies*: brutal tools for achieving military, political, or material goals rather than acts of blind rage or unthinking madness. Such violence is predictably encouraged by certain conditions that generate threats or opportunities for perpetrating groups, and is implemented according to comprehensible functional logics. Yet without incorporating ideology, such rationalist explanations fail to explain mass killings, for two key reasons.[15]

First, while mass killings may be a strategy for achieving military, political, or material goals, the claim that they are actually *the most rational strategy* to achieve such goals is implausible in most cases. Most studies of large-scale violence against civilians suggest that it is usually ineffective or of unclear utility.[16] While it might be used to coerce and repress communities, it can just as easily alienate them and provoke more determined and widespread resistance.[17] Killings in Argentina's 'Dirty War' between 1976 and 1983, massacres of civilians by the Armed Islamic Group (GIA) of Algeria, and state terror in the face of growing insurgency in Guatemala in the late 1970s, for example, all backfired in this fashion.[18] Mass killings may also prove ruinous for perpetrators by provoking outside military intervention, as in East Pakistan in 1971, Cambodia in 1979, or Kosovo and East Timor in 1999. They often, moreover, consume material and human resources that could more effectively promote security on the frontlines of combat—as in the Rwandan Genocide—or directly undermine the state's capacity to defend itself—as with Stalin's Great Terror, which included a purge of the Red Army's officers that weakened it against Nazi invasion four years later.[19]

---

[12] Esteban, Morelli, and Rohner 2015.

[13] Querido 2009.

[14] Aly 2008; Anderton and Brauer 2016a. A fourth category of rationalist theories focuses on the private benefits of violence to individuals. Macartan Humphreys and Jeremy Weinstein, for example, suggest that violence against civilians is commonly perpetrated by rebel groups with economically motivated recruits, who target civilians to extract loot. However, Humphreys and Weinstein emphasize that such an explanation probably does not apply to mass killings implemented as organized policy; see Humphreys and Weinstein 2006, esp. 433 & 444–5. See also Weinstein 2007. Criminological theories of mass killing often synthesize rationalist arguments about individual self-interest with an emphasis on situational pressures; see Karstedt, Nyseth Brehm, and Frizzell 2021.

[15] These parallel Downes' two central puzzles of violence against civilians; see Downes 2008, 2. For the reasons explained below, Downes' argument does not adequately answer either puzzle.

[16] Pape 1996; Davenport 2007, 8–10; Downes and McNabb Cochran 2010. See, however, Maat 2020, 802–4. See also Valentino 2014, 98–100.

[17] Mason and Krane 1989; Arreguín-Toft 2001; Earl 2011, 267–8.

[18] Pion-Berlin 1988; Garrard-Burnett 2009, 25–6; Hafez 2017; Thurston 2017.

[19] A particular mass killing's counterproductive consequences do not *directly render it irrational*, since subsequent consequences do not explain antecedent actions; see Elster 2015, 8 & 41. But in the

Rationalists typically argue that targeting civilians, despite this limited effectiveness, may nevertheless be the least bad gamble in extreme circumstances: a 'desperate measure'[20] that 'emerges out of frustration with conventional tactics in an effort to stave off defeat'.[21] While sometimes accurate, this argument runs up against two problems. First, the limited effectiveness of mass killings often makes them *potentially self-defeating*: they may actively worsen the perpetrators' situation. Desperation may lead actors to consider risky policies they would otherwise avoid, but it doesn't make useless or counterproductive policies rational. Second, in most mass killings it is simply not plausible to contend that no less-extreme routes to strategic advantage existed. Sometimes, indeed, groups target civilians from positions of relative strength and freedom rather than constraint.[22] Argentina's Dirty War, for example, intensified *after* the military defeat of Argentina's guerrilla insurgency (around the end of 1976),[23] mass killing in Cambodia intensified as the country *moved out* of the civil conflicts of the early 1970s,[24] and Allied mass bombing of German civilians *expanded* from late-1944 to mid-1945, even though it was clear that German defeat was imminent and that technological developments had made alternative strategies available.[25]

My point is not that mass killing is universally counterproductive for perpetrators, but that its strategic value is almost always indeterminate. This partly reflects the highly uncertain outcomes of such radically destructive violence. But it also reflects the trade-offs mass killing typically creates between different goals. If violence destabilizes economic development but satisfies political supporters, should it be engaged in? If it secures certain territories but alienates powerful neighbours, which matters more? Such questions cannot be answered by rationality alone, since they depend on the prioritization of different goals within a decision-maker's worldview. This *strategic indeterminacy* of mass killing renders rationalist explanations that neglect ideology inadequate. It is not enough for such explanations to simply show that strategic circumstances and incentives are 'in play'. If different decision-makers facing the same strategic circumstances and incentives would not have engaged in mass killing, then those circumstances and incentives are insufficient in explaining why mass killing occurs. In all my case studies, I show that mass killings were not simply the clearly rational strategy for perpetrators, and that differently inclined decision-makers could easily have seen a different course of action as rationally superior. The perpetrators *were* concerned with military,

---

absence of strong evidence that decision-makers could only access highly distorted information on mass killing's likely consequences, counterproductive outcomes make the claim that decision-makers faced clear rational incentives to engage in such violence implausible.

[20] Downes 2008, 39
[21] Valentino, Huth, and Balch-Lindsay 2004, 377
[22] Fischer 2006, 298; Hafez 2017, 1.
[23] Pion-Berlin 1988, 385.
[24] Jackson 1989a.
[25] Davis Biddle 2002, ch.5.

political, and material goals, but they saw mass killing as the best way of pursuing those goals due to ideological beliefs and assumptions, not objective incentives that all political actors would respond to in the same way.

The second problem with rationalist explanations that neglect ideology is that individuals and groups differ not only in their preconceptions of the strategic usefulness of mass killing, but also in their stances on the *normative permissibility* of such violence. Rationalist theories generally ignore normative attitudes, as though the underlying willingness to engage in mass killing is constant across individuals.[26] This is clearly inaccurate: the full diversity of human individuals—from the Mikhail Gorbachevs of the world to the Josef Stalins and from the Nelson Mandelas to the Osama bin Ladens—are not all equally inclined to target civilians.[27] Indeed, many regimes and armed groups—such as the African National Congress in the struggle against apartheid, the Popular Army of Liberation in El Salvador, and the militaries of contemporary liberal democracies—present intentional violence against civilians as illegitimate and exhibit considerable restraint in perpetrating it.[28]

This reflects the fact that mass killing is a *normatively extreme* policy, which is not widely accepted as within the bounds of legitimate political and military action.[29] Many political decision-makers are consequently reluctant to adopt such a strategy. At the same time, however, mass killing is not an entirely non-normative form of violence. Some individuals and groups positively moralize it, perpetrating or supporting atrocities, even when harmful for their material self-interest, out of vengeful justice, machoistic valour, or loyalty to one's community.[30] Such normative variation matters. From the early stages of World War II, for example, Allied policymakers explicitly declared themselves unconcerned with the deaths of German and Japanese civilians, stating that they deserved no consideration at all if *any* military benefit could be derived from targeting them. By contrast, US policymakers in recent wars—whilst tolerating significant civilian casualties—have generally affirmed the principal of civilian immunity, not contemplated mass killings of civilian populations, and gone to some efforts to minimize civilian deaths.[31]

These two features of mass killing—*strategic indeterminacy* and *normative extremity*—are theoretically crucial.[32] Of course, no form of political violence is self-evidently strategically valuable and normatively acceptable. Consequently, rationalist theorists may also miss ideological sources of variation in phenomena

---

[26] Some rationalist theories do take the costs of norm-violations into account—see, for example, Downes 2006, 175–6—but this is distinct from variation in actual normative attitudes.
[27] Mitchell 2004; Bulutgil 2017, 175.
[28] Goodwin 2007; Thaler 2012; Gupta 2014; Hoover Green 2016, 627–9; Oppenheim and Weintraub 2017, 1131–3.
[29] Savage 2013, 153; Anderson 2017b, 43.
[30] Fiske and Rai 2014.
[31] See also Bellamy 2012b.
[32] See also Downes 2008, 2.

like interstate war, civil war, and terrorism.[33] But strategic indeterminacy and normative extremity are especially characteristic of mass killings. The threat posed by civilians is always less clear than the threat posed by military forces, and the 'benefits' and legitimacy of targeting them *en masse* are correspondingly far less obvious. Ideological preconceptions therefore play a major role in shaping varied strategic and moral perceptions of mass killing.

Again, rationalists need not deny this, and some may emphasize that their theories are simplifications which ignore many key factors while still accounting for important patterns of mass killing.[34] My contention, however, is that side-lining ideology is generally an *oversimplification*: there is too much that cannot be explained if ideology is left out of the picture, including the fundamental question of why mass killings occur. Such rationalist theories have to tacitly assume that a certain sort of ideological orientation predominates amongst perpetrators of mass killing, which in reality is never assured. To genuinely explain killing, we have to place decision-makers' ideologies under the spotlight, and show how they can make the violence appear strategically and morally justifiable.[35]

Situationist Explanations

The diverse range of perspectives I refer to as *situationist* emphasize how people's behaviour is often driven, not by their personal preferences or beliefs, but by various forms of situational social pressure: such as orders from authorities, peer-pressure generated within social groups, and bureaucratic processes and norms within political or military organizations. Such accounts often acknowledge the influence of Hannah Arendt's *Eichmann in Jerusalem: A Report on the Banality of Evil*, with its picture of Adolf Eichmann as a thoughtlessly obedient bureaucratic organizer of the Holocaust, but most draw more heavily on a renowned body of research in social psychology already mentioned in Chapter 2.[36] Soloman Asch's group-conformity experiments in the 1950s showed that individuals would agree with self-evidently false statements (about, for example, which of two straight lines was longer) if surrounded by people who affirmed that they were true.[37] Stanley Milgram's famous experiments on obedience at Yale University revealed how the majority of people would inflict (apparently) severe electrical shocks on other human beings when ordered to do so by an authoritative scientist in a white lab coat.[38] The Stanford Prison Experiment, conducted in 1971 by Philip Zimbardo, similarly

---

[33] For a similar argument applied to counterinsurgency, for example, see Long 2016.
[34] Fjelde and Hultman 2014, 1251.
[35] See also Verdeja 2012, 315.
[36] Arendt 1963/2006. See also Waller 2007, 98–106.
[37] Asch 1956.
[38] Milgram 1974/2010.

revealed how ordinary people could become highly abusive once they entered a simulated prison environment and adopted the roles of prison guards.[39] Herbert Kelman and Lee Hamilton extended such work to analyse a wide range of *Crimes of Obedience*, in which individuals engage in immoral or illegal behaviour when ordered to by authorities.[40]

While this research was mainly built on simulated experiments, several influential studies have applied such psychological insights to actual mass killings. Christopher Browning's examination of German police battalions in the Holocaust and David Chandler's study of the Khmer Rouge's 'S-21' torture and killing facility in Cambodia, for example, both emphasize the extent to which perpetrators did not generally appear to be motivated by ideological fanaticism, but nevertheless conformed to orders, group-pressure, and organizational practices of mass murder.[41] Zygmunt Bauman similarly presents the Holocaust as rooted in the bureaucratic environments of modern states and the way they socialize individuals to obediently implement policies.[42] More recent research on the Rwandan Genocide by Lee Ann Fujii and Omar McDoom also stresses the power of situational social influences, especially through interpersonal relationships and social pressure within small groups, in explaining perpetrators' participation in the violence.[43] 'Micro-sociological' research, principally associated with Randall Collins and Stefan Klusemann, emphasizes situational dynamics in the form of powerful intragroup emotions which arise in the tense, confrontational situations in which atrocities occur.[44]

There are important differences between these theorists. Many situationists do not address political elites' decisions to actually initiate mass killing, leaving a role for ideology (and rational strategic calculation) higher up the chain of command.[45] Others do extend situationist arguments to elite decision-making, arguing that mass killings emerge as an escalatory output of bureaucratic competition, compartmentalization, and radicalization rather than well-formed 'ideological' plans and goals.[46] Yet this is often compatible with an emphasis of ideology in shaping bureaucratic norms, policies, and cultures.[47] Some situationists do, however, express strong scepticism towards the relevance of ideology. Paul Roth, for example, contends that social-psychological research on situational pressures 'accounts in all essentials for the number of perpetrators and their otherwise incomprehensible

---

[39] Zimbardo 2007.
[40] Kelman and Hamilton 1989.
[41] Browning 1992/2001; Chandler 2000.
[42] Bauman 1989.
[43] Fujii 2008; Fujii 2009; McDoom 2013; McDoom 2021.
[44] Collins 2008; Collins 2009; Klusemann 2010; Klusemann 2012; Collins 2013. Situational dynamics also play a central role in many criminological theories; see Karstedt, Nyseth Brehm, and Frizzell 2021.
[45] See, for example, Klusemann 2012, 471–2.
[46] Such explanations have often been termed 'functionalist'—I classify them as a subset of situationist explanations. See Moses 1998.
[47] Browning 1980; Browning 1992/2001; Kershaw 1993.

brutality'[48] so that 'no need exists for positing "deeper" reasons'.[49] Wolfgang Sofsky asserts that: 'Great crimes do not need grand ideas behind them,' characterizes ideological claims as 'justification after the event', and presents 'emotions, habit or sheer thoughtlessness' as more central.[50] Though his argument is subtler, Bauman also often downplays ideology, suggesting that the Holocaust did not rest on active ideological belief that violence was justified, but on the bureaucratic suppression of perpetrators' thinking.[51]

Such claims ignore two central ways in which situational social pressures are entangled with, rather than divorced from, ideology. First, sincerely internalized ideologies shape people's *understandings of the situations* they find themselves in, and therefore directly influence responses to social pressure. The key insight of situationist research is that human behaviour is often *not* driven by a person's deepest preferences. But as leading social psychologists emphasize, it is a mistake to think that situational pressures therefore operate 'independently' of subjective beliefs and understandings. On the contrary, 'it is the *meaning* that people assign to various components of the situation', Zimbardo emphasizes, 'that creates its social reality'.[52] Conformity to orders or group behaviour is not automatic: whether individuals perceive social pressure as emanating from a legitimate authority and valued peer group, or from an upstart pretender or distrusted outsiders, is crucial.[53] Milgram explicitly explains how:

> The perception of a legitimate source of control within a defined social occasion is a necessary prerequisite ... But the legitimacy of the occasion itself depends on its articulation to a justifying ideology ... Ideological justification is vital in obtaining *willing* obedience, for it permits the person to see his behaviour as serving a desirable end. Only when viewed in this light, is compliance easily exacted.[54]

The suggestion that dynamics of situational conformity imply that ideology's influence is weak simply misunderstands, as such, how social psychology links situational pressures to individual behaviour.

Second, the actual *direction* of social pressure—the actual behaviour encouraged in the situation—depends on the internalized ideologies and ideological structures of those involved in any given situation. Klusemann notes, for example, how the social pressure in particular communes in the early phases of the Rwandan Genocide depended on the relative balance between extremist promoters of

---

[48] Roth 2005, 206.
[49] Roth 2004, 237.
[50] Sofsky 2002, 18–19.
[51] Bauman 1989, 175–6 & 184–92.
[52] Zimbardo 2007, 221. See also Newman 2002, 51 & 60–2.
[53] Reicher, Haslam, and Rath 2008, 1320–3.
[54] Milgram 1974/2010, 143–4. See also Kelman and Hamilton 1989, esp. 17, 24, 29–30, 40–2, 56–7; Williams 2021, 64–82.

the genocide and more moderate opponents. Where extremists dominated, social pressures obviously encouraged many to participate in violence, but where moderates dominated, 'small signs of resistance were often enough to stop [genocidal] activists'.[55] Similarly, in SS units operating in Eastern Europe, actual or apparent support for Nazi norms of brotherhood, loyalty, and brutal martialism underpinned intense group pressures *to perpetrate*.[56] But in the French town of Le Chambon-sur-Lignon, immediate social pressures—generated by the moral leadership of the local pastors André Trocmé and Edouard Theis—encouraged the villagers *to resist* Nazi and Vichy authorities.[57] Situational pressures are not just 'given' by the world, but arise from group leaders, social norms, and the apparent collective attitudes of those individuals in the situation—all factors influenced by ideology.[58]

Again, the best situationist accounts readily acknowledge these interlinkages. Whilst several scholars cite Browning's influential research on German police battalions as evidence that ideology was unimportant,[59] this misunderstands Browning's arguments.[60] He explicitly emphasizes how a 'combination of situational factors and ideological overlap that concurred on the enemy status and dehumanization of the victims was sufficient to turn "ordinary men" into "willing executioners"'.[61] McDoom similarly highlights how interpersonal mobilization in Rwanda functioned in part through ideological socialization.[62] Employing 'Situational Action Theory', Kjell Anderson likewise emphasizes ideology and propaganda as constitutive of the moral context in which perpetrators act and to which they often conform.[63]

Consequently, the neo-ideological account I advance largely incorporates situationist arguments, appropriately framed. Situationists are right to emphasize that many perpetrators do not act out of deep ideological conviction and are heavily influenced by external social influence. Situationists also focus crucial attention on the cumulative radicalization of policies and norms within bureaucracies and local communities, and the forking and somewhat unplanned trajectories through which many mass killings therefore arise. But this challenges only the most monocausal 'true believer' accounts of ideology. An infrastructural account of ideology, by contrast, highlights the interdependence of ideology and situational pressures—recognizing that liberal, communist, fascist, or nationalist

---

[55] Klusemann 2012, 473. See also Straus 2006, ch.3.
[56] Kühne 2008, 59–65 & 68–72.
[57] Rochat and Modigliani 1995.
[58] See also Zimbardo 2007, 9–10 & 226–7.
[59] E.g. Roth 2004; Fujii 2009, 187.
[60] Browning 1992/2001, 184–6 & 194. It is also important to remember the scope of Browning's study: not a study of all Nazi perpetrators but of a specific group of drafted reserve policemen—*one subset* of perpetrators from *one case* of mass killing. See also Mann 2000, 333–9.
[61] Browning 1992/2001, 216.
[62] McDoom 2013, 464.
[63] Anderson 2017a.

ideological contexts vary in large part because of the different norms, discourses, and institutions that constitute them.

3.2 The Basic Argument: Ideologies and Political Crises

The Emergence of Justificatory Narratives for Mass Killing

This discussion of rationalist and situationist explanations of mass killing already illustrates some of the central reasons for ideology's importance. Because mass killings are so strategically indeterminate and normatively extreme, and because social pressures to engage in (or resist) them partly depend on prevailing norms, discourses, and institutions, ideologies are critical in determining whether individuals and groups will initiate, participate in, and support such violence. To some extent, this argument expresses a classically 'constructivist' idea: that people's willingness to engage in collective political action varies according to both the personal and social meaning of such action. But this idea is especially applicable to mass killings, because the strategic value and normative legitimacy of mass killing is so contestable and subject to such widely varying ideological interpretation.

In a 'neo-ideological' account, ideologies are crucial, not because they provide 'special' ideological motives or goals for violence, but because they vitally shape how mass killing is perceived and understood in times of crisis.[64] The critical role for ideology, here, is to provide some kind of *justificatory narrative* for the violence: a set of ideas about the meaning, character, and context of mass killing that makes it appear like a *strategically* and *morally* justifiable course of action.[65] Consistent with the infrastructural account of ideology provided in Chapter 2, I mean justifiable in a double sense here: justificatory narratives shape both the sincere *private assessment of mass killing as justified* and the perception of mass killing as *publicly justified by prevailing ideological norms, discourses, and institutions*. Within one common narrative, mass killings look like brutal, unnecessary, and wildly disproportionate atrocities against innocent victims, which violate accepted laws and norms. But a justificatory narrative provides an alternative framework of understanding, encouraging potential perpetrators to view mass killing as a necessary, potentially advantageous, and morally legitimate measure.

Many scholars might accept that perpetrators adhere, in some sense, to a justificatory narrative of this kind. The critical claim in my argument is that such narratives are not merely by-products or rationalizations of other 'deeper' motives

---

[64] In this, my argument follows closely in the footsteps of Scott Straus's work; see Straus 2015a.
[65] I thus follow Michael Walzer in recognizing that 'strategy, like morality, is a language of justification'; see Walzer 2000, 13.

or causes. They are not simply *bound to arise* when certain objective conditions create strong self-interested incentives or social pressures for mass killing.[66] Instead, justificatory narratives are vitally rooted in pre-existing ideology. Specifically, they emerge from the interaction of two factors: first, the relative *severity and character of political crises* facing a group or society, which can create potential opportunities, rationales, and pressures for violence against civilians; and second, the prior *infrastructural strength of 'hardline' ideas about security politics*, which shape how such crises and potential responses to them are understood and represented.

Pre-existing hardline ideas about security have a dual significance here. Most obviously, they are a key factor in determining whether a justificatory narrative for mass killing will be produced at all, thereby shaping *whether mass killings occur in the first place*. But the nature and strength of pre-existing hardline ideology also influences the contents of a justificatory narrative (its component ideas) and the likely distribution of that narrative as it becomes propagated (its relative strength in different organizations, areas, and segments of the population). Such ideas thereby shape the *character of mass killings* when they occur. This includes who the killings target—are justifications for mass killing orientated around ethnic or political victim groups, for example?—what the logic of the killing is—do perpetrators attempt to genocidally eliminate groups or terrorize them into submission, for instance?—and how it unfolds in different physical and organizational spaces—do some agencies or areas dominate the violence? Again, ideology is never the sole determinant here—strategic circumstances and social pressures also play a role—but it is crucial.

The exact relationship between pre-existing hardline ideologies and specific justificatory narratives varies across mass killings. In some cases, a justificatory narrative is essentially fully formed within existing hardline ideologies, and then simply *activated during crisis*. This was the case, for example, in the British area bombing of Germany in World War II, since the key justifications for targeting civilians during wartime were a feature of British conceptions of total warfare for years prior to the war.[67] But fully formed pre-crisis justifications for mass killing are not necessary for ideology to be crucial.[68] In other cases, the specific justificatory narrative is crucially rooted in pre-existing ideologies, but only *produced during crisis*. This was the case, for example, in the Rwandan Genocide, where justifications for large-scale killings of Tutsi depended on the regime's pre-existing notions of political community and social identity, but only emerged in serious strength in the 1990–94 civil war period preceding the genocide. In a small number of cases, even more extensive hardline radicalization has to occur during crisis to generate a justificatory narrative for mass killing, because pre-existing dominant

---

[66] Cf. Fearon and Laitin 2000, 846; Weinstein 2007, 21–2; Downes 2008, 25–6.
[67] See Chapter 6.
[68] Somewhat contra Downes 2008, 28.

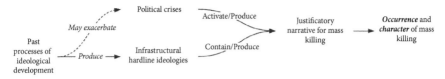

Fig. 3.1 The basic argument in brief.

ideologies are not clearly hardline. US area bombing in World War II is a possible example of such a case. Even in such cases, however, what pre-existing hardline ideas do exist plays a key role in encouraging subsequent radicalization.

This argument, summarized in Figure 3.1, is therefore *processual*—explaining mass killing as the outcome of quite complex, dynamic, and cumulative sequences of ideological radicalization rather than emerging directly from certain structural variables or static ideological plans.[69] Ideologies are constantly in motion, and the emergence of a justificatory narrative for mass killing may involve significant radicalization of pre-existing hardline ideologies during crisis. The strength and character of hardline ideologies at the outbreak of relevant crises will always be themselves, in turn, rooted in past processes of ideological developments that are to a large extent context-specific: one cannot properly explain the rise of Nazism, or Stalinism, or Maoism, or Serbian ethnonationalism without reference to specific historical contingencies of the contexts in which such ideologies emerged and evolved. Such processes of ideological development may also, themselves, be a factor in explaining the onset of the crises in which mass killing occurs.

Ideological Security Doctrines and Political Crises

What do I mean by 'hardline ideologies'? As explained in Chapter 1, I argue that the most important ideological foundations for mass killing lie, not in revolutionary ideals, utopian plans, or other 'special ideological goals', but in sets of ideas about security politics—radicalized versions of nevertheless familiar claims about threat, self-defence, punishment, war, and duty. They lie, in other words, in the prevailing *ideological security doctrines* within a particular group, organization, or society.

Different ideological worldviews do not only differ according to their preferred visions of economic policymaking, social values, or political institutions. They also incorporate different ideas about (i) the political community and political order that an ideology seeks to secure; (ii) the principal potential threats to that community and order; and (iii) the presumptive effectiveness and moral character of different forms of collective violence to defend and maintain that community and

---

[69] Mommsen 2009; Owens, Su, and Snow 2013; Murray 2015; Stanton 2016; McDoom 2021; Williams 2021. See also Pierson 2004.

Fig. 3.2 The spectrum of ideological security doctrines.

order.[70] Although not wholly unrelated to broader political goals or projects, it is these ideological security doctrines that are most important in shaping whether, and how, a justificatory narrative for mass killing emerges.

So, just as we might map an ideology's ideas about economics or social policy from left to right or liberal to conservative, I map different ideological security doctrines as straddling a spectrum between more *limitationist* and more *hardline* standpoints, as I call them (see Figure 3.2).[71] More *limitationist security doctrines*, as the name suggests, revolve around conceptions of security politics that emphasize important constraints on violence. For limitationists, the world contains specific threats linked to identifiable concentrations of violent capacities, which are fundamentally separate from civilian populations. This often reflects an inclusive vision of political order and community, in which rival identity groups or ideological factions are recognized as a constituent part rather than a suspect presence. Simultaneously, limitationalist doctrines depict violent force as costly, frequently ineffective, often avoidable, and constrained by important normative obligations.[72] The most limitationist doctrines of all, indeed, are pacifist.[73] *Hardline security doctrines*, by contrast, contain expansive conceptions of insecurity and legitimate force. For hardliners, the world contains numerous dangerous enemies which frequently operate in and through purported 'civilian' groups.[74] Often, though not always, this assumption reflects an 'exclusionary' vision of society, in which some social groups are consistently seen as lying outside the principal political community.[75] Large-scale violence, uncompromising force, and/or coercive domination, moreover, are seen as frequently effective and necessary for defeating such enemies, as deregulated from meaningful normative constraints, and as demanding the unwavering support of all 'loyal' members of the community. At the far end of this half of the spectrum, I describe certain ideologies as *ultrahardline* when they present certain civilian groups as permanent and intolerable threats. In the middle of the spectrum lie what I will call *permissive* security doctrines, which

---

[70] See also Wolfers 1952; Pion-Berlin 1988; Buzan, Wæver, and Wilde 1998; Buzan 2007.

[71] Though not mirroring their typologies, I draw on scholars of intellectual history and the philosophy of war in conceptualizing the continuum and its three categories; see, for example, Ceadel 1987; Nabulsi 1999. Chapter 4 examines the nature of hardline ideologies in more detail.

[72] Liberal ideologies tend to be limitationist, although their exact position on the security ideological spectrum can vary—for various formulations see Howard 1978; Walzer 2000; Tesón 2003; Fabre 2012. On the growth of limitationist ideas of civilian immunity see Bellamy 2012b.

[73] Fiala 2018.

[74] McDoom 2021, 163.

[75] Harff 2003; Straus 2015a.

do not assume significant normative or strategic limits on threats and warfare, but also do not possess an expansive notion of threat to the political community or order, or express high faith in large-scale violence.

Since hardline ideas about security, rather than longstanding ideological goals, are the most crucial link between ideologies and mass killing, it is the interaction between ideology and circumstances of political crisis that is crucial.[76] It is widely recognized that various forms and degrees of crisis, such as armed conflict, regime transitions, significant political unrest, serious economic downturns, and prominent terrorist attacks, generally provide necessary preconditions for mass killing.[77] Such crises are crucial in four principal and interrelated ways: first, they create a strategic context in which violence against civilians can appear potentially useful; second, they intensify psychological needs that promote support for more hardline ideas and policies; third, they generally weaken or destabilize established norms and procedures of non-violent politics; and fourth, they often shift power towards more hardline elites in control of violent capacities (sometimes even bringing new elites to power).[78]

To give an immediate sense of how I understand the interaction of ideology and crisis, Table 3.1 visualizes, in a *highly simplified* way, the expected occurrence of mass killing when limitationist, permissive, and hardline ideological security doctrines are combined with three types of crisis: *non-existential* (i.e. serious economic depressions, increased domestic political contestation and protest, escalating threat from a hostile foreign power, or limited armed conflict where there is no meaningful threat to the basic character of prevailing political order), *existential* (i.e. armed conflicts, coups, revolutions, or other forms of political instability that carry genuine potential to overturn or radically alter the prevailing political order),[79] and *supreme emergencies*[80] (i.e. scenarios involving both severe existential threats to the prevailing political order and extreme obstacles to any response that does not target civilians).

In theory, sufficiently clear and intense crisis incentives could encourage mass killings even when political elites adhere to permissive or limitationist ideologies. Hence, I theoretically anticipate mass killings in boxes #9, #8, and possibly #7 of

---

[76] In this set-up, the concept of crisis is quasi-objective—it refers to *actual observable conditions* of social instability, economic downturn, and political conflict, as distinct from perpetrators' perceptions of crisis (which are obviously shaped by ideology). This is a heuristic simplification to allow us to say something about the interaction of ideology and broader socio-political conditions. I fully recognize that what constitutes a crisis really depends on one's interests, standards, values, and expectations in any given historical context; see Buzan, Wæver, and Wilde 1998; Campbell 1998; Agamben 2005; Rousseau 2006; Buzan 2007.

[77] Fein 1993; Krain 1997; Harff 2003; Valentino, Huth, and Balch-Lindsay 2004; Mann 2005, 7; Wayman and Tago 2010; Uzonyi 2018.

[78] As well as the references in fn.77 see Staub 1989, ch.3; McDoom 2012; Straus 2012a, 546–8; Straus 2015b, 11–13; Anderson 2017a, ch.2; McDoom 2021, ch.3; Williams 2021, 178–82. On takeovers by new elites see Melson 1992; Walt 1996; Kim 2018; Tschantret 2019.

[79] This category effectively matches 'high-magnitude' conditions of 'political upheaval' as described in Harff 2003, 66, or most conditions of 'war' and 'political upheaval' as described in Williams 2016.

[80] I take this phrasing from the ethics of war literature; see Coady 2002; Bellamy 2008.

68 IDEOLOGY AND MASS KILLING

Table 3.1 The simplified interaction of ideology and crisis in mass killing

		Ideological security doctrine		
		Limitationist	Permissive	Hardline
Crisis conditions	Non-existential crisis	#1 ✘No mass killing	#2 ✘No mass killing	#3 ✘(✓) No mass killing, unless *ultrahardline* security doctrine
	Existential crisis	#4 ✘No mass killing	#5 ✘No mass killing	#6 ✓ Mass killing
	Supreme emergency	#7 ✘(✓) No mass killing, unless more limited civilian targeting impossible	#8 ✓ Mass killing	#9 ✓ Mass killing

Table 3.1. But my argument is that these 'supreme emergency' scenarios hardly ever arise *in practice*. It is highly unlikely that a regime or group will simultaneously (i) possess sufficient power to successfully engage in mass killing and (ii) face a situation where there are no plausible alternatives to using such extreme violence (this is a manifestation of what Scott Straus refers to as the 'domination–vulnerability paradox').[81] So few if any real-world cases are properly located in boxes #7, #8, and #9. Mere permissive or limitationalist ideologies are associated, at most, only with relatively selective and restrained forms of violence against civilians. Actual mass killings almost always require a prevailing hardline ideology.

The vast bulk of cases, I argue, lie in box #6—where groups or regimes with prevailing hardline ideological security doctrines encounter serious crises involving potentially existential stakes. But a major subset of cases lies in box #3, where much more limited forms of political crisis nevertheless interact with an ultrahardline security doctrine. The Great Terror under Stalin was such a case, while early phases of mass killing in Guatemala's civil war sit roughly on the boundary between box #3 and box #6. Between them, box #3 and box #6, outlined in bold, account for almost all real-world mass killings.

There are potential, although I think partial, exceptions. As suggested above, in a few cases—such as the US involvement in area bombing in World War II—it is not clear that prevailing ideologies were truly 'hardline' prior to the crisis in

---

[81] Straus 2015a, 56.

which mass killings occur. Instead, hardline ideas only come to dominate *during crisis itself*. How far such cases constitute exceptions to my argument depends on how such hardline radicalization within crisis occurs. If pre-existing ideology is essentially irrelevant, and radicalization seems purely rooted in crisis incentives and pressures, then ideology's significance does diminish. But this is generally not the case. As I show in Chapter 6, while the United States did not possess an unambiguously hardline security doctrine on entry into World War II, crucial support for hardline ideas concerning the area bombing of civilians did exist in the elite circles that most mattered, especially amongst senior figures of the US Army Air Force. The rise of those hardline ideas to dominance during the war was significantly affected by non-ideological factors—especially the limited effectiveness of alternative 'precision bombing' strategies—but this remains insufficient to explain mass killing. Area bombing was still strategically indeterminate and normatively extreme, and the existing strength of hardline ideas was key in explaining why US policymakers opted for such a course.

I re-emphasize that this argument is purely focused on *ideology's role in mass killing*. Other causal factors, beyond ideology and crisis, will be critical in three major ways. First, they are likely to play a major role in the emergence of crises themselves—so a range of existing theories focused on the origins of different kinds of existential crises must be 'plugged into' my account here. Second, they may play a role in fuelling hardline ideological radicalization (although again, ideology can never be reduced to such non-ideological factors). Third, non-ideological constraints—such as limited perpetration capacity, a low vulnerability to civilian groups, or external intervention or pressure—ensure that many scenarios in boxes #3 or #6 will still not see mass killing.[82]

I also repeat that Table 3.1 is a *highly simplified* sketch of the genesis of mass killing, whereas my actual argument is intended to foreground complexity. Neither security doctrines nor crises slot neatly into three categories—they are really matters of degree. So my basic causal contention is that *the more hardline existing ideologies are, the less serious the crisis needed to provoke mass killing; conversely, the more a political crisis creates potential pressures for mass killing, the less hardline prevailing ideologies need to be*. In addition, the exact interaction between these two factors depends on the particular ideological content and crisis dynamics of a given case—it is not enough to simply note the 'presence' of both hardline ideology and crisis. Ideologies might also take more or less hardline stances on domestic or international security, or towards different civilian groups.[83] To truly explain the occurrence and character of a given mass killing, we have to trace the interaction

---

[82] Robinson 2010; Straus 2012b; Straus 2015a, 76–7; Stanton 2016.
[83] Many historical forms of liberalism, for example, have been orientated around a limitationist security doctrine towards domestic citizens, but contained permissive or even hardline doctrines towards domestic 'aliens' or foreign civilians (especially those deemed racially 'other'). See also Gong 1984; Walker 1992; Cromartie 2015; Mills 2015, ch.6.

of context-specific hardline ideas and context-specific crisis dynamics in influencing the various actors that collectively perpetrate the violence. Throughout the remainder of the book, I seek to layer back in much of this complexity.

3.3 Ideology and Perpetrator Coalitions

Building a Perpetrator Coalition

As observed in Chapter 1, mass killings are not simply conjured into being by the diktat of political leaders 'from above' or through spontaneous outbreaks of hatred 'from below'. Instead, mass killings are complex campaigns of collective action that require contributions from three main categories of actor.[84] *Political elites* in positions of power (whether within whole societies or specific perpetrating groups) must generally instigate, authorize, and organize policies of violence. *Rank-and-file agents* mobilized from the military, paramilitary units, security agencies, or the ordinary citizenry must carry out the violence. Broader *public constituencies*, in the form of organizations, groups, and communities in society at large, must be prevented from organizing effective resistance and (to some degree) be induced to offer perpetrators their support.[85]

Again, perpetrators across these three categories are diverse—guided by various motives, and implicated in mass killing through various forms of participation.[86] In this sense, mass killings are implemented by *perpetrator coalitions*: interconnected networks of individuals guided by different motives but acting collectively to generate violence.[87] Such coalitions are not ideologically homogenous, but they are primarily created and maintained by a *dominant hardline faction*: i.e. a critical mass of elites, rank-and-file agents, and public constituents who either sincerely believe in or more opportunistically embrace a justificatory narrative for mass killing rooted in hardline ideology. These 'hardliners' readily justify mass killing privately and publicly, and work actively to orchestrate and implement it, while castigating critics and corralling non-hardliners to participate. Such hardliners organize and perpetrate much of the violence themselves, but also align

---

[84] Adapted from Mann 2005, 8–9. See also Kalyvas 2003; Kalyvas 2009, 609–11; Gerlach 2010; Owens, Su, and Snow 2013.

[85] These three categories are not entirely separate, however. Elites may participate directly in violence, blurring their distinctiveness from rank-and-file agents: in Stalin's 'Great Terror', for example, the heads of local NKVD (People's Commissariat of Internal Affairs) branches were expected to carry out official executions; see Shearer 2009, 359. Similarly, the more that the mobilization of rank-and-file perpetrators occurs directly in local communities—as in Rwanda, China's Cultural Revolution, or Nazi pogroms in the Baltic States—the more dynamics amongst rank-and-file perpetrators and local public constituencies merge; see Matthäus 2007; Fujii 2009; Su 2011.

[86] Staub 1989, 39–43; Mann 2000, 332–3; Mann 2005, 26–30; Roseman 2007, 93–5; Fujii 2008; Smeulers 2008; Gerlach 2010, 1–7; Anderson 2017a, 99–114; Williams 2021.

[87] Anderson 2017a, 99–101. See also Gerlach 2010, ch.2.

political norms, discourses, and institutions with their hardline ideological stance, drawing other (often more reluctant) participants into the perpetrator coalition. Perpetrator coalitions are, in other words, sustained by a hardline ideological infrastructure, centred around a justificatory narrative for mass killing, which enables, encourages, and shapes the violence.

Hardliners are, themselves, somewhat heterogenous. The hardline faction behind the Holocaust, for example, was not simply guided by an undifferentiated Nazism. Instead, Nazism represented an ultrahardline driving force for key elites around Hitler—themselves adhering to somewhat varying anti-Semitic perspectives[88]—which was complemented by a more traditional conservative nationalism prevalent amongst many civil servants and military elites. As Browning explains:

> Nazi Jewish policy had only to be wrapped in the flag of German patriotism to assure its vigorous defense by the professional diplomats. The state secretaries and their generation of civil servants were not motivated so much by anti-Semitism as by their desire for a nationalist revival of Germany, conceived primarily in terms of their nostalgic memories of the Imperial era ... [But] even if anti-Semitism was not the driving force behind their actions, they were still not entirely free of this affliction ...[89]

This hardline faction will never extend to a whole society or even a whole perpetrator coalition, and they often face opposition. The hardliners may even be a minority, but one able to use positions of influence, greater coordination, more intense propaganda, and coercion to dominate public political space and obtain disproportionate ideological dominance.[90] Either way, mass killings occur when such hardliners are sufficiently numerous, active, and well placed that non-hardliners find themselves sincerely outnumbered and/or structurally outflanked.

For many members of the perpetrator coalition, hardline justificatory narratives may actively *motivate* the initiation or implementation of mass killing. If internalized, such narratives may directly elicit feelings of anger, contempt, fear, or group loyalty that generate an active desire to participate in violence. Structurally, justificatory narratives may directly create normative pressures to participate in order to conform to the group or generate new instrumental reasons for participation. For other perpetrators, however, justificatory narratives may primarily serve to *legitimate* the violence—i.e. to neutralize private or public normative restraints that discourage participation. There is a tendency in some scholarship to focus narrowly on motivation and treat legitimation as relatively epiphenomenal. Yet this tendency lacks empirical or theoretical backing. It is a point of consensus in

---

[88] See some of the variation observed in Roseman 2007, 90–9.
[89] Browning 1980, 188.
[90] Bhavnani 2006.

modern social and psychological theory that the need to privately and publicly legitimate behaviour independently and powerfully shapes action.[91] As one leading survey of research on legitimacy summarizes:

> Legitimacy is crucial to impression management as well as to developing a meaningful sense of the self as a worthwhile and valid individual ... People are required ... to justify their attitudes and behaviours and to demonstrate that they are acting in a legitimate manner. Even privately, we seek to develop rationalizations for our own thoughts, feelings and actions ... the carrying out of extreme acts of exploitation, violence and evil is *socially* and *psychologically* feasible only to the extent that perpetrators are able to make their actions seem legitimate.[92]

So even when perpetrators of mass killing may be relatively lacking in ideological motives—when, for example, they are guided primarily by petty personal animosities and selfishness—the capacity to ideologically legitimate the violence typically remains crucial.[93]

Again, the emergence and triumph of a perpetrator coalition is generally the culmination of a process of ideological radicalization stretching back years or even decades. Such radicalization is partly fuelled by broader social and political changes, including the onset of crisis and, ultimately, mass killing itself. But it always also depends on significant *ideological activism*, in the form of both media propaganda and more face-to-face political agitation, including the eventual dissemination of an actual justificatory narrative. The effectiveness and importance of such efforts is a central issue in debates over ideology's role in mass killing. Certainly, the power of ideological activism is easily exaggerated.[94] 'Hypodermic needle' portrayals of propaganda, in which exposure to virulent hate speech is assumed to smoothly transform audiences into massed ranks of zealous perpetrators, are obviously too crude. Populations are not vacant recipients of propaganda, and have been found to readily reject ideas that lack psychological appeal or are highly dissonant with their existing beliefs.[95]

In rejecting exaggerated images of propaganda's power, however, some scholars veer too far to the opposite sceptical extreme, by claiming that the available evidence demonstrates that propaganda has little real impact on action.[96] This is inaccurate. Existing research on propaganda and other forms of ideological

---

[91] Skinner 1974, 292–4; Jost and Major 2001, 5; March and Olsen 2008; Malešević 2010, 9–10; Barkan, Ayal, and Ariely 2015.
[92] Jost and Major 2001, 5 (*my emphasis*).
[93] Thomas 2001; Bellamy 2012b, 27–35; Savage 2013.
[94] Straus 2007b; Benesch 2012a; Wilson 2016; Wilson 2017.
[95] Snow and Benford 1988; Petersen 2002, 251–2; Gutiérrez Sanín and Wood 2014, 222.
[96] E.g. Danning 2018; Danning 2019. More modest expressions of scepticism can be found in Straus 2007b, 614–15; Wilson 2011; McDoom 2021.

activism clearly demonstrates their real and substantial effects on audiences, including in contexts of violence. In the most systematic study thus far of the role of radio propaganda in interwar Germany, for example, Maja Adena et al. found that propaganda profoundly affected general voting patterns in parliamentary elections between 1928 and 1933—the period of the Nazi rise to power.[97] Importantly, Adena et al. leveraged a significant change in the nature of the propaganda to prove its independent impact. In elections between 1928 and 1933, radio issued pro-Weimar government (and anti-Nazi) propaganda and had a negative impact on Nazi support, whereas after the Nazis took control of most radio propaganda, this effect reversed, with exposure to radio significantly increasing Nazi support. Most significantly of all, Adena et al. found that once controlled by the Nazis, radio propaganda not only strengthened their popularity, but also increased violence and discrimination against Jews in the areas where propaganda was most intense. Historical studies likewise emphasize that 'letters to and from the eastern front betray an extraordinarily widespread internalization of official propaganda'[98] and that 'Nazi indoctrination in fact had a major and insufficiently acknowledged impact on the perception of reality of all ranks in the German army during the war'.[99] Numerous comparable studies in other contexts confirm the basic finding: that exposure to ideological activism, whether pro-violence or pro-peace, can profoundly shape actual levels of violence, feelings of fear, and/or broader support for violent institutions and policies.[100] Propaganda's impact may often be indirect, providing ideas and narratives which are then modified and disseminated 'horizontally' through peer interactions, grass roots mobilization, and rumour rather than merely being accepted 'top-down' from elite propagandists.[101] Nevertheless, in all four of my case studies, the balance of evidence suggests that ideological activism, in various forms, exerted substantial effects on rank-and-file agents and ordinary citizens.

Rather than being a crude vehicle for mass brainwashing, ideological activism operates infrastructurally. Propaganda, official statements, and face-to-face agitation can promote genuine *internalization*, to some degree, of justificatory narratives for mass killing. Three conditions make such internalization more likely. First, internalization is encouraged in contexts of high uncertainty and/or 'epistemic dependence'—i.e. when audiences are reliant on specific sources of information whose claims they cannot easily verify or falsify.[102] As noted in Chapter 1, such epistemic dependence is common in security politics, especially

---

[97] Adena et al. 2015.
[98] Fulbrook 2011, 187.
[99] Bartov 1994, 137.
[100] See, for example, Paluck 2009; Paluck and Green 2009; Collier and Vicente 2013; Yanagizawa-Drott 2014; Abrajano and Hajnal 2015, ch.5; Petrova and Yanagizawa-Drott 2016; Rigterink and Schomerus 2017; Pauwels and Hardyns 2018; Barber and Miller 2019.
[101] Das 1998; Mironko 2007, 128; Benesch 2012c; Jessee 2017, 164.
[102] See, in general, Hardwig 1985; Zaller 1992, ch.2; Hardin 2002; Baurmann 2007; Rydgren 2009, 83–7.

in crisis environments.[103] Second, audiences are more likely to accept radicalizing justificatory narratives when they are largely consistent with their pre-existing views, values, and desires—a consequence of 'motivated reasoning'.[104] This does *not* mean, however, that activism *merely reinforces* what audiences already believe.[105] Again, ordinary citizens can rarely 'self-supply' a full understanding of political events, so what interpretations are made available and plausible by prevailing social discourses is a critical variable. Third, the sheer intensity of hardline activism, so that it saturates or virtually monopolizes public discourse, can also promote real internalization even when hardline claims diverge from people's prior convictions. As Lewandowsky et al. note in their survey of research on misinformation, mere '[r]epeated exposure to a statement is known to increase its acceptance as true'.[106] Since these three factors vary even within a particular case of mass killing, the impact of ideological activism is always *uneven*. But the aggregate effect of such propaganda can be considerable, and important in sustaining a coalition willing to implement and support collective violence.

In addition, however, ideological activism is a powerful mechanism for generating or altering *structural* expectations about ideological norms, discourses, and institutions.[107] Again, this power is to a large extent a function of the sheer intensity of activism, since, as Lewandowsky et al. observe, '[r]epetition effects may create a perceived social consensus even when no consensus exists'—a form of 'pluralistic ignorance'.[108] The views of a hardline minority can come to appear dominant, incentivizing moderate but fractured majorities to stay silent and comply with that minority ideology.[109] But activism by senior political elites is also crucial in defining officially sanctioned activities, thereby leveraging the motivational and legitimating authority of the state (or other institutions) to influence public behaviour. The capture or prior control of positions of institutional authority may therefore be crucial to hardliners' influence over ideological structures, and they often enforce hardline norms in part through coercion.[110] Often, the onset of crisis, conflict, and violence is itself crucial: destabilizing existing political norms and institutions and 'endogenously' facilitating the radicalization of norms, discourses, and institutions. An unsurprising consequence is that the *most*

---

[103] This is the central problem with de Figueiredo's and Weingast's suggestion that extremist ethnic conflict requires genuinely alarming social conditions or provocative actions by other ethnic groups; see de Figueiredo and Weingast 1999. Ordinary citizens rarely experience such conditions or actions directly, and extremism often thrives against 'groups' that have engaged in no such provocative actions.
[104] Jost, Federico, and Napier 2009; Nyhan and Reifler 2010.
[105] Nyhan and Reifler 2010, 308; Adena et al. 2015.
[106] Lewandowsky et al. 2012, 113.
[107] Anderson 2017a, 69.
[108] Lewandowsky et al. 2012, 113.
[109] Noelle-Neumann 1974; Kuran 1989; Hardin 2002; Williams and Pfeiffer 2017, 78.
[110] Bhavnani 2006. Again, however, the role of coercion here is easily overstated—hardliners can rarely coercively micromanage a population into mass killing without more agential forms of participation, collaboration, and peer-enforcement of norms.

extreme activist rhetoric often appears after mass killings begin.[111] This should caution researchers against automatically fixating on the most egregious, rather than most plausibly consequential, examples of hardline activism. But it does *not* imply that broader ideological activism is unimportant—the influence and dissemination of key hardline ideas remains a visible part of the escalation to mass killing, and does not solely emerge as a 'post-hoc rationalization' once violence is underway.

Effective ideological activism is therefore generally crucial, both in pushing forward early phases of ideological radicalization and in eventually disseminating a justificatory narrative for mass killing. Such activism can consequently make the difference between a horrifyingly efficient and brutal mass killing and a fragmentary and haphazard campaign. But the role of hardline justificatory narratives in mass killing varies across political elites, rank-and-file agents, and public constituencies.

Political Elites

I define political elites as individuals in positions of authority within state agencies or non-state political organizations, and draw a distinction between *apex elites* in the senior policymaking circles of a regime or other perpetrating group and *intermediary elites* (such as mid-ranking bureaucrats, regional military commanders, and local political office holders) operating between apex elites and rank-and-file agents or public constituents. Both apex and intermediary elites tend to be *relatively* ideologically committed in mass killings. Their ideologies certainly operate alongside—and sometimes through—self-interest. Elites seek to retain power, protect their economic interests, and advance their personal careers, and ideology often serves as a tool for achieving such ends. But, given the diverse strategies that elites generally *could* pursue to advance their self-interest, and to deal with the crises they face, sincerely held hardline ideas generally play a crucial role in explaining elite decisions to pursue mass killing. Even if elites are rather ideologically non-committal, moreover, the ideological structures of prevailing political institutions and norms still matter. Without some prior structural foundation for a hardline justificatory narrative, potential organizers of mass killing will struggle to advance such a narrative, mobilize supporters, and dominate institutional decision-making in the ways necessary to prevail over more 'moderate' opponents.

The infrastructural strength of hardline ideology amongst elites affects three principal outcomes of interest. First, by generating justificatory narratives for mass killing, elite hardline ideology shapes *whether mass killings are initiated or not*. Scott Straus's analysis of occurrences and non-occurrences of genocide

---

[111] Danning 2019, 107–11; McDoom 2020b; McDoom 2021, 117–22.

in sub-Saharan Africa, for example, emphasizes how political elites do not respond to similar national crises in the same way. In Rwanda and Sudan, political elites pushed forward radical policies of violence and genocide in an attempt to eliminate political enemies or internal groups deemed hostile to those elites' vision of the national community. In Mali, Côte d'Ivoire, and Senegal, by contrast, whilst violence and repression were not absent, elites eschewed policies of genocide and mass killing despite similar crisis conditions. Straus concludes that 'to explain variation—to explain why countries with similar crises experience different outcomes—the role of ideology is essential'.[112] Similarly, Zeynep Bulutgil compares the policies of the Austro-Hungarian Empire, Tsarist Russia, and the Ottoman Empire towards ethnic minorities during World War I. Again, despite similar crisis conditions of war and tense ethnic relations, these states pursued contrasting policies: Austria–Hungary engaged in limited deportations of Italians living close to the front lines, Russia engaged in targeted massacres of Muslims, while the Ottoman government engaged in mass deportations and genocide. Bulutgil finds that political leaders' ideological priorities are crucial: leaders from parties whose ideologies emphasized ethnic cleavages were much more likely to frame ethnic minorities as dangerous threats and present lethal violence as necessary for national security.[113] Donald Bloxham similarly observes how, while Ottoman mass killing was motivated by wartime fears of nationalist Armenian uprisings, such fears were widespread amongst the imperial empires fighting in World War I, yet 'nowhere else during the First World War was revolutionary nationalism answered with total murder'.[114] The role of the Ottoman elite's ideology in exaggerating fears, promoting ambitions for ethnic homogeneity, and eliminating compunctions over the killing of overwhelmingly unrebellious Armenians was crucial.[115] Several statistical studies reach similar conclusions.[116]

Second, by generating a justificatory narrative for mass killing, elite hardline ideology shapes the *logic of violence*—in terms of both its *functions* and its *targets*. Scholars have long appreciated that violence against civilians can follow different functional logics. Many follow Kalyvas in suggesting that selective violence against civilians in civil war, for example, usually follows a *coercive logic*, aimed at controlling civilian populations.[117] By contrast, genocidal violence follows a distinct *eliminationist* logic aimed at wiping out certain population groups entirely.[118] Mass killings might also follow an *attritional* logic, which seeks to weaken specific social groups by broad collective targeting without eliminating the group as

---

[112] Straus 2015a, x.
[113] Bulutgil 2017.
[114] Bloxham 2003, 186.
[115] Ibid. 187–91.
[116] Harff 2003, 62–3 & 66; Kim 2018, 303–8; Uzonyi 2018, 483.
[117] Kalyvas 2006; Kalyvas 2012.
[118] Straus 2015a, ch.1.

such,[119] or an *overkill* logic, in which disproportionate levels of violence are unleashed on civilian populations to ensure that more specific threats within those populations are destroyed, with excess civilian deaths deemed insignificant and even a low risk of threats surviving deemed intolerable.[120] In Cambodia under the Khmer Rouge, citizens recalled: 'we were told repeatedly that in order to save the country ... it was essential to cut deep, even to destroy a few good people rather than chance one "diseased" person escaping eradication'.[121] Interlinked with this functional logic of violence is its targeting logic—individuals may be targeted based on their behaviour, location, ethnicity, religion, political allegiance, class, interpersonal associations, or a range of other attributes.

Rationalists rightly emphasize that the functional and targeting logics of violence against civilians are heavily influenced by strategic circumstances. Certain civilians, for example, may be hard to directly control but linked to real threats to perpetrators by providing a source of recruitment for insurgent movements or industrial output vital for an enemy state's military operations. Such contexts are likely to encourage attritional or overkill violence.[122] Similarly, perpetrators may employ eliminationist, attritional, or overkill logics simply because they cannot implement the kind of discriminate violence needed for effective coercion.[123] But such arguments only take explanations of mass killing so far, because there is enormous variation in how perpetrators perceive such civilian–threat linkages. Some civilians, for example, are associated with dangerous threats even when this is utterly fantastical—as with the Nazis' belief in a Jewish world conspiracy operating behind the governments of all their enemies. More commonly—as with the targeting of Tutsi in Rwanda, Armenians in Ottoman Turkey, or ethnic Vietnamese in Cambodia—such a perceived link is comprehensible but a wildly exaggerated reading of the available evidence. Even in cases where there were real links between civilians and threats to perpetrating elites—as in the latter phases of mass killing in Guatemala, or the Allied bombing of residential areas containing dispersed industrial manufacturing units in Japan—perpetrators can conceptualize the 'civilian enemy' in a wide range of ways. Are the enemy the entire civilian population of an adversary's territory? Are they particular ethnic or religious groups seen as uniformly 'suspect'? Or should the 'enemy' be confined as far as possible to those engaged in active collaboration? Elites vary considerably in how they answer such questions according to distinct ideological understandings of security.

Properly explaining mass killing often, as such, involves untangling quite complex relationships between ideological content and strategic calculation. Aristotle Kallis shows, for example, how World War II created strong strategic incentives for

---

[119] Fjelde and Hultman 2014; Esteban, Morelli, and Rohner 2015.
[120] Valentino, Huth, and Balch-Lindsay 2004.
[121] Cited in Anderson 2017a, 74.
[122] Downes 2008; Fjelde and Hultman 2014.
[123] Kalyvas 2006, ch.6.

the Nazi government to pragmatically relax a range of policies—such as restrictions on female participation in the labour force, incarceration of homosexuals, and the annihilation of Jews, Roma/Sinti, Slavs, and the disabled—which reflected Nazi ideals but undermined the war effort. In some cases, the Nazis proved responsive to such strategic considerations: relaxing restrictions on the use of female labour from 1942 and redesignating certain (Aryan) homosexual groups as admissible for military service. But with 'inferior races' the Nazis made few concessions to strategic pragmatism; indeed, violent persecution radicalized dramatically during war. As Kallis explains, this varying Nazi sensitivity to strategic incentives was itself rooted in the Nazis' varying ideological understandings of these issues. The shift in policies towards Aryan homosexuals, for example, 'was possible on the basis of both their otherwise "racial" value and the pliability that the notion of [certain forms of homosexuality as a] "curable condition" allowed—two elements that were conspicuously lacking in the case of other groups, such as the Jews and the mentally ill'.[124]

Third, the ideological orientation of different political elites, especially intermediary elites, may explain important *intracase variation* in how mass killings are implemented. Intermediaries such as military commanders, state bureaucrats, and leaders of local communities often possess considerable agency in shaping how mass killings are implemented, or subverted, in their area of political authority. Hardline intermediary elites may willingly carry out the orders of apex elites and enthusiastically drive violence forward, whereas non-hardline intermediaries can half-heartedly implement their orders, ignore directives from superiors, or actively work to obstruct policies of mass killing.

Within the Nazi bureaucracy, for example, important discontinuities in policy were sometimes generated by ideological divergences between officials. In the Interior Ministry, the 'Jewish expert' Bernhard Lösener—who had helped draft the Nuremberg Laws stripping Jews of German citizenship—dissented from the classification of 'Mischlinge' Jews (those with mixed German-Jewish ancestry) as requiring extermination, and fought to maintain Mischlinge status as a distinct legal category protected from deportation and killing. Within the Middle East desk at the Foreign Office, the diplomat Wilhelm Melchers—more wholly unsympathetic to hardline anti-Semitism—used his influence to successfully block efforts to deport and exterminate 2,400 Turkish Jews living in Paris, as well as a number of Palestinian Jews who fell into German hands.[125] During the invasion of Poland in 1939, senior German army officers similarly protested against Nazi instructions to slaughter groups of Polish civilians, with Hitler eventually rescinding the orders.[126] The contrast between this and the responses of military officers later in the war, after more prolonged radicalization, is marked.

---

[124] Kallis 2005, 14–15.
[125] Browning 1980, 193–4.
[126] Fein 1979, 4; Matthäus 2007, 220.

Geoffrey Robinson highlights similar dynamics in Indonesia's mass killings of alleged communists in 1965–66, where 'temporal and geographical variations in the pattern of mass killing corresponded closely to the varied political postures of army commanders in a given locale'.[127] Where more hardline anti-communist officers dominated, violence generally started earlier and was pursued more intensely. The deployment of hardline commanders was therefore a crucial dynamic of elite mobilization—as Robinson summarizes: 'where the army command was politically divided, faced resistance or did not have sufficient troops at its disposal, the mass killing was delayed for some time but then accelerated dramatically when the balance of forces tipped in favour of the anti-communist position'.[128] Adam Scharpf finds a similar pattern in Argentina's 1975–81 'Dirty War'. The Argentinian junta launched a campaign of political terror guided by a hardline ultraconservative-nationalist worldview which 'saw the country at the center of a communist world conspiracy'[129] so that 'teachers, students, unionists, and everybody holding liberal, Marxist or anti-Catholic values became a viable target'.[130] Yet military commanders in different parts of the country did not all subscribe to this hardline narrative. More liberal officers 'disagreed with the nationalists' solution for the subversive problem ... [and suggested] selective targeting of enemies rather than the indiscriminate re-engineering of Argentina's social fabric'.[131] Scharpf shows how the placement of officers from more liberal or nationalist branches of the army in command of territorial subunits of Argentina corresponded to significant differences in the local intensity of violence. '[V]ariation in ideological convictions', Scharpf concludes, 'explains why military officers differ in their zeal to carry out repressive orders.'[132] Bloxham and Straus reach similar conclusions for the Armenian and Rwandan Genocides respectively: with the onset and escalation of genocidal violence linked to the dominance or arrival of hardliners in specific locales.[133] In all such circumstances, the ideological orientation of intermediary elites is a key factor in determining how mass killings actually unfold.

Rank-and-File Agents

While the calculations of political elites are the most critical determinant of the occurrence and character of mass killings, they have to mobilize and organize 'rank-and-file' agents to actually carry out the violence. Following Straus, I consider 'any person who participate[s] in an attack against a civilian in order to kill or to inflict serious injury on that civilian'[134] to be a rank-and-file perpetrator of

---

[127] Robinson 2017, 466.
[128] Ibid. 468.
[129] Scharpf 2018, 2.
[130] Ibid. 7.
[131] Ibid.
[132] Ibid. 14.
[133] Bloxham 2003, 168–9, 177, & 179; Straus 2006, ch.3.
[134] Straus 2006, 102.

this kind. This includes those who physically accompany and support the direct killers but do not commit final acts of lethal violence. As noted in Chapter 1, such mobilization of rank-and-file agents is not automatic: killing is psychologically challenging for most people, and depictions of rank-and-file perpetrators as either innately murderous or as automatons who 'simply comply' with orders to kill are inaccurate.[135] Elites need to find ways to motivate rank-and-file agents to engage in killing, to legitimate the violence sufficiently to avoid large-scale resistance or evasion, and to organize the violence so that it remains tied to elite objectives.[136]

In almost all cases, some degree of coercive pressure from elites plays a role in such mobilization. But elites can rarely *coercively micromanage* rank-and-file agents, and such agents usually possess a repertoire of possible responses to elite instigation of mass killing.[137] They might enthusiastically participate and use initiative to creatively expand the scale of violence. They might reluctantly implement the violence under pressure. Or they might engage in oppositional activities of foot-dragging, evasion, non-implementation, or outright resistance. Kühne offers a vignette, for example, of the contrasting attitudes of three junior officers of a German infantry battalion involved in the Holocaust. Ordered to kill the Jewish population of an occupied Russian village, 'Lieutenant Kuhls, a member of the Nazi Party and the SS, carried out the order with his company without hesitation'. By contrast, 'Lieutenant Sibile, a teacher aged 47 ... told his superior officer than he "could not expect decent German soldiers to soil their hands with such things" [and] said his company would only shoot Jews if they were [established to be] partisans'. Captain Friedrich Nöll, meanwhile, was gravely disquieted by the order, but 'after initial evasiveness ... reacted as ordered', though passed the duty on to his company sergeant major.[138]

A huge array of factors may induce relatively willing, active implementation of violence by rank-and-file agents—Williams, for example, identifies fifteen basic motives guiding perpetrators—many of which are largely unrelated to ideology.[139] Besides coercion, private animosities and greed often encourage violence when given official sanction,[140] interpersonal relationships may draw individuals into perpetration,[141] and some perpetrators use alcohol or other drugs to ease the act of killing.[142] Many rank-and-file perpetrators more generally testify to feelings of otherworldliness, 'deindividuation', and 'derealization'[143]—what Barbara

---

[135] Collins 2008. Contra Roth 2004, 232.
[136] See also Hoover Green 2016. On the effects of distance see Waller 2007, 196–7; Littman and Paluck 2015, 91.
[137] Browder 2003, 488; Fujii 2008, 574–6.
[138] Kühne 2008, 55.
[139] Williams 2021, ch.1.
[140] See, in general, Kalyvas 2003; Humphreys and Weinstein 2006; Fujii 2009.
[141] McDoom 2013.
[142] Olusanya 2014, 97–8; Littman and Paluck 2015, 92.
[143] Olusanya 2014, 98–9.

Ehrenreich has termed 'altered states'[144]—in explaining their mindset in the act of killing itself. Consider the following testimony from a perpetrator of the My Lai massacre of civilians in a Vietnamese village by US troops in 1968:

> That day in My Lai, I was personally responsible for killing about 25 people. Personally ... I just did it. I just went. My mind just went. And I was not the only one that did it. A lot of other people did it. I just killed. Once I started the ... training, the whole programming part of killing, it just came out ... I had no feelings or no emotions or no nothing. No directions. I just killed.[145]

Such psychological processes are often encouraged by violence itself. An extensive literature demonstrates that killing brutalizes perpetrators: habituating or numbing them to violence, or leading them to identify with groups in which violence has become the social norm.[146] Such brutalization may be more crucial than ideological radicalism in explaining the horrifying forms of 'extra-lethal violence',[147] such as mutilation, inventive cruelty, and torture, that accompany much mass killing.

Because of the prevalence and strength of all these factors, I expect ideology to produce less dramatic variation in rank-and-file behaviour by comparison with political elites. Nevertheless, in the absence of coercive micromanagement, hardline ideology—especially a core justificatory narrative for mass killing—is generally crucial in leveraging various motives to participate in violence. This affects two principle aspects of mass killing.

First, the strength and distribution of hardline ideas amongst rank-and-file agents shapes *the varying intensity of both violence and resistance* within mass killing. The especially intense and early perpetration of the Holocaust in Lithuania, for example, reflected the relative abundance of enthusiastic perpetrators guided by radical anti-Semitic nationalism, and local pogroms sometimes preceded or occurred in the absence of explicit Nazi directives. As Jürgen Matthäus concludes: 'the Holocaust ... could not have evolved on Lithuanian soil if imported German violence had not harmonized with residual anti-Jewish sentiment among the local population'.[148] Su likewise shows how violence in the Chinese Cultural Revolution was often enthusiastically carried out by local hardliners even against the government's intentions and was often higher in areas of weak state control.[149] Even when units *are* under the command of central elites, the indirect nature of such control means that the ideological makeup of rank-and-file perpetrator groups is key. In Cambodia, for example, Khmer Rouge killings were often intense not because of

---

[144] Ehrenreich 1997, 12. See also Dutton, Boyanowsky, and Bond 2005, 464–9; Anderson 2017a, 19.
[145] Cited in Smeulers 2004, 244.
[146] See, in general, Valentino 2004, 55–6; Dutton 2007, 116–22; Waller 2007, 242–7; Littman and Paluck 2015; Mitton 2015, ch.6.
[147] Fujii 2013.
[148] Matthäus 2007, 230.
[149] Su 2011, 21 & 63.

the all-powerful coercive and monitoring capacities of the central party—indeed, these were comparatively weak—but because ideologically enthusiastic cadres of (often very young) rank-and-file perpetrators willingly drove forward violence. As Weitz observes:

> The violence was so extensive, the loss of life so great, both because it was decreed from the center of power *and because the regime had given completely free rein to cadres on the ground*, many of them no older than teenagers, who decided on their own who could live and who should die.[150]

Again, such enthusiastically hardline rank-and-file perpetrators may be a minority—and they need not possess especially 'intellectual' understandings of elite ideology. But their emphatic support for a justificatory narrative for violence can be crucial in creating the local norms of violence amongst rank-and-file agents that pressure more apathetic peers to participate.

For many rank-and-file perpetrators, conversely, the justificatory narrative may not constitute a central motive for violence at all, yet still be sincerely accepted as a crucial means of legitimating violence. Simultaneously, justificatory narratives reshape the ideological structures that rank-and-file agents work within—enabling and encouraging violence, often for quite self-interested reasons, while keeping that violence (largely) organized around the logics of elite ideological claims. In the former Yugoslavia, for example, many perpetrators may well have been relatively 'unideological' thugs and hooligans.[151] Yet violence still largely adhered to the ethnonationalist justificatory narrative propagated by elites, targeting civilians in opposing ethnic groups as well as moderate coethnics.[152] Similarly, in Mao's cultural revolution, many local participants had limited deep understanding of Maoist ideology, but the spread of a narrative of pervasive class enemies created structural opportunities and incentives for violence 'motivated by [perpetrators'] fear of being deemed politically lapse or by their ambition for career advancement'.[153]

Since coercive micromanagement of rank-and-file agents is so difficult, ideological security doctrines are frequently powerful because they are institutionalized in briefings, professional training, and organizational policies that socialize rank-and-file agents to comply with, understand, or sincerely accept certain norms and understandings of violent behaviour.[154] As Western militaries became increasingly aware of how few soldiers actually proved able to readily kill in warfare, for example, they increasingly used ideological indoctrination and dehumanization in

---

[150] Weitz 2003, 187 (*my emphasis*).
[151] Mueller 2000.
[152] Dragojević 2019.
[153] Su 2011, 15.
[154] See also Checkel 2017; Revkin and Wood 2020.

military training. This approach 'worked' in its intended aim of increasing soldiers' readiness to kill: US soldiers' participation rates in violence were much higher in the Vietnam War, one of the principal wars in which training of this kind was tested. A major side-effect, however, was that it also encouraged greater atrocities against civilians by US forces, of which My Lai is only the most famous example.[155] Conversely, when forms of military training and socialization are more centrally orientated around limitationist ideas, soldiers appear to more readily distinguish between civilian groups and 'the enemy' and consequently perpetrate less violence against civilians.[156]

Second, the infrastructural strength of hardline ideology amongst the rank-and-file is often crucial in indirectly shaping the *capacities and incentives of elites*. Since elites depend upon rank-and-file agents to implement violence, the pre-existing ideologies of potential perpetrating organizations can affect whether elites think mass killing is really viable, or whether it risks major backlash. Such calculations are especially important if hardline elites face significant elite opposition, since the relative abilities of hardline elites and more moderate elites to mobilize the rank-and-file may be crucial in determining which faction wins out. In Rwanda, for example, the ability of hardliners to call upon willing supporters in the Presidential Guard and militia units was key in both killing and intimidating opponents, and in nullifying potential opposition from the more ideologically ambivalent army. There were significant 'moderates' in the capital and across the country, but they were not able to mobilize similar units to defend themselves or to limit the killing campaign. Likewise in Indonesia in 1965–66, the army's capacity to contemplate and implement mass killings was generated, in part, by its long-time mobilization of politicized paramilitary militias that readily embraced the military's justificatory narrative for violently purging the nation of communists.[157] Conversely, the Indonesian air force—known to be more sympathetic to the political left—was a potential problem for the hardliners and had to be effectively side-lined.[158]

So, if scholars associate ideology with the kind of 'true believer' model critiqued in Chapter 2, it is almost inevitable that they will find few 'ideological' rank-and-file perpetrators in most mass killings. Even ideologically committed rank-and-file agents sometimes understand remarkably little of the full ideological edifice constructed by their leaders. This should be unsurprising, since it has long been appreciated that highly systematized and intellectually elaborate ideology is, in most societies, largely an elite phenomenon. Yet this leaves plenty of room for looser, more selective, and more vernacular forms of ideological internalization amongst rank-and-file perpetrators, which may critically motivate and legitimate violence. Simultaneously, rank-and-file perpetrators frequently attest to

---

[155] Turse 2013. See also Grossman 2009.
[156] Hoover Green 2016; Oppenheim and Weintraub 2017. See also Bellamy 2012b; Thaler 2012.
[157] Robinson 2018, 25 & 31.
[158] Ibid. 46, 49–50, & 60–1.

significant pressure from ideologically rooted norms and institutions. Variation in such infrastructural ideology amongst the rank-and-file will rarely prove sufficient to determine whether or not mass killings occur in the first place. But it remains significant in both shaping the intensity and character of the violence and in emboldening or restraining hardline elites in pursuing mass killing in the first place.

Public Constituencies

By public constituencies, I mean all those social groups that do not directly participate in the perpetration of mass killing, but which perpetrating states or organizations control, represent, and draw support from. The centrality of such groups in mass killing is easily overstated. Outside observers are often too quick to assume that atrocities are rooted in widespread societal animosities, rather than question whether the bulk of the population actually supports violence at all.[159] Mass killings are not instigated by public ballot, and it is difficult and risky for ordinary members of the public to effectively resist them.

Nevertheless, as Hugh Trevor-Roper observed of sixteenth- and seventeenth-century witch hunts: 'no ruler has every carried out a policy of wholesale expulsion or destruction without the cooperation of society'.[160] Mass killing requires some mobilization of active collaboration from the public, and more extensive neutralization of potential opposition. All ruling elites depend on some sort of supportive coalition partly rooted in influential public constituencies, such as business elites, religious organizations, party memberships, media elites, and portions of the mass population.[161] When such a coalition fragments, even entrenched regimes can rapidly crumble—as in Slobodan Milošević's Yugoslavia in 2000 and Hosni Mubarak's Egypt in 2011. Even if this does not happen, failures to adequately legitimate policies of violence—especially amongst the most influential constituencies—risk provoking public ire and thereby creating costs in political capital, social disruption, and the erosion of an elite's strength and cohesion. Effective ideological justification to public constituencies is thus always highly functional, and often a genuine precondition, for any campaign of large-scale political violence.

As with rank-and-file agents, the infrastructural strength of hardline ideology amongst public constituencies can affect, first, the *structural incentives and capacities of elites* to implement (or resist) mass killing. In the Holocaust, the ways European states acted ranged from full implementation of Nazi plans, through

---

[159] For prominent critiques of such views see Gagnon 2004; Brass 2005.
[160] Cited in Getty and Naumov 1999, 7.
[161] See, in general, Mayer 2001; Brady 2009; Svolik 2012; Wedeen 2019.

various forms of partial or minimal implementation, to active obstructionism and rescue. Semelin observes that: 'governments' attitudes were generally determined by the degree of apparent anti-Semitism in local opinion—that is, the extent of social division on the "Jewish problem" within the countries'.[162] Sometimes the intensity of direct Nazi control mitigated the relevance of the local government's posture.[163] But where direct control was weaker—in Bulgaria, for example— limited support for the killing of local Jews (though not 'foreign' Jews) resulted in only partial implementation.[164] Even powerful hardline elites, moreover, sometimes make surprising concessions to public opposition, as when the Nazi regime caved in to the *Rosenstrasse Protest* of February 1943 by the non-Jewish wives of Jewish men targeted by the regime.[165] Similarly, as Rowan Savage observes, '[t]he partially successful resistance in Nazi Germany to the T4 programme (killing disabled people) is evidence that sometimes mass killing is impeded when discursive preparations have failed to convince'.[166]

Conversely, public support for ideological justifications of violence might positively incentivize policymakers to consider mass killing. Colonial genocides against the Native Americans in California or the Herero and Namaqua people of German South West Africa, for example, were actively lobbied for by local settler populations who had been rallied by hardline racist and nationalist narratives that cast the indigenous populations as a dangerous, parasitic, and criminal population.[167] In Indonesia, the army was empowered, in its mass killings of communists, by a supportive 'middle-class coalition' of strongly anti-communist groups, including the leading Islamic and Catholic parties, certain student groups, and an array of liberal intellectuals.[168] In the Wars in Yugoslavia, while most of the population did not appear guided by longstanding 'ancient hatreds' against other groups, residents of mixed communities in the war zones testified to significant popular belief in ethnonationalists' narratives of atrocities committed by the other side, empowering local hardliners while ostracizing moderate or critical voices.[169]

Second, members of the public, like rank-and-file perpetrators, tend to retain meaningful although constrained agency within a campaign of mass killing, so the infrastructural strength of ideological justifications can significantly shape *the intensity of mass collaboration and resistance*. The coordinated rescue of over a thousand Jews by the residents of Le Chambon-sur-Lignon in Vichy France, for example, rested on the determined rejection of fascist justifications of violence by

---

[162] Semelin 1993, 137. See also Fein 1979; Fein 1990, 63–6. For a more fine-grained analysis of such variation across communities in Romanian participation in the Holocaust see Solonari 2010, 193–9.
[163] Semelin 1993, 141–2.
[164] Segal 2018, 112–15. See also Dawidowicz 1987, Appendix B.
[165] Stoltzfus 2001.
[166] Savage 2013, 146.
[167] Madley 2005; Haas 2008; Madley 2016.
[168] Robinson 2018, 62–3.
[169] See the numerous interviews and anecdotes throughout Silber and Little 1997; Dragojević 2019.

the village pastors, but also the way 'their response resonated strongly with the beliefs of many people of Le Chambon'.[170] Such ideological opposition is rarely sufficient to entirely prevent violence. But when differences in popular ideological attitudes are strongly patterned across areas, they can generate significant regional contrasts in violence. For example, Diana Dumitru and Carter Johnson analyse a 'natural experiment' involving two Romanian-controlled territories in World War II, Bessarabia and Transnistria, characterized by profound differences in civilian attitudes towards the Holocaust. While ordinary civilians in Bessarabia displayed antagonistic attitudes towards Jews and collaborated in widespread violence, in Transnistria, the population displayed more cooperative attitudes towards Jews and avoided anti-Jewish pogroms. The crucial explanation of this contrast lies in the fact that Bessarabia had been controlled by highly anti-Semitic and increasingly fascist Romania from 1918 onwards (bar a brief Soviet occupation from 1940 to 1941), whereas Transnistria had been controlled from 1918 until Romanian occupation in 1941 by a Soviet government ideologically committed to Jewish integration.[171] As Dumitru and Johnson summarize:

> It was [the] constant vilification, blame, and construction of the Jew as an enemy of the Romanian people that, in combination with a wartime regime that sought their destruction, opened up the possibility for local gentiles to unleash a wave of violence against their Jewish neighbours. The Soviet Union, meanwhile, had encouraged locals to view their Jewish neighbors not as Jews, but as compatriots, as neighbours, as equals.[172]

As with rank-and-file perpetrators, few if any mass publics will be found in which 'true believers' in hardline ideology are a majority. But, when propagated effectively, justificatory narratives for mass killing tend to receive significant uptake from public constituencies—even if they do not accept the entire elite ideology that underpins them. As with rank-and-file perpetrators, justificatory narratives also generate important structural pressures to collaborate in and support mass killing. The consequence of decisions by individual members of public constituencies is limited in isolation. But collectively the level of public support for hardline justifications of mass killing can shape the relative attractiveness of different policy courses for elites and shape the extent of mass collusion in or resistance to the violence.

---

[170] Rochat and Modigliani 1995, 202.
[171] Dumitru and Johnson 2011, 11–31.
[172] Ibid. 35. See also Solonari 2010, 193–9.

3.4 The Roots of Hardline Ideology

If the behaviour of political elites, rank-and-file agents, and public constituents is profoundly shaped by the infrastructural strength of hardline ideologies, it is crucial to finally ask why such ideologies become dominant in particular contexts. Traditional-ideological perspectives often implicitly or explicitly answer this question through reference to *longstanding cultural traditions*, sometimes coupled with more immediate propaganda campaigns through which age-old stereotypes, animosities, and myths are revived. At its rather infamous extreme, this explanation is often associated with two works: Robert Kaplan's *Balkan Ghosts*, routinely cited by scholars for wrongly suggesting that 'ancient hatreds' underpinned 1990s ethnic conflict in Yugoslavia,[173] and Daniel Goldhagen's *Hitler's Willing Executioners*, often criticized for presenting Nazism as merely a reflection of longstanding mass enthusiasm amongst ordinary Germans for 'eliminationist antisemitism'.[174] There are, however, many more moderate versions of such a focus. John Weiss, while critical of Goldhagen, presents the Holocaust as deeply rooted in legacies of German and European anti-Semitism dating back to the origins of Christianity.[175] In their discussion of the psychosocial roots of genocide, Linda Woolf and Michael Hulsizer similarly suggest that 'a cultural history that values violence ... provides the foundation for future movement along the path of violence'.[176] Many case-specific studies of mass killing ground their analysis, in part, in an analysis of cultural roots of the violence stretching back centuries.[177]

Yet the explanatory power of such longstanding cultural traditions is too often assumed rather than demonstrated.[178] Evidence of widespread pre-existing animosities amongst the populations that come to engage in mass killing is relatively rare—often such hatreds seem to emerge only during or after the escalation of violence. As observed in Chapter 2, assertions that pre-existing ethnic hatreds drove violence in the fall of Yugoslavia in the 1990s fit poorly with evidence that three quarters of Yugoslavians described ethnic relations as good, and that rates of ethnic intermarriage were high over the preceding decades.[179] Regarding the Armenian Genocide, Gerlach likewise reports how: 'a number of observers held that local Muslim–Armenian relations were good and changed only under the circumstances, or never, and many credible sources testify to local opposition to the persecution and killing'.[180] Even where longstanding cultural foundations

---

[173] Kaplan 2005.
[174] Goldhagen 1996.
[175] Weiss 1997.
[176] Woolf and Hulsizer 2005, 114.
[177] E.g. Prunier 1995; Taylor 1999; Tuckwood 2010.
[178] For prominent critiques see Fearon and Laitin 2000; Mueller 2000; Gagnon 2004; Brass 2005; Fujii 2009.
[179] Malešević 2006, 176.
[180] Gerlach 2010, 115.

for violence do exist, it's not clear how far they actually constrain the key elite choices that bring about mass killing. After all, the regimes that perpetrate mass killing often *radically overturn* established cultural traditions—the Soviet Union under Lenin and Stalin felt able to abolish almost every formal institution of traditional Tsarism across Russia, the Nazi regime systematically dismantled many more liberal aspects of German society, and the Young Turks who orchestrated the Armenian Genocide willingly abandoned much of an Ottoman political structure that was over six centuries old. Since cultures are vast and heterogenous, it is always possible to look backwards through history and find certain cultural 'precedents' for later violence. But the mere existence of such precedents does not demonstrate a clear causal link between the two.

Consequently, while it is always important to place hardline ideologies in their broader historical and cultural context, I place primary emphasis on key ideological changes in the decade or two preceding mass killing. As already emphasized, ideologies are dynamic systems of ideas, not static cultural monoliths. The protofascist ideologies of the German far-right in the early 1920s were not the same as Nazi ideology in the later 1920s, which in turn was not identical to Nazism as it existed in 1941. To fully explain mass killings, we therefore need to trace processes of *ideological radicalization*, defined in the context of political violence as *people's increasing perception of more expansive levels or forms of violence as privately or publicly justifiable*. This definition is intended to permit radicalization along three main dimensions: more individuals may perceive such justifiability, individuals may have greater levels of conviction in such justifiability, or individuals may see more extreme levels or forms of violence as justifiable.[181]

This emphasis of ideological dynamism and processes of radicalization could return us to a sceptical worry, however. How can we be sure that hardline ideologies, and the justificatory narratives for mass killing they produce, are not just *symptoms* of more fundamental material or social causes of mass killing? Many rationalists or situationists may suspect, in other words, that ideological radicalization is always likely to occur if certain material or social conditions – such as a desperate armed conflict or a large-scale challenge to an authoritarian regime – are in place. If so, then such underlying non-ideological conditions may once again start to look like the *real* roots of mass killing.[182]

---

[181] The term *radicalization* has its drawbacks—what seems 'radical' is subjective, and the concept is often associated with crude, linear models of ideological change or deterministic assumptions that radical beliefs automatically produce radical behaviour. Nevertheless, I have found no better concept, and such problems arise from crude portrayals of radicalization rather than the concept itself. For a deeper discussion that accords with my own views see Neumann 2013. For a more critical perspective see Schuurman and Taylor 2018.

[182] Danning 2019, 108–10. Ideology would therefore be largely 'endogenous' to non-ideological causes and lack *causal priority* and/or *causal depth*; see George and Bennett 2005, 185–6. This has been a dominant assumption about ideology in, for example, the study of civil wars; for discussion and critique see Hafez, Gade, and Gabbay 2022.

I take this sceptical concern seriously. Ideological radicalization *is* powerfully fuelled by broader material and social changes, without which the trajectory of radicalization can rarely be explained. The appeal of extremist ideologies is, for example, enhanced by experiences of material frustration, insecurity, or loss.[183] Experimental psychological research also finds that threatening environments increase psychological desires for cognitive certainty, positive in-group identity, and assertive leadership, and can thereby increase the attraction of relatively right-wing or authoritarian ideological positions.[184] Major economic crises or wars often represent 'shocks'[185] or 'anomalies'[186] that trigger ideological change. The Great Depression profoundly challenged the claims of conventional liberal ideologies in the late 1920s, for example, lending plausibility to fascist and communist contentions that liberal democratic systems were failing and that radical alternatives were necessary. Conversely, the decline and growing dysfunctionality of the Soviet economy in the 1980s was critical in prompting Mikhail Gorbachev's reforms and the eventual collapse of Communism.[187] Crises may also bring hardline ideological elites into power. The progressive collapse of the Russian war effort allowed Vladimir Lenin's Bolsheviks to topple the provisional government of Russia in 1917, just as the Cambodian Civil War, coupled with devastation created by massive US bombing, permitted the Khmer Rouge to take power in 1975. Obviously, then, 'non-ideological' material and social factors matter in explaining ideological change, including the emergence and radicalization of hardline ideas.

Yet hardline ideologies cannot be *reduced* to such factors, for three reasons. First, the way people interpret and respond to material and social changes is *itself influenced by pre-existing ideological worldviews and structures*. In other words, the character of any ideology is itself a key determinant of how it may evolve in response to broader social and material developments. People cannot examine all the information available in the world simultaneously from scratch, so they must use either their existing beliefs and assumptions or prominent ideological discourses to interpret new information.[188] Motivated reasoning also encourages individuals to interpret new information and ideas in ways that maintain or reinforce their existing ideological worldviews.[189] Indeed, contemporary political psychology strongly emphasizes how individuals in the same material or social circumstances will be drawn towards different ideological orientations according to their different underlying personalities and psychological

---

[183] Staub 1989, ch.3; Midlarsky 2005; Midlarsky 2011; Anderson 2017a, 19–23.
[184] Jost et al. 2003; Landau et al. 2004; Jost and Hunyady 2005; Haller and Hogg 2014; Jost 2017. See also McDoom 2012.
[185] Kowert and Legro 1996, 473–4; Legro 2005; Anderson 2017a, 27.
[186] Owen 2010.
[187] Brooks and Wohlforth 2000/2001. But see also English 2002.
[188] Jervis 1976, ch.4; Hardwig 1985; Nyhan and Reifler 2010; Swire et al. 2017.
[189] Jervis 1976, ch.4; Woolf and Hulsizer 2005, 109–10; Lewandowsky et al. 2012, 118–19.

dispositions.[190] Thus, psychologists find, for example, that individuals' receptiveness to harsh perceptions of intergroup threat; preferences for violent, coercive, or hierarchically dominant social relations; tendencies towards prejudice and stereotyping; and propensity to demand authoritarian in-group solidarity and out-group denigration in response to threat, are generally predicted by a range of personality orientations.[191] Simultaneously, existing ideological structures may incentivize or constrain both elite and non-elite radicalizing activities.[192] The propagation of new norms is most effective when they are consistent with or even 'grafted' onto existing norms, for example.[193] The creation of more exclusionary or discriminatory policies and laws typically facilitates more radical policies down the timeline.[194] In all these ways, as Jeffrey Legro observes: 'Collective ideas fundamentally shape their own continuity or transformation.'[195]

Second, while material or social conditions and crises shape ideology, the reverse is also true: *ideologies shape material and social conditions and crises*.[196] It goes without saying that the 'crisis' in which Nazi Germany began to implement the Holocaust was very much a crisis of its own making. The same is true in other cases. The Russian Civil War helped radicalize Lenin's Bolsheviks, for example, but the character of that civil war was *itself* a consequence of the Bolsheviks' ideological radicalism. A clear opportunity for Lenin to build a broad-based united socialist government existed in late 1917—which would have either avoided civil war or hugely strengthened the government's position in it. But this possibility was quashed, as discussed in Chapter 5, by Bolshevik hardliners who framed cooperation with other socialist groups as a betrayal of the working class.[197] Likewise, in Indonesia, while mass killings in 1965–66 did emerge out of an internal power struggle, a central cause of that struggle was Indonesian hardliners' ideological antipathy to what they saw as unacceptable overtures to communism made by the incumbent President Sukarno.[198] In colonial genocides, similarly, the crises in which colonial empires slaughtered indigenous populations were direct products of the progressive dispossession and persecution of those populations under the doctrines of colonialist ideologies.[199] In all of the cases examined

---

[190] Jost, Federico, and Napier 2009. For related perspectives see Zaller 1992, 22–8; Sidanius and Pratto 1999; Caprara and Zimbardo 2004; Block and Block 2006; Jost 2006; Caprara et al. 2010; Haidt 2012; Gries 2014.
[191] See, for example, Sidanius and Pratto 1999; Pratto, Sidanius, and Levin 2006; Hodson, Hogg, and MacInnis 2009; Haidt 2016. For broader perspectives on the psychological needs driving what I conceptualize as hardline ideological belief see Covington et al. 2006; Bar-Tal 2007; Sjoberg and Via 2010; Kruglanski and Orehek 2011; Baugher and Gazmararian 2015; Covington 2017, esp. chs.6 & 8.
[192] See, in general, Skinner 1974; Katzenstein 1996b; Finnemore and Sikkink 1998; Morrow 2015.
[193] Morrow 2020, 165.
[194] Mommsen 2009; Murray 2015; Stanton 2016.
[195] Legro 2005, 13.
[196] See also Mann 2004, 78–90.
[197] Ryan 2012. See also Chapter 5.
[198] Robinson 2018.
[199] Madley 2004.

in Chapters 5–8, pre-existing hardline ideology played a notable role in deepening the crises in which elites initiated mass killing.

Third, and in part because of the previous two reasons, *the dynamics of ideological change are too complex and indeterminate* for material and social conditions to be simply substituted for ideological variables.[200] Ideological changes often unfold through sudden 'tipping point' shifts that are infamously unpredictable,[201] and significant medium- or long-run developments often result from contingent events that occur at certain 'critical junctures'.[202] Given the centrality of Hitler to Nazism, for example, the ideological development of interwar Germany may have been radically different if he had been killed in a gas attack on the German front line in 1918. Equally, the entire ideological history of the Soviet Union in the twentieth century was shaped by what most historians recognize as an unlikely eventuality in 1917–20: the successful takeover and maintenance of government by Vladimir Lenin's ultraradical faction of Russian socialism. Since such events exert enormous influence over ideological developments, yet cannot be reliably predicted by appealing to material or social factors, ideologies always possess a significant degree of explanatory 'autonomy'. Even in very similar material and social conditions, individuals, groups, and organizations frequently diverge dramatically in their ideological orientations and doctrines.

So the emergence and radicalization of hardline ideologies is shaped by *both* broader material and social conditions and the nature of prevailing ideologies themselves, with the exact trajectory of radicalization being highly context-dependent. Some macro-historical generalizations may be possible. Existing research has, for example, collectively emphasized four main (overlapping) settings in which hardline ideologies commonly appear: first, when social systems involving salient ethnic or religious social categories lack cross-cutting forms of political mobilization and intergroup coalition building;[203] second, when revolutionary movements take power with a conception of political rule that involves monopolistic control by the revolutionary vanguard in at least the medium term;[204] third, when entrenched conservative authoritarian governments anticipate major future challenges to their political control in the absence of intra-regime demands for reform;[205] and fourth, when states anticipate future warfare in which their likely opponents depend on civilian mobilization, especially when such warfare is conducted in a colonial or other overseas context where one's own civilian

---

[200] Straus 2015a, 65. Mann's detailed cross-national analysis of fascists, for example, finds no straightforward relationship between certain material interests and adherents to (or the broader rise of) fascist ideology, instead emphasizing the complex interaction of material and ideational factors; see Mann 2000; Mann 2004.
[201] Granovetter 1978; Jervis 1997; Goldstein 1999; Mahoney 2000; Pierson 2004.
[202] See also Capoccia and Kelemen 2007.
[203] See, in general, Mann 2005; McMahon 2007; Bulutgil 2016; Nyseth Brehm 2016.
[204] See, in general, Melson 1992; Kim 2018.
[205] See, in general, Rummel 1994; Davenport 2007.

population is relatively insulated from reprisal.[206] Yet such associations are loose and accompanied by many exceptions. In each of my case studies, I show that hardline ideology was profoundly rooted in pathways of ideological development that were effectively historically 'contingent'—rooted in the particular course of cultural and political developments of the society in question—and not simply a natural or predictable consequence of certain features of the material or social conditions of that case.[207] Ideology is, in short, of explanatory significance in its own right, and not simply a manifestation of supposedly 'deeper' unideological causes.

3.5 Conclusion

There is considerable sympathy within scholarship on mass killing for the claim that ideology has some important role to play. Yet many scholars nevertheless side-line ideology from their explanations of the violence, in favour of a focus on either strategic circumstances and rational incentives or situational pressures and dynamics within bureaucratic structures, local communities, or small groups of perpetrators. This chapter has sought to reaffirm ideology's central role in mass killing, but in synthesis with strategic decision-making and situational social pressure. Mass killings are indeed strategic instruments utilized in response to perceived crisis, and they are initiated and sustained in large part through dynamics of conformity, bureaucratization, and endogenous radicalization. But strategic circumstances and situational social pressures explain mass killing only in tandem with ideology—specifically, with the strength and content of certain 'hardline' ideas about threat, community, order, and violence. In each of my case studies in Chapters 5-8, I therefore stress the way in which mass killing emerged out of case-specific ideological contexts and took radically different forms according to the contextually dominant ideological infrastructures.

This is a fairly complex picture of mass killings in which, to quote Jeffrey Checkel, 'explanatory richness is being purchased at the expense of theoretical parsimony'.[208] The trade-off is real. More parsimonious accounts offer more precise predictions and can still identify important causes of mass killings without reference to ideology. But a large quantity of evidence shows that side-lining ideology leaves an unacceptably large proportion of the key decisions and processes that produce mass killings unexplained. Mass killings do not emerge out of monolithic ideological systems, which uniformly motivate perpetrators via deep commitments to a preformed ideological plan for extermination. But ideologies do

---
[206] See, in general, Valentino, Huth, and Balch-Lindsay 2004; Downes 2008; Cromartie 2015.
[207] Straus 2015a, 330-1.
[208] Checkel 1997, xi

influence the constellation of choices made by elites, rank-and-file agents, and public constituents in times of crisis, through varying degrees of sincere internalization combined with the structural pressures of ideological norms and institutions. This influence is critical, profoundly affecting the occurrence and character of mass killings.

In offering a synthesis of ideology, strategic calculation, and situational pressure, this 'neo-ideological' explanation of mass killing has therefore moved away, in important respects, from many traditional-ideological accounts. Chapter 2 focused, in particular, on explicating how ideologies can crucially influence perpetrators even in the absence of deep ideological convictions. In this chapter, I have coupled this with a greater emphasis of ideological heterogeneity within mass killings, ideology's diverse roles amongst different types of perpetrator, and the central importance of hardline ideas about security politics that shape interpretations of crisis, rather than ultimate ideological goals. On this last issue, however, more needs to be said. Why are hardline ideas about security politics more important, in mass killings, than radical ideological goals? What are these hardline ideas, and how do they work to encourage violence? It is to these questions that I turn in Chapter 4.

4
The Hardline Justification of Mass Killing

An increasing number of scholars agree that ideologies shape the occurrence and character of mass killings. However, as Scott Straus observes, 'less consensus exists on *which type of ideology* prompts elites to commit genocide and mass atrocity'.[1] What are the key ideological foundations of mass killing? How do they make certain civilians look like strategically and morally justifiable targets of violence? What do the critical justificatory narratives for mass killing actually look like?

To some extent, the answers to these questions vary across different mass killings. As emphasized throughout my four case studies, such context-specific ideological content is crucial. An authoritarian and racially supremacist state at war, such as Nazi Germany, may carry high general risks of engaging in mass killing. But one cannot explain why the Nazis attempted to annihilate the Jews specifically, why they also felt it necessary to eliminate disabled people or the Sinti and Roma, or why they pursued such campaigns even when doing so undermined their war effort, without understanding the distinctive biologized and apocalyptic manner in which the Nazis understood racial categories and racial conflict.[2] Similarly, one cannot understand why the Khmer Rouge felt it necessary to empty Cambodia's cities and target urban and bourgeois society without appreciating the specific diagnoses of Cambodia's economic and social problems advanced by the party's leading ideologues.[3]

Nevertheless, scholars generally agree that important ideological similarities exist across mass killings. Indeed, numerous typologies of common ideological arguments, principles, or themes have been offered to capture such similarities.[4] These typologies inform my analysis in this chapter, but they collectively suffer from two shortcomings. First, most employ a relatively descriptive 'theme-spotting' approach, identifying recurring clusters of ideas in the discourse of perpetrators across cases. This is an important first step, especially when based on broad and systematic cross-case comparisons.[5] But the mere recurrence of

---

[1] Straus 2016, 57.
[2] Kallis 2005; Kallis 2009.
[3] Jackson 1989b.
[4] Examples include Kiernan 2003; Slim 2007; Alvarez 2008; Murray 2015; Richter, Markus, and Tait 2018; Saucier and Akers 2018.
[5] Saucier and Akers 2018, 82.

such themes does not itself tell us what impact they have on mass killing: it does not reveal their *causal powers* to actually encourage and shape violence. Second, these typologies have tended to 'pile on top' of each other rather than theoretically accumulate. Each identifies different themes that encourage the violence, but it is not clear what is really at stake when we compare claims that mass killings are fundamentally caused by 'dehumanization', 'moral exclusion', 'nationalism', or 'purity'. This is a natural consequence, to some extent, of a 'theme-spotting' approach. There are, after all, countless themes in the ideologies that underpin mass killing, and countless ways they can be described.

I do not want to simply add another typology to the top of the pile. This chapter therefore attempts to clarify what is really at stake in competing portrayals of the ideological themes or elements that encourage mass killing. The key controversy, I argue, is over the extent of continuity between the key ideological foundations of large-scale violence against civilians and 'ordinary' politics. To what extent, in short, do mass killings depend on a radical break with conventional strategic and moral ideas about political action? The dominant view (typical of traditional-ideological perspectives) associates ideology with certain 'extraordinary' ideological foundations—presenting (ideological) mass killings as rooted in radical political projects that demand the violent purification of society and invert conventional moral standards that prohibit such violence. An alternative view, by contrast, downplays such extraordinary ideological projects, stressing a vital degree of continuity between the ideological foundations of mass killing and 'ordinary' political, strategic, and moral ideas. It is this alternative view that I aim to defend and develop in this chapter. I do so in two stages. Section 4.1 delves deeper into this core debate over *what makes* certain ideologies liable to encourage mass killing, emphasizing the centrality of 'hardline' ideas about security politics. The bulk of the chapter, in section 4.2, then analyses *how* hardline ideologies promote mass killing: identifying six key 'justificatory mechanisms' that can generate support for, and participation in, such extreme violence; and that contrast with alternative limitationist ideas that tend to discourage mass killing.

4.1 What Kinds of Ideologies Matter?

The Traditional Focus: Revolutionary Transformations and Moral Disengagement

There are two main versions of the traditional-ideological argument linking mass killings to 'extraordinary' ideology.[6] First, scholars often focus on the role of *radical revolutionary projects to transform society*. The most infamous

---

[6] The two versions are not necessarily incompatible, and many scholars aver to elements of both.

ideologies associated with mass killing—such as communism, fascism, and extremist ethnonationalism—often contain ambitions to implement such transformational projects, which are generally theorized as encouraging mass killing through three main mechanisms. First, they either generate or express hatreds against certain social groups, who are identified as incompatible with the new society perpetrators seek to create and must therefore be violently purged. Second, they create a utopian calculus in which the ideological 'ends' justify any 'means', no matter how violent. Finally, they encourage a dogmatic commitment to such ideological goals that can tolerate no dissent, promoting a reliance on dictatorial, coercive, and violent methods of rule.[7]

This perspective originates in early genocide research, which often emphasized the role of totalitarian ideologies that advance utopian visions of the ideal society, denigrate the value of individual human life, and inculcate hatred for political, religious, or racial opponents.[8] While recent scholars have shifted the focus away from totalitarianism, the emphasis on revolutionary transformational goals has remained dominant. Robert Melson contends, for example, that '[t]o respond to the revolutionary situation and to implement their ideological desires [revolutionaries] may destroy not only the old regime's institutions but also their political opponents and the communal groups and classes that they identify with their enemies'.[9] Eric Weitz presents genocides as generally the product of: 'ideologies of race and nation, revolutionary regimes with vast utopian ambitions, [and] moments of crisis generated by war and domestic upheaval'.[10] Patricia Marchak stresses 'absolutist and exclusionary' ideologies in which 'anyone who fails to meet the test of purity—however it may be defined—is less than human, to be degraded, spat upon, or killed in the name of a higher good'.[11] Valentino argues that: 'radical communist regimes have proven such prodigious killers primarily because the social changes they sought to bring about have resulted in the sudden and nearly complete material and political dispossession of millions of people',[12] while racist/ethnonationalist mass killings likewise 'result from the effort to fundamentally reorganize society at the expense of certain groups'.[13]

There is a lot of truth in these claims. The contention that mass killing is generally ideologically justified by perpetrators as a brutal but necessary way of achieving particular political goals, and that the perception of certain groups as enemies is fundamentally embedded in particular ideological worldviews, has been an important corrective to approaches that portray mass killing as a product of

---

[7] The same basic argument has been made about religious ideologies—for examples and critique see Cavanaugh 2009.
[8] Fein 1993, 83. See also Kirkpatrick 1979; Kuper 1981, ch.5; Courtois et al. 1999; Richter, Markus, and Tait 2018.
[9] Melson 1992, 260.
[10] Weitz 2003, 15.
[11] Marchak 2003, 88. On purification see also Volkan 2009, 212.
[12] Valentino 2004, 93.
[13] Ibid. 153. See also Alvarez 2010, 62–73; Midlarsky 2011, 12.

thoughtless conformity, mindless bureaucratic inhumanity, or self-evident rational incentives.[14] There is, moreover, evidence that radical, revolutionary projects are associated with the most intense campaigns of mass killing.[15] Yet there are two key problems with this portrayal of the ideological foundations of mass killing.

First, the correlation between transformational ideological goals and mass killing only appears to exist in a few—albeit especially well-known—cases. The most infamously destructive regimes in world history such as Nazi Germany, Stalinist Russia, Mao's China, and Khmer Rouge Cambodia may appear to be guided by such goals. But mass killing has also been employed by many non-revolutionary states, including liberal democracies in World War II, the Korean War, and the Vietnam War, and relatively reactionary authoritarian regimes, as in Indonesia in 1965–66 and Guatemala in 1978–83. Some theorists might simply conclude that ideology matters in the former cases but not the latter, but this risks circularity. If we treat the way ideologies encourage mass killings as an open question, rather than assuming it *must* reflect transformational ideological goals, then why are liberal democracies or conservative autocracies assumed to act independent of ideology? Such states do have ideologies, and like revolutionary regimes, they conduct mass killing guided by distinctive sets of ideas—just not transformational goals. Conversely, many regimes that do adhere to transformational goals rooted in revolutionary ideology, such as communist Nicaragua, Vietnam, and Cuba, have largely avoided mass killings in favour of more limited repression, despite also undergoing periods of war or severe crises.

Second, even in those cases where relatively revolutionary or utopian regimes do engage in mass killing, transformational goals often appear surprisingly peripheral to the violence. As shown in Chapter 5, Stalin's Great Terror of 1937–38 was only very indirectly linked to utopian ambitions, and was primarily guided by concerns with security and the maintenance of political order as understood through Stalinist ideology. Similarly, violence in China's Cultural Revolution was primarily rooted in Maoist narratives of counter-revolutionary threats and criminality, and the patterns of escalatory denunciation these generated in local parties and communities.[16] In both cases, groups *were* identified as suspect on ideological grounds, but neither Stalin nor Mao initially saw the physical annihilation of such groups as a direct or necessary corollary of their revolutionary projects. Indeed, they sometimes assumed that such groups would be integrated into or contained within revolutionary society without large-scale violence. A focus on longstanding transformational goals also struggles to explain the timing of mass killing in such cases. If a regime's core ideological goals directly require the elimination of certain groups, then that regime should initiate such eliminationist policies as soon as it is feasible

---

[14] See also Reicher, Haslam, and Rath 2008.
[15] Kim 2018. See also Tschantret 2019.
[16] Walder 1994; Xiuyuan 1994; Su 2011.

to do so.[17] Yet this is not the typical pattern—instead, mass killings typically emerge from quite complex processes of escalation during crisis.[18] Part of the issue, here, is that traditional-ideological perspectives remain tacitly indebted to the image of ideologies as providing longstanding 'blueprints' for action. But even communist regimes have generally lacked such detailed blueprints, and vacillated significantly in their conception of exactly what the transition to utopia requires in practical terms. In only two major cases—Nazi genocide and Khmer Rouge mass killing in Cambodia—is it clear that longstanding transformational goals directly implied ideological rationales for mass killing. Even here, such notions were closely tied to conceptions of (in)security and radicalized under contextual crisis.

A second kind of traditional-ideological argument linking ideology to mass killing focuses on processes of *disengagement from conventional morality*, which remove important constraints on violence.[19] Rather than focusing on transformational goals, this argument stresses the moral exclusion or dehumanization of victims, the inversion of conventional moral values, and the promotion of doctrines of unthinking obedience in the ideologies of perpetrators of mass killing.[20] Herbert Kelman, for example, contends that the central task in explaining massacres of civilians is 'to ask how the voice of conscience is subdued',[21] since 'neither the reason for the violence nor its purpose is of the kind that people would normally consider justifiable ... [such violence] is entirely outside of the realm of moral discourse'.[22] In his study of the Holocaust, Zygmunt Bauman likewise claims that 'the Holocaust could be accomplished only on the condition of neutralizing the impact of primeval moral drives'.[23] Kjell Anderson also emphasizes the need for perpetrators to employ 'techniques of neutralization' to get around the fact that '[s]ocieties impose a general prohibition on killing',[24] while Adam Lankford presents atrocities as committed by 'a sufficiently indoctrinated workforce with little concern for traditional moral values'.[25] Ervin Staub concludes that in atrocities: 'Behavior toward the victims that would have been inconceivable becomes accepted and "normal." Institutions are changed or created to serve violence. The society is transformed. In the end, there may be a *reversal of morality*.'[26]

---

[17] Maat 2020, 778.
[18] Again, see Valentino 2004; Mann 2005; Gerlach and Werth 2009; Mommsen 2009.
[19] Bandura 1999; Slovic et al. 2013. In this chapter, I always mean 'moral' in the subjective, descriptive sense of people's beliefs or assumptions about moral standards and right action, as opposed to the normative sense of actually justified moral standards; see Gert and Gert 2016.
[20] Such arguments are not always explicitly linked to ideology, and can appear in more situationist theories of violence—but the causal explanation involved is almost identical.
[21] Kelman 1973, 43.
[22] Ibid. 33.
[23] Bauman 1989, 188.
[24] Anderson 2017b. See also Littman and Paluck 2015, 79. Anderson acknowledges that 'many acts of perpetration ... are rooted in ties to conventional values', but sees this as standing in essential contrast to ideological influence, which he associates with moral disengagement; see Anderson 2017a, 109–10.
[25] Lankford 2009, 3.
[26] Staub 1999, 183 (*emphasis in original*). This view has precedents in Arendt—who suggests that totalitarian ideologies encourage a kind of 'thoughtlessness' in which responsibility for action is completely disavowed; see and compare Arendt 1951/1976; Arendt 1963/2006. See also Covington 2012.

Again, such scholarship has advanced our understanding of mass killing. There is evidence of certain forms of moral disengagement in mass killings, and experimental psychology has convincingly demonstrated the power of such processes to encourage aggression in laboratory settings.[27] But this causal story is also misleading, in two ways.[28]

First, moral disengagement arguments excessively idealize the 'usual standards of morality', conventional 'voice of conscience', or 'primeval moral drives' found in real-world societies. Such theorists often seem to take it as self-evident that conventional moral standards are incompatible with mass killing—implying, as Rowan Savage critically observes, that 'everyone has a conscience that must be thwarted or obstructed in order to do "wrong", which, on some level, the perpetrator must realise is wrong'.[29] But real-world 'conventional moral standards' are not so pacific, and should not be assumed to match the specific humanitarian and cosmopolitan values held by contemporary scholars.[30] It is clearly not true, for example, that human societies place general prohibitions on killing. They typically prohibit *private killings by individuals*. But most social orders affirm that political authorities are both permitted and required to kill in certain circumstances, and potentially even kill civilians, especially in defence of the community and punishment of serious criminals.[31] Large proportions of people in surveys and experiments, moreover, have been found to openly believe that killing 'enemy' civilians can be justified if it is ordered by state authorities or framed as protecting 'our troops'.[32]

Kelman, indeed, observes this. In survey research he conducted, 51% of respondents said they would follow a superior officer's orders to shoot civilians, compared to 33% who said they would refuse the order. Most people appear to be 'prepared, in principle, to engage in mass violence if faced with authoritative orders to do so, [and] are certainly prepared to condone such actions'.[33] Recent research by Daryl Press, Scott Sagan, Ben Valentino, and Janina Dill extends such findings. They show, for example, that a majority of the American public would approve of a nuclear strike which killed *two million* Iranian civilians if this avoided a ground invasion of Iran in which 20,000 American soldiers would be expected to die.[34] Mass killings are clearly a normatively extreme course, but commonplace moral attitudes and norms in real societies imply that such killing *could* be justifiable under extreme conditions. For most people, moreover, these 'extreme conditions'

---

[27] Bandura, Underwood, and Fromson 1975; Bandura 1999; Slovic et al. 2013. See also Monroe 2011.
[28] This discussion expands my comments in Leader Maynard 2022.
[29] Savage 2013, 145.
[30] See also Frazer and Hutchings 2019; Rathbun and Stein 2020.
[31] Savage 2013, 144.
[32] See also Millar 2019.
[33] Kelman 1973, 41.
[34] Sagan and Valentino 2017, 59.

are vague. When security apparatuses and legal authorities declare that large-scale threats or criminals have been identified—claims that may fit with pre-existing social prejudices or narratives—it is often not clear to ordinary citizens that such claims are false. In such circumstances, there is no need for people to radically disengage from conventional moral norms in order to support the violence. Indeed, conventional moral norms surrounding war, self-defence, group loyalty, and duties to obey state authorities may *enjoin them to do so*.

Indeed, the second problem with moral disengagement arguments is that mass killings provide abundant evidence of strong moral *engagement* with violence on the part of many perpetrators.[35] Hardliners who champion violence against civilians righteously portray targets of the violence as criminal evildoers, while denouncing opponents of the violence as traitors. Rank-and-file perpetrators testify to feelings of duty to follow the directives of higher authorities, as well as loyalty to their comrades and a desire to live up to the manly virtues expected of good soldiers, security officers, or patriotic citizens. The ideological narratives that surround the violence aggressively assert, moreover, that atrocities have been committed *against the perpetrators* and their kin, rendering violence necessary self-defence and just retribution. To an outsider, such claims may appear outlandish, but they are unquestionably *moralized*—appealing to, rather than disregarding, familiar moral values and principles about self-defence, justice, and duty.[36]

The Neo-Ideological Focus: Radicalized Security Politics

So while transformational ideological goals and moral disengagement play a role in mass killings, that role has been overstated. Our abhorrence at mass killings has encouraged a misplaced assumption that the violence is alien to conventional political and moral concerns. In truth, killing is a conventional part of practices of war-waging, group protection, geopolitical struggle, and policing and punishment in most human societies. It is from these conventional elements of security politics, and their ideological underpinnings, that perpetrators of mass killings draw both sincere and instrumental justifications for their actions. The core ideological roots of mass killing lie in hardline visions of security: in which the world is presented as highly threatening, such threats are not limited to identifiable armed forces but instead associated with (certain) civilian groups, and proper responses are thought

---

[35] This argument has been pressed strongly by Reicher, Haslam, and Rath 2008; Powell 2011; Powell 2012; Fiske and Rai 2014; Lang 2020. See also Midlarsky 2005, 107–10 & 180–3; Campbell 2009, 151 & 155–8; Morrow 2020, 54. Though famous for his work on moral disengagement, the psychologist Albert Bandura also affirms this; see Bandura 1999, 195.

[36] Indeed, when violence is meant to serve these sorts of moralized goals, forms of moral disengagement such as dehumanization often appear to *undermine* violence; see Rai, Valdesolo, and Graham 2017.

to often require uncompromising violence and intense group loyalty, all to protect a political community and political order whose rigid preservation is an absolute imperative.

Such hardline visions are not wildly unusual, but nor are they typical. They represent a distinctive *ideological* position, with sympathizers in most societies, that stands in contrast to more limitationist ways to understand security politics. Like other ideological positions, this hardline orientation has an elective affinity with certain common value orientations and psychological dispositions, while also gaining influence through ideological activism by hardline movements and authorities.[37] Hardline ideas are not 'exceptions' to some sort of default cosmopolitan and humanitarian ideals, but powerful rivals to those ideals which at times achieve ascendency.[38] Mass killing is fundamentally a manifestation of *ideologically radicalized security politics* more than revolutionary utopianism or disengaged thoughtlessness. Justifications of the violence capitalize on arguments, norms, and institutions that are familiar to the discourses and practices of security in modern societies, to make large-scale violence against civilians appear like a necessary form of collective protection. Mass killing is, as Martin Shaw puts it, a form of 'degenerate war', in which harsh moral and strategic logics of warfare are ideologically transferred onto civilian groups.[39] Recognizing this rootedness of mass killings in certain radical interpretations of conventional strategic and moral ideas is crucial to understand the mentality of perpetrators and the power of justificatory narratives to draw ordinary people into mass killing.

Some hardline doctrines go even further: presenting certain civilian groups as not merely suspect, but as locked in an ongoing existential conflict, involving terrible crimes and atrocities, against the political community. Such *ultrahardline* security doctrines increase the chance of mass killing occurring even in limited crises, and increase the likelihood that perpetrators will resort to a genocidal logic of violence. If certain civilian groups are seen as constituting irredeemable and existential threats to the political community—a 'toxic' population, in Rhiannon Neilsen's depiction, or kind of 'anti-nation', as Elisabeth Hope Murray puts it—perpetrators are likely to conclude that such groups need to be eliminated *as such* for the political community to be secure.[40] More esoteric ideological elements—such as racist pseudoscience—may play a role in underpinning such understandings of threat. But the principal roots of genocidal violence still lie

---

[37] See, again, Sidanius and Pratto 1999; Jost, Federico, and Napier 2009.
[38] For parallel observations about racist ideas see Mills 2015, 133–4.
[39] Shaw 2002; Shaw 2003. See also David Moshman's contention that genocides are 'an extreme result of normal identity processes', or Kathleen Taylor's suggestion that atrocities are rooted in processes of 'otherisation' which are 'part of natural human behaviour'; see Taylor 2006, 231; Moshman 2007, 115. For the conventional, security-orientated justifications of authoritarian repression more generally see Edel and Josua 2018; Tschantret 2019.
[40] Murray 2014; Murray 2015; Neilsen 2015.

in radical perversions of familiar ideas about insecurity, national self-defence, patriotic loyalty, and the punishment of those who wrong the community.

Again, this neo-ideological perspective represents a targeted revision rather than a wholesale rejection of traditional-ideological arguments. I strongly emphasize that a regime's or group's understanding of 'security politics' always remains connected to its broader ideological goals, since *what constitutes a threat* to security depends on the values and interests one seeks to promote and preserve. Security politics, including war, is never merely a neutral defence of a self-evident status quo, but involves the prioritization of certain political goods over others, and always depends on value-laden processes of reordering social life to some extent.[41] Moreover, since mass killings do remain normatively extreme courses of action, *selective* moral disengagement typically exists alongside hardline moral justification,[42] mutually 'warping the moral landscape', as Stephen de Wijze puts it, in which mass killing takes place.[43]

Nevertheless, placing primary emphasis on ideologically radicalized security politics, rather than transformational goals or moral disengagement, has important theoretical implications. For a start, it foregrounds the interdependence, rather than opposition, of ideology, strategy, and morality. Rationalist insights on the strategic logics of repression and war can therefore be combined with an emphasis of the ideological constructions of threat, strategic necessity, and group identity within which those logics operate.[44] Simultaneously, strategic security politics is typically *moralized*[45] in mass killing, rather than being purely driven by cold rationality. Strategic calculation is overlain with appeals to vengeance, loyalty, duty, heroism, and valour, which can be essential in explaining the willing participation and support it is given by elites, rank-and-file agents, and pubic constituencies, who sometimes have little to gain and much to lose from the violence.[46]

In addition, the neo-ideological perspective emphasizes that mass killings emerge out of shifting ideological diagnoses and narratives *in response to perceived crisis*, rather than emerging directly from longstanding goals and preformed

---

[41] See, in general, Foucault 1977; Buzan, Wæver, and Wilde 1998; Campbell 1998; Hagenloh 2009, 324–6. While I downplay the importance of radical projects of social re-engineering, my account thus has affinities with scholars who emphasize the expansion of state aspirations to regulate and control society; see Scott 1998; Naimark 2001, 6–11; Powell 2011; Hagenloh 2009.

[42] Somewhat contra Lang 2010; Lang 2020.

[43] Wijze 2019. See also Fujii 2004; Morrow 2020, 45–54.

[44] Straus 2015a.

[45] I conceptualize moralized action broadly here, in the sense that it is motivated or legitimated with reference to other-regarding or symbolic values, interests, sentiments, rules, and obligations, as opposed to private self-interest. This recognizes that many moral concerns are consequentialist, avoiding the error, common amongst social scientists, of treating only 'deontological' or 'non-consequentialist' concerns as 'moral', e.g. Ginges and Atran 2011, 2930. For a critique of this error see Rathbun and Stein 2020.

[46] Snow and Benford 1988; Jabri 1996; Kaufman 2006; Bar-Tal 2007; Ginges and Atran 2011; Tausch et al. 2011; Fiske and Rai 2014; Costalli and Ruggeri 2015; Kaufman 2015; Costalli and Ruggeri 2017.

'plans' for extermination. In a few cases—such as the Holocaust—some exterminatory ambitions can be found to significantly predate violence, although the path from such ambitions to mass killing generally remains forking and uneven. But in general, scholarship has come to strongly emphasize the complex, crisis-driven, escalatory paths of radicalization to mass killing, in which perpetrators often lack a clear idea of the form of or even need for mass killing at the outset.[47]

Finally, the neo-ideological perspective more plausibly accounts for the apparent power of hardline ideas to influence and radicalize 'ordinary' perpetrators. There is no need to assume that ordinary perpetrators are culturally preprogrammed with intense hatreds, or that techniques of extremist propaganda can rapidly 'brainwash' ordinary citizens *en masse* to accept esoteric transformational goals or abandon established moral norms. Instead, hardline justificatory narratives work because they draw on claims, norms, institutions, and practices that are familiar to and accepted by most ordinary people, but which are vulnerable to radicalization via their extension to civilian groups. Commonplace ideas about security politics are often strategically and morally elastic, and can be made to look applicable to violence against civilians if integrated into the right ideological narratives. It is to the power of such narratives that I now turn.

4.2 The Power of Justificatory Narratives

Hardline ideologies are crucial parts of the explanation of mass killing because they provide, in times of political crisis, a justificatory narrative which makes such extreme violence appear strategically and morally justifiable. While the exact content of these narratives varies in important ways, the basic elements are relatively consistent across cases. I term such consistent elements the *recurring justificatory mechanisms* for mass killing.[48]

There is no single 'correct' way of typologizing these justificatory mechanisms, but I try to advance beyond existing studies by more systematically following three principles of typological design. First, I limit myself to mechanisms that have strong evidence for being *causally relevant*, based on leading research from both psychological science and comparative studies of mass killing. Second, I focus on mechanisms that are close to being *empirically universal* across cases of mass killings. This is not entirely clear-cut. Justificatory narratives are powerful

---

[47] Valentino 2004; Gerlach 2010; Owens, Su, and Snow 2013.
[48] This notion draws on the extensive literature on causal mechanisms; see Gerring 2007; Hedström and Ylikoski 2010; Waldner 2012. As ideational mechanisms, the justificatory mechanisms are cognitive and discursive: ultimately exerting causal effects via cognition but communicated between individuals via discourse. Each justificatory mechanism describes a commonly recurring configuration of certain ideas that, when invoked, tends to produce certain similar kinds of causal outcome (specifically, increasing the private or public justifiability of violence).

precisely because they provide multiple means of justification, which matter to different extents for different perpetrators. Consequently, the 'presence' of a given mechanism in an overall case is a matter of degree, and assessing it involves some substantive analytical judgement. Finally, I condense identifiable sets of justificatory ideas into the minimal number of mechanisms that, though overlapping, are nevertheless not subsumable by each other.[49] Again, psychological research plays a role in this claim of *non-subsumability*. For example, while there is strong linkage between the representation of victims as threatening and as guilty, there are important differences in the psychological responses to perceptions of threat and perceptions of guilt.

On this basis, I identify six recurring justificatory mechanisms of mass killing which I label, with varying degrees of originality: (i) threat construction; (ii) guilt attribution; (iii) deidentification; (iv) futurization; (v) valorization; and (vi) the destruction of alternatives. Human beings are able to sincerely support mass killing when they internalize the ideas involved in such mechanisms, and this is the core way of getting inside most perpetrators' understandings of their own actions. Equally, when these justificatory mechanisms become embedded in structural norms, institutions, and discourses, all individuals face social pressures, incentives, and opportunities to reproduce them and enact the violence they justify, even if they do not privately approve of such violence. The mechanisms are interrelated and mutually reinforcing. But the first three mechanisms primarily shape perceptions of civilian targets of violence, while the latter three mechanisms primarily shape perceptions of the violence itself and those who perpetrate it.

The rest of this section unpacks and analyses these six justificatory mechanisms. My analytical approach here is eclectic, blending comparative political and historical research, psychological science, and critical political theory. This analysis does not produce a tightly predictive, quantifiable, or parsimonious model, but it substantiates the causal link between ideology and mass killing outlined in Chapter 3. Rather than merely observing the association of certain ideas and violence, it identifies sociopsychological processes through which those ideas have real causal power to encourage violence. Unpacking the mechanisms also demystifies the perpetrator mindsets and social structures involved in mass killing. Many observers find it hard to comprehend how perpetrators could possibly see appalling atrocities against civilians as privately or publicly justifiable. The justificatory mechanisms show how this becomes possible—how the formulation and propagation of certain ideas allow many perpetrators to 'believe that mass killing is the right thing to do,'[50] while also creating powerful norms and institutions that promote violence.[51]

---

[49] Cf. Saucier and Akers 2018, 86.
[50] Chirot and McCauley 2006, 5.
[51] See also Morrow 2020.

Portraying the Victims: Threat, Guilt, and Exclusion

The way that perpetrators of mass killing portray their victims represents one of the most intensely studied issues in existing research. Those processes that are present across cases can be condensed into three key justificatory mechanisms: (i) the presentation of civilians as threatening, (ii) the presentations of civilians as guilty of wrongdoing that makes them proper targets of violence, and (iii) the denial of normatively relevant links of identity between perpetrators and victims that would inhibit such violence. These mechanisms interrelate but are non-subsumable, while incorporating many other concepts analysed in existing scholarship.

Threat Construction

The most ubiquitously recognized justification for violence is individual or collective self-defence. The assertion that killing is a necessary form of (perhaps pre-emptive) self-defence is central in all cases of mass killing, and stressed by virtually all scholars. Those who initiate, carry out, and support mass killing perceive and/or portray themselves as in extreme peril from dangers emanating from or fundamentally linked to civilian victim groups. Such threat perceptions are *somewhat* linked to a range of predictable objective circumstances, as rationalist scholars suggest, but remain heavily underdetermined by them. As Paul Staniland argues: 'it is impossible to make sense of threat perception without paying attention to the symbols and cleavages that regimes and their dominant security institutions view as salient, acceptable, and intolerable'.[52] Ideological worldviews shape the understanding and relative prioritization of different values, interests, and identities that might be 'threatened', and encourage different factual assumptions and perceptions about the likely dangers posed to those values and interests.

While relevant to other forms of political violence, ideological threat construction is especially crucial in mass killings, because the strategic indeterminacy of violence against civilians is partly rooted in the fact that unarmed civilian groups are, by their very nature, not obviously threatening by comparison with armed forces. Indeed, the threats ascribed to victims of mass killing are often hugely exaggerated or even fantastical—and look absurd to those operating outside the perpetrators' ideology.[53] The violence depends, as Shaw comments, on a 'construction of civilian groups as enemies, not only in a social or political but also in a military sense, to be destroyed'.[54]

---

[52] Staniland 2015, 777.
[53] Brubaker and Laitin 1998, 441–3; Hiebert 2017.
[54] Shaw 2007, 111. I therefore dissent somewhat from Laura Sjoberg's argument that gendered notions of civilians as feminized and in need of protection creates incentives for targeting civilians as the enemy's vulnerable 'centre of gravity'; see Sjoberg 2013, 196–202. This may apply to some degree

Threat construction reconstrues the perpetrators of violence as well as their victims, transforming the meaning of killing itself from mass-murder to self-defence, from a criminal act to a heroic one, so that 'massacre takes on the appearance of an act of war'.[55] As Kristen Renwick Monroe observes of her interviews with rescuers, supporters, bystanders, and perpetrators in the Holocaust in the Netherlands:

> the strongest Nazi supporters' self-images are—ironically—those of victims, of people besieged by threats to their well-being, who must strike pre-emptively to protect their ontological security and that of their community against Jewish threats.[56]

Threat construction is itself, in turn, often reinforced by other justificatory mechanisms. As discussed below, threats may be rooted in dehumanized conceptions of civilian groups which present them as insidious, biological dangers—as Neilsen emphasizes, civilians are frequently construed as intolerable toxic threats requiring elimination.[57] The Khmer Rouge, for example, asserted great threats from a 'concealed enemy boring from within'[58] and promised to 'clean up hidden enemies burrowing from within … so that they are completely gone, cleansed from inside the ranks of our revolution'.[59] Again, when threats are presented as inhering in the fundamental *essence* of the civilian group—implying that future coexistence with them is impossible—mass killing is liable to take its most intensive, genocidal form.[60]

As shown across my four case studies, the construction of civilians as threats often appears to sincerely persuade elites, rank-and-file agents, and public constituents. The power of threat construction is rooted in a range of human cognitive responses to danger, which our capacity for complex language allows to be triggered not just by immediate physical threats, but ideologically imagined ones too. As neuroscientist Kathleen Taylor observes:

> In humans … the ability to use symbols—words and images—to trigger emotions has decoupled threat responses from the necessity to have an actual threat to hand. Using symbols … [one] can conjure up the strong emotions of a threat response—hatred, fear, disgust and anger—and link them to prior beliefs about

---

in sexual atrocities, and also in Allied area bombing in World War II; see Chapter 6. But it relies on perpetrators perceiving their victims as civilians, and in almost all cases of mass killing they do not. Ideological justifications of mass killing are gendered, but primarily through a different mechanism emphasized by Sjoberg—the valorization of male perpetrators as tough, heroic defenders against (typically male, often sexualized) enemies.

[55] Semelin 2007, 145 See also Straus 2012a, 353.
[56] Monroe 2011, 197.
[57] Neilsen 2015.
[58] Cited in Anderson 2017a, 82.
[59] Hinton 2005, 34.
[60] Hiebert 2008; Straus 2015a, ch.1.

a target group, prompting an audience to act on those beliefs without pausing to check their accuracy.[61]

Experimental evidence correspondingly suggests that threat construction has significant psychological impacts: even momentary reminders of threat and death can significantly increase prejudice, stereotyping, and support for violence against 'outgroups',[62] while also encouraging 'an increased feeling of [ingroup] togetherness ... increased respect for leaders, increased idealization of ingroup values, and increased readiness to punish deviates from ingroup norms'.[63]

Again, however, threat constructions are consequential even for perpetrators who do not sincerely accept them, because they are typically institutionalized: embedded in the discourses and administrative practices of state security agencies in ways that can structure violence independently of individuals' private beliefs.[64] Such agencies typically produce a stable securitized lexicon for describing targets of mass killing: Suharto's anticommunist policies in Indonesia targeted 'gangs of security disruptors',[65] the Nazis fought 'Judeo-Bolsheviks'[66] or 'International World Jewry',[67] while Stalinists targeted 'socially harmful elements'.[68] Such institutional processes of categorization and classification are crucial to the process of constructing threats,[69] and are spread and reinforced through security practices. Michael Wagner observes how in Rwanda:

> Civil defense gave people the experience of conducting roadblocks, house searches, security meetings, and night patrols. It also developed the shared vocabulary, as well as the techniques, for identifying and seeking out 'enemies of the people' and their 'accomplices'.[70]

Such categories are never conjured out of thin air, but draw upon ideological underpinnings that give them meaning and plausibility.[71] Critical to the lethal potential of such categories, moreover, is their typical elasticity.[72] It was hard for citizens in Stalin's Soviet Union to know where the category of 'kulaks' really stopped, for example, allowing it to be used to legitimate ever-expanding circles

---

[61] Taylor 2006, 239. See also Zimbardo 2007, 11.
[62] Hirschberger, Pyszczynski, and Ein-Dor 2015, 4–8; Littman and Paluck 2015, 89–90.
[63] Chirot and McCauley 2006, 65. See also Kelman 2007, 83–6; Hammack 2008, 224 & 233; Cohrs 2012, 60; Lickel 2012, 99.
[64] du Preez 1994, 48–9 & 52; Chirot and McCauley 2006, 82–3. See also Berger and Luckmann 1967, 139.
[65] Kiernan 2003, 47.
[66] Weitz 2003, 105, 107–8, 125, & 139. See also Chalk and Jonassohn 1990, 332; Mann 2005, 301.
[67] Chalk and Jonassohn 1990, 337.
[68] Hagenloh 2000; Werth 2003, 219. See also Chandler 2000, 6 & 93–4; Harff 2003, 61.
[69] Browning and Siegelbaum 2009.
[70] Cited in Fujii 2004, 107–8.
[71] Fujii 2004, 102–7; Browning and Siegelbaum 2009, 231–2.
[72] Hagenloh 2000, 289.

of violence. Similarly, at the height of the Argentinian 'Dirty War', the military regime relied on an extremely general category of 'subversion'—defined as 'anyone who opposes the Argentine way of life'.[73] As David Pion-Berlin observes: 'With such a vague notion of threat, the junta was more likely to lash out against a broad cross-section of the population ... [The military regime] showed little precision in its work, instead casting its coercive net broadly and unpredictably such that few could tell when or why they would be at risk.'[74]

Even when individuals may be privately apathetic or sceptical about such notions of threat, being associated with the individuals identified with threat categories risks being labelled an 'accomplice', while failure to participate in 'defence' against such threats connotes shameful treason to the ingroup. Categorized as dangers or enemies rather than citizens or innocents, civilians become apparently proper objects of police and military powers, while being stripped of the kinds of institutionalized social protections that may deter more self-interested abuses and exploitation. Pervasive ideological threat construction of certain groups therefore opens the doors even for ideologically disinterested individuals to target victims for conformist or opportunistic reasons.

Guilt Attribution

Like the presumption that violence is justified in self-defence, the notion that criminals and heinous wrongdoers can and should suffer violent punishment is commonplace.[75] Like threat, however, guilt can rarely be inferred directly from objective facts accessible to ordinary citizens—instead, the politics of blame is driven by ideological narratives about purported wrongdoers which are often widely accepted even when they bear little relation to hard facts. In mass killings, such claims frequently appear to inform both elite-level diagnoses of national problems and mass support for discrimination and violence.

Research on political violence—partly due to the influence of rationalist theories—has not traditionally emphasized perceptions of guilt by comparison with the role of threat and material self-interest. Even in genocide studies, there has been some uncertainty: early 'scapegoating' theories of genocide proposed that victims are killed because they are (wrongly) blamed for perpetrators' real losses and frustrations,[76] but many scholars now see this causal narrative as inaccurate. Research on genocide and atrocities has increasingly highlighted how perpetrators of violence have rarely suffered the kind of *personal* frustration or loss that might motivate scapegoating, and that the accusations levelled at victims are often very indirectly linked, at best, to actual experiences of social suffering.[77]

---

[73] Pion-Berlin 1988, 401.
[74] Ibid. 400.
[75] Lickel et al. 2006, 373.
[76] E.g. Staub 1989; Staub 1999. See also Midlarsky 2005; Midlarsky 2011.
[77] Straus 2007a, 482–3; Straus 2012a, 549. See also Fischer 2006, 298.

Yet recent scholarship has seen a resurgence of interest in dynamics of blame and revenge.[78] Scapegoating models may suggest too close a relation between objective losses and violence, but mass killings are always accompanied by pervasive *ideological representations* of victims as guilty of great crimes, and therefore legitimate targets of violence. Such claims can play different roles for different perpetrators. For some, they may primarily serve to legitimate violence by placing victims in a category of 'criminals' for whom, under quite conventional moral and social norms, violent punishment is permitted as a form of 'reprisal'. Perpetrators frequently suggest genuine belief in such portrayals, but, again, deep convictions are not necessary. Vague acceptance of guilt-attributing categories used by authorities to define and refer to victims can nevertheless induce framing effects: victims are presented as 'criminals', 'enemies of the people', and 'traitors', who can be treated as such with varying degrees of (un)reflective behaviour. Even where sincere belief in such assertions is entirely lacking, guilt attribution, like threat construction, strips away victims' social and institutional protections against violence and can generate social pressure to participate in their persecution. As Lickel et al. observe:

> those who retaliate against an outgroup ... will often go unpunished by other ingroup members. In fact, such retaliatory actions might confer advantages in the form of increased respect and status within the group ... [Conversely], group members who fail to retaliate ... will be viewed as deviants and as having insufficient commitment to the group ... Thus, to avoid ostracism, people may retaliate on behalf of their group even when they do not want to.[79]

But guilt attribution goes beyond legitimating categorizations, since revenge frequently *motivates* violence.[80] Isabel Hull says of the Herrero genocide that 'the most lethal factor ... was the desire to punish',[81] and Véronique Nahoum-Grappe remarks on the way that ideological mobilization of past massacres is frequently used to evoke 'the desire, the obligation, the *duty toward the dead* to seek revenge'.[82] Sometimes, the egregious accusations involved in guilt attribution may whip up genuine hatred, but this is not necessary for punitive motivations to operate.[83] The desire to punish is psychologically rooted and probably evolutionarily selected: experimental set-ups demonstrate that individuals will often incur significant personal costs simply to impose punishments on others, in ways that are normally functional for social cooperation.[84] But a consequence is that the pure desire to

---

[78] See, in general, Steele 2013; Souleimanov and Aliyev 2015; Balcells 2017.
[79] Lickel et al. 2006, 378.
[80] Chirot and McCauley 2006, 25–31 & 66–71; Dutton 2007, 73–84.
[81] Hull 2003, 158.
[82] Semelin 2007, 43 (*emphasis in original*).
[83] Bar-Tal 2007, 1440.
[84] Carlsmith, Darley, and Robinson 2002; Lickel et al. 2006, 377–8; Hirschberger, Pyszczynski, and Ein-Dor 2015, 2. See also Midlarsky 2005, 107–10. The prevalence of revenge motives is well recognized

make sure perceived wrongdoers get their 'just deserts' carries considerable motivational power. In contrast to 'scapegoating' theories, individuals often feel this desire even when they have suffered no personal loss from the purported crimes in question. The mere perception that other members of one's ingroup have suffered at the hands of members of an outgroup appears to be capable of generating intense desires for 'vicarious retribution'.[85]

Again, conditions of 'epistemic dependence' are often key here, with rumours and unsubstantiated assertions by authorities, and the incessant repetition of isolated anecdotal cases, often sufficient to create a social perception of victim guilt.[86] Members of the Lord's Resistance Army—involved in widespread atrocities in Uganda, the Democratic Republic of the Congo, and the Central African Republic—reported how: 'The commanders … sat every evening and told us, that … Ugandan soldiers are sleeping with their mothers, abusing their children, and causing a lot of atrocities. That is the reason why we must fight tirelessly.'[87] Serb media in the Yugoslav Wars repeatedly propagated rumours of Croat and Bosnian atrocities as established facts. One broadcast reported that 'It seems that Muslim extremists invented the most horrific crime on the planet. Last night they fed Serb children to the lion at the Sarajevo Zoo',[88] while in another, a Serbian interviewee stated:

INTERVIEWEE: I didn't see such things, but I heard from others that there was torturing … they were cutting off fingers, pulling finger-nails off children … we have found children in pots ready to be baked. We discovered beheaded children.
REPORTER: They have no mercy for anyone, do they?[89]

It is now widely recognized that many exposed to such 'fake news'—with the exception of those strongly predisposed to view such claims with suspicion—often accept such stories when numerously repeated.[90] The well-documented tendency of people to engage in 'just-world thinking' and assume that individuals tend to deserve what they get reinforces this tendency.[91]

Desires to blame and enact revenge, as well as the common attribution errors that accompany such processes, are to some extent *pre-ideological*: they are widespread reactions that transcend ideological differences between individuals.

in research on domestic homicides, roughly 20% of which appear linked to revenge; see McCullough, Kurzban, and Tabak 2011, 221.
[85] Lickel et al. 2006.
[86] Brubaker and Laitin 1998, 439–40; Osborn 2008.
[87] Cited in Anderson 2017a, 75.
[88] Cited in ibid. 84.
[89] Cited in ibid.
[90] Lewandowsky et al. 2012.
[91] Furnham 2003; Jost and Hunyady 2005; Bénabou and Tirole 2006. See also Sagan and Valentino 2017, 66.

But the direction and intensity of such processes is substantially affected by prevailing ideological narratives. Most individuals do not lash out at literally *anyone* in the attempt to find a blameworthy culprit for crimes—victims are selected according to claims and stereotypes that vary across ideological worldviews.[92] Moreover, such ideological attributions of guilt are typically collective in character. Civilians are often deemed guilty as *a homogenous group entity*—in the language of the psychological sciences, guilty groups are accorded high 'entitativity'—seemingly justifying collective punishment for any crimes individual members purportedly committed.[93] Once crimes are repeatedly attributed to 'Jews', 'Tutsi', 'Muslims', or 'Japs', many people stop differentiating individual members of such groups and instead conceive of all members of the referenced category as equally and uniformly complicit and thus proper reciprocal targets of revenge. As Hugo Slim points out: 'The tragedy, of course, is that such revenge is not reciprocity at all. It seldom inflicts suffering upon the perpetrators who committed the original offence. Instead, it selects yet another group of civilians to die as representatives of the perpetrators and punishes them instead.'[94]

Deidentification
Human beings have an evolved need to relate to and identify with groups, and to orientate their thinking and behaviour around collective categories.[95] This sustains remarkable levels of uncoerced human cooperation, including people's willingness to sacrifice their own lives in service to the group. But it can also sustain people's willingness to brutally harm 'outgroups' in the name of their 'ingroup'. The most powerful forms of group-identification involve close kinship and personal relationships. But over human history, the cultural construction of large-group identities has extended identification beyond small kinship groups to various kinds of 'imagined community'—such as nations, religions, or transnational political movements—in which most members never meet each other and yet feelings of identification remain deep.[96]

Such large-group identities are implicated in all facets of politics, and it is hardly surprising that exclusionary identity processes like 'othering' and 'dehumanization' are amongst the most widely recognized foundations for atrocities against civilians.[97] As Leonie Huddy observes, however: 'Each of us has many potential identities derived from diverse group memberships, but relatively few of these

---

[92] Lickel et al. 2006, 375–80.
[93] Haslam 2006, 259 & 261–2; Lickel et al. 2006, 378–80; Slim 2007, 143–51 & 175–6; Lickel 2012, 96–8.
[94] Slim 2007, 142.
[95] Taylor 2006, 231–2; Hammack 2008; Hogg 2014.
[96] Anderson 1983/2006; Malešević 2006.
[97] Fearon and Laitin 2000; Suny 2004; Taylor 2006; Volkan 2006; Moshman 2007; Hammack 2008; Alvarez 2015, 230.

identities develop or become politically consequential.'[98] Again, basic measures of identity divides, such as the level of ethnic or religious diversity in society, are poor predictors of violent conflict or mass killing. What matters is how identity categories are ideologically interpreted, mobilized, and institutionalized to encourage (or discourage) violence.[99]

In explaining the justification of mass killing, I focus on processes of *deidentification*: the denial or suppression of bonds of common identity between perpetrators and victims. Deidentification goes beyond the mere categorization of individuals into groups. Individuals sorted into different groups often still identify across group boundaries, through some higher-level shared identity such as a common nationality which transcends ethnic differences or a common humanity which transcends nationality. Deidentification goes further, suppressing emotional and moral concern for the welfare of certain individuals by emphasizing their separateness from the relevant moral community of the ingroup.[100] As the US commander of the brutal American occupation of the Philippines told subordinates, for example: 'All consideration and regard for the inhabitants of this place cease from the day I become commander.'[101]

Deidentification is a form of what Susan Opotow influentially labels 'moral exclusion'—the elimination of individuals from a society's domain of moral consideration or 'universe of obligations', in Helen Fein's language.[102] How the universe of obligations is delimited varies ideologically: while ethnonationalists demand exclusive loyalty to the ethnic nation, for instance, cosmopolitans assert the equal moral importance of all human beings. Importantly, victims of mass killing are often from groups that *have* traditionally been members of a society's universe of obligations (as with Jews in Germany, Bosnians, Croats, and Serbs in Yugoslavia, Hutu and Tutsi in Rwanda, or the many victims of Soviet violence).[103] Even when victims may never have been established parts of perpetrators' universe of obligations (as with Native Americans in North America, or the Herero people of German South West Africa), their excluded status is, again, not mandated by the mere fact of difference. So while certain 'objective' conditions may facilitate ideological deidentification, they do not require it. In Cambodia under the Khmer Rouge, for example, it was hardly surprising that supporters of the preceding regimes were viewed with suspicion and excluded from power. But nothing beyond ideology mandated that they be seen as intrinsically incompatible with the new order and therefore requiring extermination. It was due to Khmer Rouge

[98] Huddy 2001, 137.
[99] Hammack 2008.
[100] Taylor 2006, 233.
[101] Valentino 2004, 202.
[102] Fein 1979, 4; Opotow 1990. Threat construction and guilt attribution also, however, involve moral exclusion.
[103] Hinton 2002, 10–11; Valentino 2004, 2; Midlarsky 2005, 165; Waller 2007, 252 & 254; Straus 2015a, 84.

ideology, as Karl Jackson observes, that such individuals 'were *not* perceived as fellow countrymen who had made mistakes and could be remolded to perform useful functions in the new society'.[104] Instead, the Khmer Rouge declared: 'These counterrevolutionary elements which betray and try to sabotage the revolution *are not to be regarded as being our people*. They are to be regarded as enemies of Democratic Cambodia, of the Cambodian revolution, and of the Cambodian people.'[105]

The most extreme and commented on form of deidentification is dehumanization: the representations of certain outgroups as lacking common humanity.[106] Dehumanization is a widespread feature of atrocities. As Ben Kiernan notes:

> Democratic Kampuchea referred to its enemies as 'microbes', 'pests buried within', and traitors 'boring in'. The Germans had talked of 'vermin and lice' … they prefigured biological depictions by Bosnian Serbs of the 'malignant disease' of Islam threatening to 'infect' Europe.[107]

Such dehumanizing conceptions are frequently endorsed by senior political, military, and intellectual authorities, including, remarkably, the medical professions. As one German doctor at Auschwitz wrote: 'Of course I am a doctor and I want to preserve life. And out of respect for human life, I would remove a gangrenous appendix from a diseased body. The Jew is the gangrenous appendix in the body of mankind.'[108] A senior Ottoman doctor under the Young Turks likewise stated: 'Armenian traitors had found a niche for themselves in the bosom of the fatherland; they were dangerous microbes. Isn't it the duty of a doctor to destroy these microbes?'[109] In his work on American soldiers involved in massacres of civilians in Vietnam, psychiatrist William Barry Gault observed how:

> Orientals are regularly referred to as 'gooks' and 'dinks'; are said to be 'like children' … And above all, Orientals are held to be inscrutable, strange, profoundly and irreconcilably different from and incomprehensible to Western man. These attitudes serve to psychologically soften the experience of killing Orientals, so that some soldiers feel that the individual dead enemy was 'not like you and me, but more like a Martian or something.'[110]

---

[104] Jackson 1989b, 56.
[105] Cited in ibid. (*Jackson's emphasis*).
[106] There is debate as to how complete this denial of humanity must be and what kind of humanity is denied. Since all kinds of dehumanization are sub-strands of deidentification, my stance here is neutral in that debate. See Haslam 2006; Smith 2011; Steizinger 2018; Owen 2021.
[107] Kiernan 2003, 33. See also Fleming 2003, 109.
[108] Goldhagen 1996, 269. See also Gellately 2003, 246; Haas 2008.
[109] Mann 2005, 172. For examples from the conflict in Darfur see Hagan and Rymond-Richmond 2008, 882.
[110] Gault 1971, 451–2.

Many other perpetrators of atrocities subsequently highlight such dehumanization in explaining their own behaviour. 'We thought of them as things, not people like us,'[111] reported one Japanese General regarding Chinese victims in the 'Rape of Nanking', whilst a junior Japanese soldier stated: 'perhaps when we were raping her, we looked at her as a woman, but when we killed her, we just thought of her as something like a pig'.[112] A guard at the Nazis' Chelmno death camp told a superior in 1943 that he could kill easily because: 'Little men or little women, it was all the same, just like stepping on a beetle.'[113]

Dehumanization is clearly a powerful psychological mechanism in promoting violence. Yet its centrality is often exaggerated, and it is not—at least in any explicit and widespread form—a *necessary* foundation for mass killing. By contrast with cases like the Holocaust, many mass killings do not depend on substantive conceptions of victims as inhuman or subhuman. Policymakers in Britain and America during World War II may have considered Germans contemptible, but they rarely denied their humanity, and while the Soviet Union was willing to rhetorically cast the victims of Stalinist purges in dehumanizing terms, there was no deep sense in Soviet ideology that the victims were subhuman. Dehumanization is not irrelevant in such cases, and may facilitate violence for some.[114] But what matters most is not that the victims of violence are consciously seen as subhuman, but that they are placed into identity categories to which the perpetrating governments cognitively and institutionally attach little or no moral concern. Dehumanization is best understood, in other words, as an extreme form of deidentification. This resolves a purported puzzle in the literature on dehumanization: namely that the treatment of victims often involves justifications and cruel techniques of violence that make little sense unless the victims have something like human emotions.[115] This is only a 'puzzle' under a needlessly extreme conception of dehumanization as necessitating the complete denial of *any* human qualities to victims.[116] In real mass killings, it is clear that perpetrators do not need to see their victims as *entirely lacking* in human characteristics in order to commit atrocities against them. It is sufficient to portray the individuals as inferior outsiders, whose welfare is of no concern to perpetrators.

Deidentification's power to encourage violence is rooted in the way it ties together basic cognitive processes of categorization and stereotyping in ways that increase the perceived 'social distance' between perpetrators and victims, eroding empathy and other psychological barriers to violence.[117] Albert Bandura's

---

[111] Cited in Zimbardo 2007, 307.
[112] Cited in Dutton 2007, 65.
[113] Cited in Hiebert 2008, 22.
[114] See Owen 2021.
[115] E.g. Rai, Valdesolo, and Graham 2017; Lang 2020, 18–19.
[116] Rai et al. acknowledge the possibility that perpetrators might engage in selective dehumanization; see Rai, Valdesolo, and Graham 2017, 19.
[117] Littman and Paluck 2015, 91–2. See also Hogg 2014.

pioneering research on dehumanization found that when experimental subjects overheard even a single utterance of a dehumanizing characterization of a group, this would significantly increase the severity of pain they would subsequently inflict on that group.[118] It should be unsurprising that the far more extensive deidentification that accompanies atrocities can have lethal consequences. In particular, strong deidentification is involved in emotional states of *contempt*, in which high denigration of an outgroup is accompanied by a lack of any desire to preserve future social relationships, encouraging support for violence.[119]

But as with the other justificatory mechanisms, deidentification also generates structural pressure on individuals to collude in mass killing. When identities have become highly polarized between ingroups and outgroups, even those who do not sincerely internalize such distinctions face significant costs to positively defending or associating with the outgroup.[120] Ideological redefinitions of identity categories are especially powerful when they are legally institutionalized—as in the Nazi government's Nuremberg Laws, which legally defined Jews by heredity and excluded them from German citizenship and marriage rights. The death camp survivor (and secular, non-identifying 'Jew') Jean Amery recalled his awareness that in the Nuremberg Laws:

> Society concretized in the national representatives of the German people, had just made me formally and beyond any question a Jew ... [with those deemed Jews facing] the social reality of the wall of rejection that arose before us everywhere.[121]

As this illustrates, ideological constructions of identity are crucial not just as objects of genuine attachment, but as the conceptual coordinates around which social norms and institutions revolve. Again, deidentification serves to strip individuals of protective social norms and institutions, while allowing them to be targeted by ideological policies that specify their outgroup status as a basis for violence.

Portraying the Violence: Virtue, Future Goods, and Inevitability

Many scholars suggest that the common ideological roots of mass killings purely revolve around the portrayal of victims. Rowan Savage, for instance, claims that '[t]he *central* and the *only common* aspect of ideological motivation for genocide ... is the construction of the Other'.[122] Omar McDoom defines radicalization in the context of genocides and mass killings purely around negative beliefs and attitudes

---

[118] Bandura, Underwood, and Fromson 1975, 253–9. See also Bandura 1999; Staub 2002, 15–16; Tileagă 2007, 720; Zimbardo 2007, 13–14 & 307–13.
[119] Tausch et al. 2011.
[120] Moshman 2007, 119.
[121] Hiebert 2008, 14.
[122] Savage 2013, 146–7 (*emphasis in original*).

'towards a perceived outgroup'.[123] In their survey of a large number of themes in the mindsets behind mass killing, Saucier and Akers likewise observe how:

> All themes represent distortive *oversimplifying beliefs that facilitate the inculpating of an outgroup*, and not just the outgroup loosely or abstractly but *with every member of that outgroup thereby inculpated*. That is, we see a group portrayed as threatening, impure, absolutely bad, and less than human, all in the context of a worldview that not only essentializes groups but treats group-membership as the essence of every individual.[124]

Such portrayals are certainly key, but this victim-centric picture is lopsided. Ideological justifications of mass killing never focus purely on the victims, but must also ideologically construct the perpetrators and the violence they commit—as unavoidable, valorous, and ultimately efficacious.[125] Many perpetrators, indeed, display limited animosity towards their victims—for them, representations of perpetration as a necessary and patriotic duty often appear more important. I argue that three depictions of mass killing are crucial here: as valorous and praiseworthy, as delivering immense future benefits, and as an unavoidable course of action without meaningful alternatives.

Valorization

While many analyses of political violence purely focus on self-interested rationality, it is widely recognized that, in reality, moral sentiments and emotions play vital roles in generating support for violence and a willingness to participate in it.[126] People obviously violate moral codes, on a regular basis, that they do not meaningfully internalize. But this should not obscure how human beings are often willing to incur enormous personal costs for broader values and goals that they do identify with, while also expending significant effort to demonstrate their adherence to prominent social norms.[127] These values and norms are often intuitively felt and understood rather than rooted in elaborate ethical reasoning, but they are one of the critical ways in which different ideological standpoints diverge and conflict.[128]

Such divergence is especially clear in evaluations of violence, with different ideologies promoting approval or condemnation of the *intrinsic moral quality of*

---

[123] McDoom 2020b, 125.
[124] Saucier and Akers 2018, 91. See also Hiebert 2008; Hiebert 2017.
[125] Reicher, Haslam, and Rath 2008, 1327; Lang 2020, 18.
[126] The relevant literature is immense; see, in general, Bar-Tal 2007; Smith 2009; Ginges and Atran 2011; Atran and Ginges 2012; Powell 2012; Fiske and Rai 2014.
[127] Atran and Axelrod 2008; Elster 2009; Ginges and Atran 2011; Atran and Ginges 2012; Morrow 2020.
[128] Lakoff 2002; Graham, Haidt, and Nosek 2009; Haidt, Graham, and Joseph 2009; Haidt 2012.

certain acts of violence.[129] Within more limitationist ideologies, the act of committing violence against civilians is morally tainted, dubious, stigmatized, or taboo. Even if potentially justified in emergency circumstances, it is a shameful and regrettable activity. But this is not how most perpetrators of mass killing understand their actions. Hardline security doctrines encourage a reconstruction of the moral meaning of killing civilians in ways that can legitimate and even actively motivate many perpetrators to participate. In other words, justificatory narratives for mass killing *valorize* violence against civilians: framing it as an admirable, glorious, and praiseworthy act indicative of the virtuous, rather than evil, character of those who carry it out.

How is this achieved? In general, justificatory narratives for mass killing appeal, not to esoteric moral arguments, but to culturally established and widely admired virtues, such as patriotism, duty, loyalty, discipline, vigilance, and intelligence,[130] while denigrating opposition to violence as treasonous, sentimental, and foolish.[131] Much of this 'virtuetalk'[132] is intensely gendered, drawing on familiar hierarchical portrayals of masculine/feminine stereotypes and representing violence as an embodiment of heroic 'militarized masculinity'.[133] Hardliners emphatically demand 'toughness', 'loyalty', and, of course, 'hardness', while dismissing values of compassion, restraint, or critical reflection as 'effeminate', 'soft', and 'weak'. 'My pedagogy', Hitler stated, 'is hard. What is weak must be hammered away ... I want the young to be violent, domineering, undismayed, cruel ... They must be able to bear pain. There must be nothing weak or gentle about them.'[134]

Valorization leverages both powerful psychological needs and social pressures to motivate and legitimate participation in mass killing. At a minimum, valorization promotes a mental representation or social definition of violence which disassociates it from morally problematic categories of murder, barbarism, or atrocity. Such implicit moral reframing of actions is known to affect individuals' propensity to support or oppose particular courses of action.[135] Whether by weakening expectations of social condemnation, convincing individuals that violence against civilians is genuinely praiseworthy, or simply sowing moral confusion about the violence, valorization can thereby legitimate perpetration, facilitating violence when other motives for it are present.[136]

---

[129] There is a moral dimension, indeed, to all six justificatory mechanisms—but here I am concerned with the most direct valorization of violence itself.
[130] du Preez 1994, 130; Markusen and Kopf 1995, 75; Hinton 1998, 96–7; Mann 2005, 200; Slim 2007, 25–6; Cohrs 2012, 66.
[131] Hinton 1998, 116; Mann 2005, 200; Semelin 2007, 242 & 263.
[132] In earlier work I used the label 'virtuetalk' in place of 'valorization'; see Leader Maynard 2014, 832. I now understand virtuetalk as merely the commonest form of valorization.
[133] See, broadly, Elshtain 1987; Theweleit 1989; Das 1998; Peterson 2010; Sjoberg 2010.
[134] Miller 1983, 142.
[135] E.g. Petrinovich and O'Neill 1996. See also, in general, Jost and Major 2001.
[136] Littman and Paluck 2015, 91.

Yet valorization can also engage active motives to perpetrate. Most individuals feel intense motives to establish both positive moral self-identity and social self-esteem, and valorization links participation in violence to both.[137] Such needs and anxieties are especially visible amongst relatively young men organized into groups—a constituency often, though not always, overrepresented amongst atrocity perpetrators. Again, militarized portrayals of masculinity seem to matter here: we need further empirical studies of the causal impact of gendered language on political violence, but there is established evidence that male 'gender role stress' can promote aggression.[138] As an aspect of ideological structure, valorized norms of violence attach social esteem and possibly indirect material rewards for those willing to prove their 'hardness' and 'loyalty' via participation in mass killing, while threatening shame and reputational damage for those who refuse to participate.[139] Former Nazi *Einsatzgruppen* members testified that refusing to participate in killing risked being shamed for 'cowardice', an inability to conduct 'tough action', or for being 'not as hard as an SS-Mann ought to have been'.[140] The Commandant of Auschwitz, Rudolf Hoess, similarly emphasized how: 'I wished to appear hard, lest I be regarded as weak'.[141] But valorization may also be internalized, allowing individuals to derive positive self-satisfaction from acting out what they take to be highly positive moral traits. Omar Bartov observes of German soldiers that 'as their own letters and diaries, frontline journals, and memoirs clearly indicate, they were strongly motivated by an image of battle as a site of glory precisely because it was harsh, pitiless, and deadly'.[142]

In stressing this exploitation of *conventional* virtues, I am not denying that justificatory narratives for mass killings significantly radicalize prevailing moral sentiments and norms. Hardliners espouse an explicit 'ethos of violence',[143] built around the exaltation of machoistic brutality, and the repudiation of humanitarian or compassionate moral reflection.[144] As Burleigh observes of Nazi Germany:

> Nazism represented a sustained assault on fundamental Christian values, regardless of any tactical obeisance to the purchase it had on most Germans. Compassion, humility or love of one's neighbour were dismissed as humanitarian weakness by an organisation which regarded hardness, sacrifice and self-overcoming as positive virtues.[145]

---

[137] See also Monroe 2011, ch.9; Littman and Paluck 2015, 83.
[138] Baugher and Gazmararian 2015. See also Mann 2004, 26; Karstedt, Nyseth Brehm, and Frizzell 2021, 10.9–10.
[139] Littman and Paluck 2015, 90.
[140] Mann 2005, 269.
[141] Hoess 1959, 82.
[142] Bartov 2003, 129.
[143] Mann 2005, 253. See also Weitz 2003, 52.
[144] Hinton 1998, 91, 112, 115, & 117. See also Gault 1971, 452–3.
[145] Burleigh 2001, 196.

Similarly, for many communist regimes, as James Ryan observes: '"humanist" and "compassion" could become hidden insults, "destruction" could become a term of immense praise'.[146] The Khmer Rouge idealized strict military harshness and denigrated humanitarian feelings,[147] issuing demands for the people 'to be clean, to be good, to be rough, to be strong'.[148] As Hinton has written, such discourse was part of an overarching glorification of brutal hardness, in which 'the ideal new communist citizen would be able to "cut off his or her heart" from the enemy who was "not real Khmer"'.[149] The same can be said of fascist ideologies, in which war and violence were adulated as noble activities essential for a healthy society. In Mussolini's words: 'War alone brings up to their highest tension all human energies and puts the stamp of nobility upon the peoples who have the courage to meet it.'[150]

But this selective rejection of humanitarian values is accomplished, not through complete moral disengagement or inversion, but by asserting the overriding importance of *other conventional moral sentiments*. Rather than inventing new radical moral frameworks anew, hardline ideologies shuffle, reorder, and reinterpret conventional values in distinct ways. Valorization of mass killing consistently emphasizes what Raul Hilberg labelled 'the doctrine of superior orders'—the need to unquestioningly obey the commands of authorities—which, as Hilberg notes, offers 'the oldest, the simplest, and therefore the most effective' means of justifying participation in violence.[151] More generally, hardliners draw on familiar images of militarized masculinity and soldierly virtue rooted in ideologies and practices of war across the world. Thus, Nazi leaders called on ordinary Germans to demonstrate their 'duty' to the Fatherland and their 'hardness' in implementing violence to defend the *Volksgemeinschaft*. German police regiments were told by the Police Commander of the Lublin District, where they had been shooting large numbers of Jews: 'I feel bound sincerely to thank you all … for your indefatigable work, as well as your proven loyalty to me and willingness to sacrifice. You have all given your best for Führer, Volk and Fatherland in the tenacious, hard, and bloody partisan fighting.'[152] The SS leader of Warsaw likewise wrote privately to his superiors, after the killing or capture of 56,065 Jews in the destruction of the city's ghetto, to suggest that: 'High credit should be given [to the Waffen SS] for the pluck, courage, and devotion to duty which they showed'.[153] As Michael Mann observes, such appeals 'resonated amid commonplace virtues like loyalty, obedience, comradeship,

---

[146] Ryan 2012, 22.
[147] Hinton 1998, 112. See also Chandler 2000, 101.
[148] Weitz 2003, 155.
[149] Hinton 2002, 15.
[150] Cited in Bell 1986, 58.
[151] Hilberg 1985, 288. I diverge from Hilberg, however, in his implication that this tends to be merely a 'rationalization' for violence.
[152] Goldhagen 1996, 262.
[153] Doc.6.13, Stackelberg and Winkle 2002, 368.

dutifulness, honour and patriotism, which were especially strong among the SS core recruitment constituency of ex-soldiers, policemen, civil servants, and educated professionals'.[154] Valorization generally appeals to ordinary virtues, elevating and stretching them to override and outmanoeuvre moral obstacles to violence, and to thereby reconstruct mass killing as socially legitimate and praiseworthy action.

Futurization

As just emphasized, support for violence is rarely a purely instrumental calculation. Yet, all other things being equal, human beings are unsurprisingly more inclined to perpetrate and support violence if they think it will yield beneficial outcomes.[155] In rationalist analyses of mass killing, such anticipated benefits are the central driver of violence, and are a function of the incentives created by a decision-maker's external environment. People are expected to support mass killing if their objective circumstances make it appear comparatively beneficial for their interests.

As rationalists are fully aware, this picture radically simplifies actual decision-making. Besides the fact that people often deviate from rational calculations in both predictable and unpredictable ways,[156] available information on the likely effects of mass violence is, as observed in Chapter 3, generally highly ambiguous. In consequence, the link between the *actual* likely effects of violence and the *perceived* likely effects of violence is often tenuous, and highly affected both by individuals' ideological preconceptions and by ideological representations of violence in public political discourse. At a more basic level, perceptions of violence as beneficial will also vary ideologically, since 'what overall consequences are valuable, satisfactory or good can be very different depending on the actor's ideology'.[157]

This is true to some degree of all political violence.[158] But with many forms of violence, such strategic indeterminacy should recede over time, as decision-makers receive increasing feedback on the actual consequences of their actions. If a military offensive is not succeeding, decision-makers should be expected to gradually accumulate information on its inefficacy, and eventually explore other options irrespective of their initial ideological convictions. In the four mass killing campaigns I examine in this book, however, such processes of feedback appeared to have little to no effect in three cases (Stalinist repression, Allied area bombing, and the Rwandan Genocide) and only very modest effects in the fourth (the Guatemalan Civil War). This reflects the fact that mass killings are not normally justified by immediate, short-run benefits which could be verified or falsified by such feedback. Instead, they are justified by what I term *futurization*: the assertion

---

[154] Mann 2005, 200.
[155] Saab et al. 2016. See, however, Ginges and Atran 2011.
[156] Elster 1982; Sunstein 2000; Ariely 2009; Kahneman 2012.
[157] Aguilar, Brussino, and Fernández-Dols 2013, 449–50.
[158] Frazer and Hutchings 2019, 18–19.

that there are ultimate long-run benefits to violence which outweigh the apparent costs and suffering it causes in the short term. Futurized violence is justified not by the 'tactical' utility it immediately yields, but because of a hypothesized eventual 'strategic' utility: ultimate military victory, elimination of threats from the body of the nation, salvation of the revolution, and so forth. Futurization is deeply ideological, because such long-term consequentialist calculations are highly underdetermined by the available evidence and primarily rooted in preconceived beliefs and value-judgements.

There is nothing inherently invalid about pursuing such long-term consequences. But scrutiny on the epistemic status of such benefits—how sure we are that they will arise, and what the foundation for that certainty is—is often weak when the costs are borne by others.[159] Consequently, the futurization of political violence is commonly fallacious: cost–benefit calculations are skewed towards the future without significantly discounting future benefits for their uncertainty, so that the *definite* deaths of present victims are weighed up against *hypothetical*, and often totally fictitious, future benefits as if these were outcomes of an equivalent kind. Perpetrators thus justify mass killing through a kind of consequentialist calculus, but the calculus is loaded—creating a moral posture of speculative and unsubstantiated consequentialism which serves to insulate mass killings against available evidence of their ineffectiveness.

In some cases, futurization does have a 'utopian' quality, reflecting certain transformational ideological goals. Communists used the prophesized 'certain' revolution and advent of Communist society to collapse the ethical regulation of revolutionary violence in the here and now—anything framed as 'defending the revolution' might appear justifiable since the future well-being of all humanity appeared to rest on advancing the revolution in the present.[160] In colonial atrocities, sweeping discourses about 'historical development' and 'progress' towards a prosperous future were utilized to justify mass killings of 'backwards' and 'primitive' peoples.[161] In Nazism, the vast prosperity of the thousand-year Reich rested on the eliminations of racial parasites and 'life unworthy of life'.

In general, though, futurization revolves around more familiar security concerns and the protection of a political order's future safety. Indeed, consistent with the influential findings of 'prospect theory'—which shows that people tend to take greater risks to avoid outcomes framed as 'losses' than to acquire 'gains'— futurization in mass killing tends to revolve around the prevention of terrible losses more than the achievement of utopian goals, and is therefore closely interwoven with threat construction.[162] As one Kemalist member of the Turkish parliament

---

[159] Reliance on abstract consequential calculations also appears to increase as more 'psychological distance' is created between the decision-maker and potential victims, linking futurization to deidentification. See Aguilar, Brussino, and Fernández-Dols 2013.
[160] Lukes 1985; Glover 1999, 252–6; Lukes 2001; Figes 2002, 92; Weitz 2003, 54–6.
[161] Bodley 2002; Waller 2007, xv.
[162] See also Midlarsky 2005, 103–7.

said of mass killing at the end of World War I: 'We acted thusly simply to ensure the future of our fatherland that we consider to be dearer and more sacred to us than our own lives.'[163] Most commonly, futurization manifests in elastic assertions about 'military necessity'. Military operations that kill civilians are frequently justified in terms of promised military benefits that, in reality, never arise. Justifiers do not usually need to offer a detailed consequentialist proof of the net worth of killing, but merely a tokenistic 'lining-up' of individual acts of killing with assertions of military benefits.

Even in Nazism, it was the future *security* of the German and Aryan race that dominated justifications of mass killing—not complex political utopias. Himmler warned his SS Group Leaders in 1938, 'if we are the losers in the struggle, not even a reservation of Germans will remain but rather everyone will be starved out and slaughtered'.[164] Hoess reported Eichmann likewise telling him that 'any compromise, even the slightest [in the extermination of the Jews] would have to be paid for bitterly at a later date'.[165] On 30 March 1941, Hitler briefed 250 senior officers, combining all the justificatory mechanisms with a futurized presentation of mass killing as ensuring ultimate security:

> Communism is a tremendous danger for the future. We must get away from the standpoint of soldierly comradeship. The Communist is from first to last no comrade. It is a war of extermination. If we do not regard it as such, we may defeat the enemy, but in thirty years' time we will again be confronted by the Communist enemy. We are not fighting a war in order to conserve the enemy ... The troops must defend themselves with the methods with which they are attacked. Commissars and the GPU people are criminals and must be treated as such ... In the East toughness now means mildness in the future. The leaders must make the sacrifice of overcoming their scruples.[166]

As historian Michael Burleigh observes: 'Mere links in big biological time or cosmic space, the SS had [within Nazi ideology] been granted a one-off opportunity to save future generations of their nation and race from chaos, subversion and oblivion ....'[167]

A prominent form of futurization, blended with threat construction, characterizes violence as a preventive measure against future retaliation by the victims. This is often critical in explaining the otherwise puzzling targeting of children in mass killings. In Indonesian massacres of alleged communists in 1965–66, for example, confident predictions that the offspring of those killed would grow up seeking

---

[163] Cited in Marchak 2003, 95.
[164] Burleigh 2001, 336.
[165] Mann 2005, 244.
[166] Cited in Burleigh 2001, 518.
[167] Ibid. 196.

vengeance was used to justify the expansive mass murder of women and children. An Australian Embassy official in Indonesia reported being told by locals that they wished to exterminate communists 'thoroughly (thoroughly meaning wives and children as well), as some sort of guarantee against future reprisals', and as late as 1969 (over three years after the violence), an Indonesian businessman lobbied the Nixon government to support the Indonesian military on the grounds that 'the offspring of the executed hard core communists will grow up within the next ten to fifteen years. They will just be the same like their fathers were.'[168] Likewise, in Guatemala, soldiers engaged in mass killings of Maya in the early 1980s explained that they were targeting male children 'because those wretches are going to come some day and screw us over'.[169]

That cost–benefit calculations can be influenced by inflated expectations of future gains is well established. Research on 'positive illusions' suggests a common tendency for elite instigators of war and political violence to adhere to highly overconfident beliefs in violence's likely outcomes.[170] Amongst rank-and-file agents and public constituencies, moreover, expectations of long-run future gains are often weakly anchored in reality simply because such individuals are rarely able to confidently estimate the long-term effects of violence. The aforementioned evidence that most American citizens would support a nuclear attack that would kill two million Iranian citizens if it promised to save 20,000 US soldiers' lives shouldn't obscure the fact that most American citizens are never in any position to tell whether such an act would *actually* save 20,000 US soldiers. Rank-and-file agents and public constituencies frequently accept the future validation of their actions as assured by their superiors. Nurses accepted the health benefits asserted by doctors accruing from the Nazi 'euthanasia' of the mentally disabled, while soldiers accepted their commanders' assertions of the military need for 'anti-partisan operations' against Jews in Polish towns, Vietnamese villages, and Guatemalan provinces. Such military appeals might seem absurd when those targeted are unarmed civilians, but, as conflict scholars have long appreciated, civilian populations *can* be important bases for partisan recruitment and support in wartime. Soldiers are rarely in a position to effectively challenge their superiors on whether specific assertions against civilian populations of this kind have any basis in fact. So once a formula of justification has been established in which the costs of killing are weighed against future benefits undiscounted for uncertainty, moral legitimacy can be hijacked by whoever can articulate the relevant categories of costs and benefits with most authority.

Indeed, futurization should often not be understood as a serious form of cost–benefit calculation at all, but a mechanism of *framing* violence—powerfully

---

[168] Cited in Gerlach 2010, 41.
[169] Cited in Goldhagen 2010, 469.
[170] Johnson 2004.

legitimating it by reconstructing the private or public meaning of violence around its purported purposes. Participation in mass killing becomes understood as simply *part of* the 'victorious war effort', the 'building of communism', the 'defence of the revolution', or the 'war to end all wars'. Futurizing frames can thus be used to neutralize potential moral criticism—in the Soviet Union, futurization was embedded in trivializing metaphors that legitimated the violence: 'you can't make an omelette without breaking eggs', Bolsheviks would ritualistically retort, or 'you can't plane down wood without losing a few chips'.[171] More powerfully, futurization socially redefines violence as part of a seemingly unimpeachable 'glorious mission', as Anderson labels it, towards immense future gains.[172] This glorious mission may be genuinely identified with by perpetrators. But appealing to it can also offer a means of public justification and a source of social power, with hardliners seeking rhetorical leverage over their opponents by presenting themselves as the most committed and effective servants of that mission.

Destruction of Alternatives

One of the central puzzles of mass killing is why regimes and groups, even if convinced that they face extreme threats, would resort to such a radical and risky course of action rather than employ more limited and selective forms of coercion or violence. Given the high costs and dubious effectiveness of mass killing, many rationalist scholars plausibly suggest that decision-makers come to favour it because more limited forms of violence are unfeasible. As Downes puts it: 'civilian victimization is often driven by perceived strategic necessity: leaders may see themselves as having little choice but to target noncombatants'.[173] Valentino likewise notes how the 'politics of extermination ... usually emerge only after leaders have concluded that other options for achieving their ends ... are ineffective or impractical'.[174] Understanding this perception—that there are *no acceptable alternatives*—is critical to the explanation of mass killings. In almost all cases, perpetrators assert that they have little choice, and that mass killing is effectively unavoidable.

Where I depart from rationalist theories, however, is in their tendency to treat this perception of unavoidability as rational—as though perpetrators of mass killing have good reasons, rooted in genuine constraints and circumstances, for deeming alternatives to mass killing unavailable. Genuine constraints are certainly *relevant*: perpetrating states and organizations often face real difficulties in engaging in more limited and selective violence.[175] They may lack good information on who to target, have technologies of violence which make careful targeting difficult,

---

[171] Glover 1999, 254–6.
[172] Anderson 2017a, 87.
[173] Downes 2008, 39.
[174] Valentino 2004, 3.
[175] Kalyvas 2006, ch.6.

or have rational concerns about the benign intentions and commitments of their political opponents in ways that make negotiated resolutions of crises difficult. In the four mass killings I consider in detail, such constraints on more targeted violence were clearly a factor in Allied area bombing, and may have played some role in Stalinist repression, Guatemalan mass killings, and the Rwandan Genocide. In none of these cases, however, was the perception of mass killing as 'unavoidable' primarily rooted in genuine constraints. There were always alternatives and, given the immense exaggeration of civilian threat, simply not killing civilians—whatever other course was pursued instead—would have been a plausibly rational choice. This is, as I argued in Chapter 3, almost always true in mass killings: groups and regimes with the power to successfully implement mass killing are rarely if ever so circumstantially constrained that alternatives to such a strategy are unavailable.

Instead, the perception of mass killing as unavoidable is rooted in strong ideological biases. Decision-makers can rarely give serious consideration to *all* possible courses of action in any given scenario—they consider only the 'field of possibilities'[176] that is prejudged to be relevant. Justificatory narratives for mass killing promote the *destruction of alternatives* to extreme violence within that contemplated field of possibilities, intensely denigrating potential alternatives so that decision-makers feel as though only one 'realistic' course of action is open to them.[177] Often this overlaps with valorization, as alternatives are removed from consideration because they represent 'weakness', 'sentimental humanitarianism', 'naivety', 'rotten liberalism', or some other denigrating frame.[178] Unsubstantiated assertions that such policies would be 'unfeasible', 'risky', 'expensive', or 'impractical' likewise often prove sufficient to eliminate them from discussion, particularly within bureaucratic policymaking in which groupthink and shared ideological assumptions are common.[179]

This ideological destruction of alternatives may directly shape strategic assessments, with people's preconceptions about the viability of less violent courses of action shaping their support for mass killing.[180] But it also serves to naturalize and 'deagentify' violence: presenting mass killing as an inevitable result of irresistible forces or unavoidable circumstances beyond the control of human agency, and thus eroding perpetrators' sense of responsibility in bringing it about. Providence, the nature of war, the laws of class or racial struggle, technological progress, or simply the actions of others are held up as 'the real cause' of the killing, rather than the deliberate decisions made by elites, rank-and-file killers, and public supporters. Perpetrators appear to just respond to the dictates of the situations or forces they find themselves subject to—doing as 'necessity' requires of them.[181] Such loss

---

[176] Foucault 1982, 788.
[177] I take this phrase from Gordy 1999.
[178] Getty and Naumov 1999, 34. Noakes and Pridham 2001, 462.
[179] See also Janis 1982.
[180] Tausch et al. 2011.
[181] Slim 2007, 151.

of perceived agency has been linked to increased tendencies to conform to social pressure and to morally disengage from one's actions.[182]

The destruction of alternatives can take two broad forms. In a more theoretically elaborate variety, quasi-deterministic conceptions of eternal and irresistible forces underpin the presentation of violence as unavoidable. Nazism held that the extermination of inferior races was simply a requirement of the natural and eternal laws of racial struggle—unavoidable if any race wished to ensure its own survival. As the Nazi Party Chairman Martin Bormann wrote in private correspondence:

> What is it we Nazis want? We want to adapt our people to the laws of nature ... we want to make it fit for the ineluctable struggle for existence. This struggle exists, whether we like it or not, whether we reject or accept it ... Just as the individual— as every individual creature, be it animal or plant—must assert and maintain his existence, so must the nation as a whole.[183]

A German army instruction manual correspondingly instructed soldiers:

> [The Jew] wants to live among us as a parasite who can suck us dry ... Who can believe it possible to reform or convert a parasite (a louse for example)? Who can believe in a compromise with the parasite? We are left with one choice only, either to be devoured by the parasite or to exterminate it.[184]

In this respect, Nazism was preceded by colonialist ideologies, which similarly held that the destruction of inferior races was, as perpetrators put it, 'as ultimately beneficial as it was inevitable',[185] 'just a natural process',[186] and an 'unavoidable'[187] part of the necessary spread of Europeans around the world in their providential exploitation of new lands.[188] Under such quasi-deterministic ideologies, as Hannah Arendt wrote, 'the liquidation is fitted into a historical process in which man only does or suffers what, according to immutable laws, is bound to happen anyway'.[189] Perpetrators thereby evade moral regulation:

> All concerned are subjectively innocent ... the murderers because they do not really murder but execute a death sentence pronounced by some higher tribunal. The rulers themselves do not claim to be just or wise, but only to execute historical or natural laws ....[190]

---

[182] Bandura 1999; Monroe 2011.
[183] Cited in: Trevor-Roper 1954, 54
[184] Cited in: Aronsfeld 1985, 48
[185] Gellately and Kiernan 2003, 24.
[186] Madley 2004, 169
[187] Ibid.
[188] For an extended example see Cass 1830.
[189] Arendt 1951/1976, 349.
[190] Ibid. 465.

More often, however, the destruction of alternatives is more mundane—justificatory ideologies portray violence as unavoidable, not because of mysterious historical forces, but by the circumstances in which perpetrators find themselves—often those of war. Again, such moves are common in discourses of 'military necessity'.[191] General William Sherman thus justified his decision to employ scorched earth policies in the American Civil War with the sweeping assertion: 'War is cruelty and you cannot refine it ... You might as well appeal against the thunderstorm as against these terrible hardships of war. They are inevitable.'[192] The relevant features of the 'circumstances' that mandate mass killing are, moreover, often locked in the ideological preconception that the victims 'only understand violence'. Thus General von Trotha, orchestrator of the genocide against the Herero and Namaqua people in German South West Africa from 1904 to 1908, asserted that: 'My knowledge of many central African peoples, Bantu and others, convinces me that the Negro will never submit to a treaty but only to naked force ... The uprising is and remains the beginnings of a racial war.'[193] As should be expected, such assumptions are often woven in with futurization—mass killings are presented by perpetrators as the only viable path to future security. An SS soldier involved in killing Russian Jews was thus able to believe that: 'the more one thinks about the whole business the more one comes to the conclusion that it's the only thing we can do to safeguard unconditionally the security of our people and our future ... what we are doing is necessary'.[194]

The Limitationist Alternative

Hardline justificatory narratives of crisis, built from these six justificatory mechanisms, therefore appear to have real causal power to encourage people to initiate, implement, or support mass killings. I have argued, moreover, that such ideological narratives are essential, because broader material and social conditions alone never straightforwardly encourage a policy of mass killing over other plausible and less extreme courses of action. Indeed, however tragically numerous they are in world history, mass killings remain relatively rare events—most wars and crises do not involve them. Even when mass killings do occur, moreover, substantial numbers of people do not endorse them and may engage in efforts to resist them and rescue potential victims.

This reflects the fact that strongly hardline visions of security politics are rarely *dominant* in modern societies.[195] In most crises, elites never develop a radical

---

[191] Weitz 2003, 61; Bellamy 2012a; Bellamy 2012b.
[192] Cited in Slim 2007, 25.
[193] Cited in Kiernan 2003, 30.
[194] Cited in Valentino 2004, 47.
[195] On the general spread of ideas of limited violence see Pinker 2011; Bellamy 2012b.

justificatory narrative for mass killing, instead espousing more limitationist or, at worst, permissive security narratives that suggest more constrained forms of violence. After the end of the Vietnam War, for example, the victorious Vietnamese Communist Party could well have resorted to large-scale mass killings as a means for securing its rule and eliminating enemies—as the Khmer Rouge did in Cambodia and Stalin did in the Soviet Union. By contrast with such intensely hardline communist regimes, Maureen Hiebert observes how, in Vietnam: 'Former opponents were constructed as individuals who could be rehabilitated and integrated into a unified socialist Vietnam.'[196] This was consistent with the inclusive pre-existing 'united-front' strategy the Vietnamese Communist Party relied on, in which 'southerners as a group were not conceptualized as enemies, let alone as a lethal threat. Southerners who sided with the US-backed GVN [Government of Viet Nam]—the real enemy, described as "cruel" and "dictatorial" fascists—were portrayed as wayward brothers who had yet to see the socialist light.'[197]

The absence of intensely hardline ideology does not necessarily equate to a benign political environment. Many regimes that do not endorse a justificatory narrative for mass killing may nevertheless engage in significant human rights violations, involving political repression, large-scale incarceration, and even limited killings. All of these were present in communist Vietnam.[198] But without strongly hardline security doctrines, regimes and armed groups prove highly reluctant to escalate beyond such measures to outright mass killing. Limitationist ideas reduce the motive and opportunity for hardline radicalization, and discourage or counteract each of the two interwoven portrayals on which justificatory narratives for mass killing rely.

First, limitationist ideologies advance *ideas of civilian protection* in which civilians, or at least particular relevant groups of civilians, are construed as non-threatening, innocent, and sharing critical ties of identity with those the ideology articulates itself to. Such doctrines may rest on universalist notions of a common humanity, as expressed in human rights protections and standards of civilian immunity.[199] Or, somewhat more commonly, they may involve inclusive or pluralist constructions of national, religious, or communal identity, in which potential minorities or 'out-groups' are presented as full in-group members.[200] By discouraging the association of threat-based or guilt-based fear and anger with civilian groups, by promoting potential solidarity, compassion, and loyalty towards them, and by sustaining (where institutionalized) particular social and legal protections on civilians, such limitationist ideas discourage the motivation or legitimation of violence against them.

---

[196] Hiebert 2017, 192.
[197] Ibid. 197.
[198] Canh 1983; Vo 2003.
[199] Bellamy 2012b; Fabre 2012.
[200] Straus 2015a; Hiebert 2017. See also Saucier and Akers 2018, 89.

Thus, for example, Straus's analysis of three 'negative cases' in which large-scale genocidal mass killing was avoided—Senegal, Mali, and Côte d'Ivoire—places central emphasis on the inclusive visions of the national community promoted by government leaders, which prevented a narrative of civilian groups as dangerous, guilty enemies from emerging. By contrast with such a hardline narrative, Straus explains how:

> Léopold Senghor and Alpha Oumar Konaré ... created, defended, defended again, and practiced a pluralist vision of their nation, one rooted in the local realities and cultures of Senegal and Mali, respectively. Felix Houphouët-Boigny was more instrumentally pluralist. He also favoured his own group as superior, but nonetheless he practiced and preached a vision of a tolerant, unified, peaceful, multiethnic Côte d'Ivoire.[201]

Stephen McLoughlin similarly shows how in Botswana, Zambia, and Tanzania, large-scale atrocities against civilians were avoided in part because 'Seretse Khama, Kenneth Kaunda and Julius Nyerere all steered their countries towards independence with ideologies that aimed to transcend tribal, ethnic and religious division, while forging new national identities that incorporated diversity'.[202] Alex Bellamy, meanwhile, demonstrates how the decline of 'anti-civilian ideologies' and the spread, acceptance, and institutionalization of broader norms of civilian immunity have worked to obstruct and limit mass killings since the end of the Cold War and even promote significant but inconsistent international efforts to prevent or halt them.[203]

Second, limitationist ideologies advance *ideas of restricted violence*, in which large-scale violence is preconceptualized as frequently counterproductive, as typically strategically and morally inferior to alternative strategies of limited violence or non-violence (which may be actively valorized), and as a stigmatized and morally dubious activity to be avoided if at all possible.[204] When internalized, such ideas create predispositions to emphasize the costs and risks of mass killing, and may create strong attachments to virtues and practices of non-violent politics.[205] But even as elements of ideological structure, such ideas can shape prevailing norms of warfare and violence within military and police organizations or mass political movements. By shaping strategic doctrines or procurement and recruitment processes, they may also imbue such organizations with an

---

[201] Straus 2015a, 324.
[202] McLoughlin 2020, 1562. See also Goodwin 2007.
[203] Bellamy 2012b. Bellamy emphasizes the fragility of civilian immunity norms, however, and the capacity of regimes to use uncertainty and misinformation as an alternative to ideological justifications to facilitate atrocities against civilians. See also Thomas 2001, ch.5.
[204] Thomas 2001; Chirot and McCauley 2006, 111–16; Slim 2007, 260–6.
[205] On pacifist and non-violent movements see Ceadel 1980; Karagiannis and McCauley 2006; Stephan and Chenoweth 2008; Chenoweth and Cunningham 2013.

operational orientation towards, and capacity for, more limited and discriminate violence.

In negative cases, such ideas of restricted violence typically appear influential. McLoughlin emphasizes, for example, Nyerere's 'preference for non-violence ... informed by the failure of past uprisings and inspired by the teachings of Gandhi'.[206] In South Africa, while the anti-apartheid movements were not pacifist and did employ limited strategies of violence, strong preferences for non-violent action characterized the African National Congress's strategy of resistance and the nature of the post-apartheid transition.[207] Notions of restricted violence are especially evident in the strategic evolution of the militaries of the United States and its principal allies. As shown in Chapter 6, in the mid-twentieth century, the Western Allies in World War II embraced a total war doctrine in which the massive and direct bombing of civilian populations was seen as a strategically effective, necessary, and hard-headed way to prosecute modern war, forcing the enemy to surrender by attacking their 'morale' and 'social fabric'. But this doctrine was gradually abandoned from the 1950s onwards. Technology, specifically the development of more precise aerial weaponry, was a factor here, but insufficient, since the strategic theory of area bombing was built on the claim that killing civilians was actively effective, not just difficult to avoid. Equally critical was the discrediting of massive bombing in the Vietnam War (itself a comparatively restricted war for the time), growing concerns about public hostility to mass bombing, and the parallel development of counter-insurgency doctrines, in which military strategists emphasized the superiority of 'hearts and minds' strategies that sought to win over civilian populations rather than attempting to coerce them with mass violence.[208]

Again, limitationist ideas are not a panacea, and the fact that states do not perpetrate mass killing should not obscure their engagement in other forms of ethically dubious violence. But in the absence of more radical hardline ideas, states and non-state armed groups generally avoid more extreme violence against civilians, even when significant political crises hit.[209] I re-emphasize that the relative balance between hardline and limitationist ideas in any social context is a matter of degrees. Since all societies are ideologically heterogenous, and hardline and limitationist ideological doctrines vary in their intensity and specific content, binary

---

[206] McLoughlin 2020, 1560.

[207] Zunes 1999; Goodwin 2007; Gupta 2014. Though not focused on mass killing, Paul Collier's and Pedro Vicente's study of a campaign of ideological activism to discredit electoral violence by ActionAid International Nigeria prior to the 2007 Nigerian elections also highlights the power such ideas can have within specific cases. Collier and Vicente found 'clear and statistically significant effects of the campaign on diminishing perceptions of political violence and increasing empowerment of the population against political violence' and, most importantly, a reduction in the actual likelihood of violence by 47% in the areas targeted; see Collier and Vicente 2013, F344 & F350–1.

[208] See, in general, Clodfelter 1989; Pape 1996; Thomas 2001; Krcmaric 2018, 21–2; Schulzke 2019, ch.5. Such developments were obviously closely interwoven with notions of civilian immunity.

[209] On ideology and restraint by non-state armed groups see Thaler 2012; Oppenheim and Weintraub 2017.

classifications of whole societies as 'hardline' or 'limitationist' should be avoided. To reach genuine explanations of mass killings and the violence involved, it is necessary to trace the influence of specific sets of hardline (or limitationist) ideas in key organizations, policymaking circles, and segments of the public. This is what I seek to do in the four case studies that form the rest of the book.

4.3 Conclusion

During the first phases of the Armenian Genocide in 1915, the leading Ittihadist ideologue Dr Nazim Bey told the Ottoman regime's Central Committee:

> If we remain satisfied with … local massacres … if this purge is not general and final, it will inevitably lead to problems. Therefore it is absolutely necessary to eliminate the Armenian people in its entirety … Perhaps there are those among you who feel it is bestial to go so far. You may ask, for instance, what harm could children, the elderly or the sick do to us that we feel compelled to work for their elimination. Or you may feel that only those guilty should be punished … I beg you, gentlemen, don't be weak. Control your feelings of pity. Otherwise these very feelings will bring about your own demise.[210]

Such statements abound in the preparation and commission of mass killings and are, I have argued, critical in explaining such violence. When civilians are categorized as dangerous, guilty, and external to the primary political community, and when violence against them is represented as valorous, ultimately generative of huge benefits, and as effectively unavoidable, mass killing can appear a logical, laudable, and justified political or military strategy. Such *justificatory narratives* for mass killing do not depend on especially esoteric ideological goals, nor need they involve sweeping moral disengagement. Yet they are also far from an obvious, natural, or universal response to objective material or social conditions.

Instead, such narratives are fundamentally rooted in distinctive ideological conceptions of security, and of the politics and policies through which it should be pursued. They look plausible within certain preconceptions of potential civilian threat, group identities, and the virtuousness and efficacy of violence. They depend on a *selective* dismissiveness towards certain moral values and principles—of compassion, tolerance, and critical reflection—but a strong emphasis of others—including duty to authorities, loyalty to the community, and masculine strength and toughness. The justification of mass killing, in short, depends on *hardline*

---

[210] Cited in Valentino 2004, 165.

ideological foundations: expansive in their conceptions of danger, narrow in their conceptions of moral community, allured by uncompromising violence, and contemptuous of restraint. Our best evidence suggests that such distinctive ideological conceptions make a difference. The psychological evidence that such ideas can encourage support for violence is extensive, and such ideas can frequently be linked to specific patterns and logics of violence in mass killings that are otherwise hard to explain.

The six justificatory mechanisms at the heart of such narratives operate *infrastructurally*. Different mechanisms are likely to resonate with different individuals to different extents, while their extensive institutionalization within the operational norms and organizational practices of militaries and security agencies creates diffuse social pressures, incentives, and opportunities for violence. These narratives tend to sincerely guide the political elites who initiate mass killings, to at least a significant degree, as well as many of their supporters. But they also facilitate participation by much broader segments of the population, while creating powerful obstacles to effective resistance and opposition. Without becoming objects of universal enthusiasm or agreement, dominant justificatory narratives for mass killing can thereby profoundly reshape the psychological and social terrain on which violence takes place.

As emphasized in Chapter 3, my explanation of ideology's role in mass killing is thus *processual* in nature. We can rarely identify a static ideological doctrine, long predating the onset of mass killing, that contains a fully formed 'blueprint' for exterminatory violence. But mass killing emerges out of a process of ideological radicalization in which pre-existing hardline ideas and worsening political crises interact to fuel the more gradual or rapid development of a justificatory narrative for extreme violence. That process tends to be contested. Hardliners espouse a justificatory narrative for mass killing and push for policies to implement such violence, while more moderate figures express scepticism about such justifications and seek to pursue alternatives. This is, in important respects, a *power-struggle*, but one in which the propagation of a justificatory narrative is a critical form of power: it is by widely disseminating and institutionalizing hardline claims that hardliners are able to generate support for mass killing and pacify potential opposition. An infrastructural justificatory narrative is thus both means and motive: it both constitutes many elites' sincere understanding of why mass killing should be pursued, and allows them to mobilize the necessarily perpetrating coalition to implement it.

In the remaining chapters of the book I show how such processes of ideological radicalization, in which prior hardline ideology and contexts of political crisis interact, are essential to explain the occurrence of mass killing in Stalin's Soviet Union, Allied area bombing in World War II, the Guatemalan Civil War, and the Rwandan Genocide. In each case, mass killing was not simply a logical or even probable consequence of broader social and strategic circumstances. Nor, however, were longstanding projects of societal transformation or disengagement from conventional morality the central foundations for the violence. These mass killings were primarily rooted in hardline conceptions of security politics, which, in contexts of crisis, generated a justificatory narrative for extreme violence that guided elite perceptions and choices, helped mobilize rank-and-file perpetrators, and promoted collaboration amongst broader public constituencies.

5
Stalinist Repression

5.1 Overview

The Soviet Union under Josef Stalin is infamous as one of the most violent regimes in history. Between Stalin's consolidation of power around 1928 and his death in 1953, a bare minimum of 1.8 million Soviet citizens died directly at the hands of Soviet state security, with a further 1.5 million dying through forced labour and deprivation in the gulag prison camp system.[1] Millions more died from Stalinist economic policies, through forced relocation, famine, and accompanying violence. In this chapter I focus on direct state killing in the especially violent 1930s, and primarily on the 'Great Terror' of 1937–38, in which around 750,000 Soviet citizens were executed (including many members of Stalin's Bolshevik party)[2] and thousands more killed in the gulags.[3] Stalinist mass violence was not, however, confined to a single period of intense radicalization, but was a persistent and fluctuating feature of the Stalinist political system, as shown in Figure 5.1.[4] It comprised 'a number of interrelated repressive lines and policies, divergent in scope, character and intensity; implemented through legal and extralegal means; and aimed at different categories of "enemies".'[5] Indeed, the USSR was arguably engaged in mass killing for almost the entire period of Stalin's political hegemony.[6]

Stalin's Great Terror is puzzling for established theories of mass killing. Early explanations—formulated in the first decades of the Cold War—bore some resemblance to rationalist theories, portraying the Terror as a brutal campaign to

---

[1] Nove 1993; Overy 2004, 196; Viola 2007, 3; Gerlach and Werth 2009, 176. Much higher estimates are now largely debunked.

[2] The 'Bolsheviks' were Lenin's 'majority' wing of the Russian Social Democratic Labour Party, as opposed to the 'Menshevik' minority. After reorganizing as a separate party in 1912, the Bolsheviks took control of Russia in November 1917, forming the Soviet Union. From 1925 to 1952, they were officially the 'All-Union Communist Party (Bolsheviks)'.

[3] Whitewood 2015, 1. I do not focus on the mass famine, primarily of the Ukrainian peasantry, in a campaign now referred to as the *Holodomor* (roughly 'hunger-extermination'). State-induced famine might be viewed as a form of mass killing, but differs in several ways from the kind of direct and systematic violence focused on in this book. I do not dispute the atrocious character of Stalinist famines, which several scholars describe as genocidal. See Bilinsky 1999; Naimark 2010; Graziosi 2017.

[4] Naimark 2010, 3.

[5] Gerlach and Werth 2009, 135.

[6] Figure 5.1 does not show true death counts—it includes labour camp sentences (which were frequently but not always deadly), while excluding extrajudicial killings—but it roughly depicts the trends of direct state violence.

Ideology and Mass Killing: The Radicalized Security Politics of Genocides and Deadly Atrocities. Jonathan Leader Maynard, Oxford University Press. © Jonathan Leader Maynard (2022). DOI: 10.1093/oso/9780198776796.003.0005

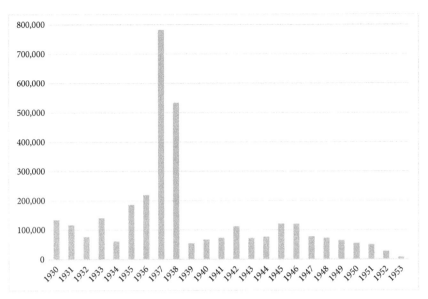

Fig. 5.1 Official Soviet executions and forced labour camp sentences, 1930–53.
Source: Overy 2004: 195.

consolidate Stalin's personal power, for which ideology was mere rationalization.[7] This portrayal is now widely rejected by historians. By the time the Great Terror started, in 1937, the Stalinist leadership already completely dominated Soviet politics. Far from consolidating power, the Terror was hugely counterproductive, causing chaos, weakening state structures, retarding industrial output, and undermining regime efforts at counter-intelligence and cultural control.[8] One of its most infamous elements—the purge of 35,000 military officers from the Red Army—was a remarkable act of self-harm, with ruinous consequences when the Nazis invaded the USSR in 1941.[9] As David Priestland observes:

> [T]he Terror still remains difficult to understand because it seems to be so irrational. Stalin's victims were of no real threat to him, and many of them were members of the industrial bureaucracies and the military high command, precisely the people whom Stalin needed to achieve some of his central objectives ... The Terror also caused enormous political and economic disruption that proved highly counterproductive on the eve of war.[10]

---

[7] For a summary and critique of such perspectives see Whitewood 2015, 3–12.
[8] Kotkin 1995, 348; Getty 2002, 116–17; Shearer 2009, 369–70; Shearer and Khaustov 2015, 13.
[9] Whitewood 2015.
[10] Priestland 2007, 2. See also Harris 2016, 1–2 & 188.

It is not for nothing that John Arch Getty and Oleg Naumov's leading study of the Great Terror is subtitled *Stalin and the Self-Destruction of the Bolsheviks*.[11] The timing of the Terror is also curious from rationalist perspectives, since it was not a reaction to war or any sudden challenge to the regime's territory, status, or power.[12] Indeed, the years immediately preceding the Great Terror were ones of relative prosperity and security, yet Stalinist terror in these peacetime years vastly exceeded the violence of the civil war period of 1918–21 under Lenin, when the regime faced real existential threats.[13] 'The paradoxical fact remains', as Igal Halfin observes, 'that killings increased as the internal security of the Stalinist court improved.'[14]

To resolve this paradox, we must focus on the Stalinist regime's ideological character. Historians generally agree that Stalinist ideology was significantly, but unevenly, internalized by both the elite and the mass population. At the elite level, archival evidence suggests that in ideological terms 'the Stalinists said the same things to each other behind closed doors that they said to the public'.[15] Early historical portrayals of Stalin as merely a cynical power-hoarder have been replaced by accounts stressing his ideological worldview.[16] As Peter Whitewood emphasizes: 'Stalin frequently used Marxist language outside of his public speeches and in his private correspondence. It is now clear that he did not just cloak a base desire for power in Marxist rhetoric. Stalin viewed the world through a Marxist lens and appears to have been ideologically committed.'[17] Below the elites, most Soviet citizens understood relatively little of Marxist 'high doctrine', but accepted many elements of the regime's worldview. In particular, 'Soviet citizens of the epoch were inclined to lend credit to the regime's propaganda about the subversive activities of plotters and foreign agents'.[18] At first glance, the Stalinist case may therefore appear to support a traditional-ideological perspective on mass killing. Stalinists clearly pursued transformational revolutionary projects, and many scholars have therefore emphasized Marxist utopianism and zeal in the Terror,[19] or more broadly assumed that communist mass killings are rooted in totalitarian ambitions to radically restructure society.[20]

---

[11] Getty and Naumov 1999.
[12] Midlarsky does attempt to apply his theory, focused around ephemeral gains and loss/fear of loss (especially of territory), to Stalin's extremism; see Midlarsky 2009.
[13] On Leninist terror see Ryan 2012.
[14] Halfin 2009, 15.
[15] Getty and Naumov 1999, 22. See also Gould-Davies 1999, 92; Priestland 2007, 4.
[16] Ramet 2006.
[17] Whitewood 2015, 12–13.
[18] Rittersporn 1993, 100. See also Inkeles and Bauer 1959.
[19] Rolf Binner and Marc Junge offer one such interpretation, summarized in Bernstein 2016, xxvi. Kotkin's landmark analysis of the city of Magnitogorsk under Stalin carries shades of this interpretation, in emphasizing Stalinism as a 'theocracy' seeking to combat 'heresy'; see Kotkin 1995, 333 & 351. Peter Holquist's and David Hoffman's analyses of Bolshevik violence also foreground utopianism and revolutionary politics; see Hoffman 2003, 72; Holquist 2003, 650–2. Overall, however, Kotkin's, Holquist's, and Hoffman's portrayals are more multifaceted.
[20] Kołakowski 1978/2008; Kuper 1981, ch.5; Lukes 1985; Valentino 2004, ch.4.

I argue, however, that Stalinist Terror is more consistent with a neo-ideological perspective. At the heart of Stalinist mass killing lay not ideological goals to restructure society, but an obsession with state security understood through a particular ideological worldview.[21] Violence was strategic, but strategic calculations were made under the influence of an ultrahardline security doctrine, in which existential threats were assumed to operate, hidden, within society as an inevitable consequence of the global class struggle. Stalinist ideology was also 'indispensable, regardless of whether people believed in it'.[22] Stalinism constituted a powerful set of institutional ideologies, political norms, and ideological discourses, which created intense structural pressures on Soviet citizens to participate in ideologically justified state violence. For this reason, situationist theories do offer significant purchase on Stalinist repression,[23] and Stalinist perpetrators were guided by many 'non-ideological' concerns: careerist ambitions and incentives, petty local animosities, and self-protection.[24] But all these motives became vitally clothed in the ideological language of the Stalinist regime: a discursive structure which Soviet citizens *had to use* to participate meaningfully in public life, and which could prompt the deployment of violent force by the state.[25] In most contexts, local political rivalries or disputes between workers and managers do not trigger violent state repression. Stalinist ideology changed this, creating an environment in which everyone understood that being accused of 'wrecking' or a 'lack of revolutionary vigilance' could warrant lethal violence by state agents.[26]

As with all mass killings, then, Stalinist repression cannot be explained by ideology alone. Operational weaknesses in the security forces hampered more selective repression,[27] top-down pressure for arrests created self-interested incentives for essentially arbitrary violence,[28] and a worsening international environment in the later 1930s was a major concern for Soviet elites.[29] But none of these factors rendered mass killing necessary or even likely for what was, by the 1930s, an internally secure Soviet regime. To understand why the state lashed out so destructively, Stalinism's ideological influence on elite perceptions, public attitudes, institutional orientations, and political norms must be accorded a central focus. I first, in section 5.2, examine the relatively contingent path of ideological development

---

[21] Hagenloh 2009, 16.
[22] Kołakowski 1978/2008, v.
[23] See, for example, Getty 1985/2008; Rittersporn 1991; Getty 2002.
[24] Viola 1993, 97–8; Viola 2013, 12–13.
[25] Kotkin 1995, ch.5.
[26] See also ibid. 284–6 & 539–41, fn.19–23.
[27] See also Kalyvas 2006, ch.6.
[28] Hagenloh 2009, 273; Shearer 2009, 345–9; Harris 2016, 159.
[29] See, in general, Harris 2016.

that generated Stalinism and placed it in a position of ideological dominance in the 1930s Soviet Union. I then, in section 5.3, analyse the role of that ideology in generating and shaping Stalinist mass killing.

5.2 Explaining the Rise of Stalinism

Marxism, Leninism, and the Early Soviet Union

I use 'Stalinism' to refer to the form of Marxism-Leninism that dominated the Soviet regime from around 1928 until Stalin's death in 1953. Like most ideologies, it represented an identifiable but somewhat sprawling set of ideas about politics rather than a tightly defined belief-system, and was not entirely homogenous. Priestland, for example, identifies five major ideological currents within the Soviet elite in the 1920s and 1930s—'elitist revivalism', 'populist revivalism', 'elitist technicism', 'populist technicism', and 'neo-traditionalism'—which all (but especially the revivalist strands) influenced Stalinism.[30]

Stalinist ideology obviously emerged out of the development of Marxist revolutionary thought in the nineteenth and early twentieth century. Marx famously justified revolutionary violence, writing in response to the revolutions of 1848, for example: 'There is only one way in which the murderous death agonies of the old society and the bloody birth throes of the new society can be shortened, simplified, and concentrated, and that way is revolutionary terror.'[31] Several scholars therefore suggest that communist mass killings are rooted in intrinsic features of Marxist thought.[32] This often reflects an interpretation consistent with the traditional-ideological perspective. Leszek Kołakowski suggests, for example, that mass repression by communist regimes 'to be sure … was not Marx's intention, but it was an inevitable effect of the glorious and final benevolent utopia he devised'.[33] While there are merits to these accounts, such a deterministic claim is implausible. Overall, Marx's theory of revolution was ambiguous, encouraging huge divergences between his followers—who range from some most violent regimes of the twentieth century to numerous peaceful, and even pacifist, movements.[34] Marxist ideology had no inevitable momentum in any one direction. What mattered was how it was understood and elaborated by subsequent thinkers and activists.

---

[30] Priestland 2007, 35–55.
[31] McLellan 2000, 297–8. See also Shorten 2012, 214–17.
[32] Kołakowski 1978/2008; Lukes 1985; Malia 1999.
[33] Kołakowski 1978/2008.
[34] Schaff 1973, 266–7; Cohen-Almagor 1991, 4; Ryan 2012, 17; Frazer and Hutchings 2020, 15–18. Contra Malia's assertion that all Communist parties adhere to 'the same' ideological worldview: Malia 1999, xiv–xv.

Effective explanations of Stalinism therefore focus more specifically on the political movement from which it emerged: Vladimir Lenin's Bolshevik party, one of the most hardline factions of Russian and international socialism. As James Ryan summarizes: 'Though Marxists were certainly not averse to using violence, Lenin was the first and most significant Marxist theorist to dramatically elevate the role of violence as revolutionary instrument.'[35] Lenin's thought, in this respect, highlights the problems of contrasting ideology with strategic pragmatism. Lenin was both fanatically dogmatic in his Marxist convictions and relentlessly obsessed with the pragmatically best tactics for advancing the revolution. This did not mean that, in practice, ideology fell out of the picture. Lenin arrived at different strategic diagnoses from other revolutionary groups *because* such diagnoses were shaped by his ideological beliefs.[36] By contrast with more moderate Russian socialists, Lenin held that massive revolutionary violence was unavoidable given the brutality of socialism's reactionary opponents.[37] As early as 1906 he argued that: 'We would be deceiving ourselves and the people ... if we concealed from the masses the necessity of a desperate bloody war of extermination, as the immediate task of the coming revolutionary action.'[38] Two factors marked out the Leninist position as especially hardline. First, Lenin's defence of revolutionary violence was far more sweeping in scale by comparison with other socialists. Lenin assumed that historical laws meant that *reactionary classes as a whole*, probably including the 'petit-bourgeois' peasantry, would fight the revolution bitterly—consequently, he anticipated a full-fledged 'civil war' against broad categories of class enemies. Second, Lenin's imaginative space for non-violence was particularly narrow, because he saw all other socialist parties as tending (intentionally or not) to serve counter-revolutionary interests. *Only the Bolsheviks*, in Lenin's eyes, correctly understood the needs of the proletarian revolution, so even cooperating with other socialist parties risked undermining the revolution.

On 7 November 1917,[39] the Bolsheviks seized power in Russia, toppling the short-lived 'provisional government' that had ruled since the abdication of the Tsar in March. As Ryan observes: 'The Bolsheviks came to power with no clear "blueprint" as such for employing physical violence.'[40] But as the 'revolution' transitioned into the Russian civil war (1917–21), pitting 'Red' Bolshevik forces against their 'White' opponents, the Bolsheviks readily resorted to large-scale violence

---

[35] Ryan 2012, 3. See also Shorten 2012, 217–24.
[36] By contrast, Orlando Figes contends that: 'In everything he did, Lenin's ultimate purpose was the pursuit of power. Power for him was not a means—it was the end in itself ... he did not establish a dictatorship to safeguard the revolution, he made a revolution to establish the dictatorship.' See Figes 1996, 504. Such an interpretation flies in the face of Lenin's long devotion to the revolutionary cause before he had any chance of obtaining power, and when it exposed him to immense personal cost. Lenin *was* ruthless in centralizing power, but this does not imply that he lacked revolutionary commitment.
[37] Bar a very brief exception in mid-1917; see Ryan 2012, ch.3.
[38] Cited in ibid. 40.
[39] Or 25 October, under the old Russian Calendar.
[40] Ryan 2012, 9.

against civilians. As rationalists would emphasize, the context of civil war was crucial in radicalizing the Bolsheviks, of which Stalin was already a senior member. Any new government faced incentives in these circumstances to resort to violent coercion to consolidate its control.

Yet crisis circumstances alone do not explain Bolshevik violence. For a start, many difficulties facing the Bolsheviks were a *consequence* rather than a cause of their ideological radicalism. The Bolsheviks had taken power in a narrow coup, but had an opportunity in late 1917 to build a broad socialist ruling coalition which could have strengthened the new 'Soviet' regime. This was blocked, however, by Lenin and Leon Trotsky, who (falsely) characterized other socialist factions as aligned to right-wing counter-revolutionaries, and established an essentially one-party government that was widely perceived as illegal.[41] This was hardly a rationally incentivized strategy—indeed, at the time, it looked potentially suicidal.[42] Even Lenin recognized the danger, but adopted such a strategy due to his erroneous ideological conviction that the Bolsheviks' revolution in Russia would soon be accompanied by other socialist revolutions across Europe.[43]

As this suggests, moreover, Lenin's Bolsheviks consistently interpreted the crises they faced through idiosyncratic ideological lenses.[44] Grain shortages to the cities were seen, not as a consequence of wartime devastation that could be remedied by building alliances with peasant parties, but as counter-revolutionary machinations by wealthy peasants or 'kulaks' (closely foreshadowing the ideological narrative underpinning Stalin's later campaigns of 'dekulakization').[45] Such narratives were intensified by the Bolshevik's optimistic revolutionary predictions failing to materialize. Having assumed that handing over factory control to workers would benefit productivity,[46] the Bolsheviks blamed the resulting economic chaos on bourgeois infiltration and workers' ill-discipline. Common criminals and 'idlers' became classified as enemies of the people, with Lenin calling for: 'war on the rogues, the idlers and the rowdies! All of them are the same brood—the spawn of capitalism, the offspring of aristocratic and bourgeois society.'[47] He continued to reiterate that all Bolshevik violence was historically unavoidable: 'a long period of "birth-pangs"; lies between capitalism and socialism ... violence is always the midwife of the old society ... What dictatorship implies and means is a state of simmering war, a state of military measures of struggle against the enemies of the proletarian power.'[48]

---

[41] Swain 1996, 53–69; Figes 1996, 500–2. The Bolsheviks did initially cooperate with the left-wing faction of the Socialist Revolutionary Party.
[42] Ryan 2012, 101–2 & 137–8.
[43] Ibid. 72 & 81–2.
[44] Ibid. 9.
[45] Ibid. 100.
[46] Priestland 2007, 80.
[47] Cited in Ryan 2012, 85. See also ibid. 98.
[48] Cited in ibid. 86–7. See also ibid. 90–5 & ch.5.

These hardline ideological foundations also shaped the institutional contours of the new regime. In late 1917, the Bolsheviks created a new political police—the Cheka. Initially intended as a temporary measure, it would rapidly take on broad powers and become a central instrument of the Soviet state, eventually becoming the feared NKVD (People's Commissariat for Internal Affairs) under Stalin. The Cheka embodied and institutionalized a hardline security doctrine from its outset. Its first leader, Felix Dzerzhinskii, declared: 'Do not think that I seek forms of revolutionary justice; we are not in need of justice. It is war now—face-to-face, a fight to the finish.'[49] He called for Chekists (i.e. secret police operatives) to be 'the most determined, hard, and solid comrades, without feelings of pity, ready to sacrifice for the safety of the revolution'.[50] Alongside the expansion of political police powers, article 23 of the 1918 Constitution removed rights from individuals and groups who 'use them to harm the interests of the socialist revolution'.[51] While the Bolshevik party was not uniformly so hardline, Lenin drew similarly inclined figures such as Trotsky, Dzerzhinskii, Stalin, and Nikolai Bukharin to fill leading roles within the new revolutionary government.

With this ideological infrastructure, Lenin's Bolsheviks employed mass killing as part of their civil war strategy from 1918 onwards. Imposing a 'food dictatorship' on the countryside, Lenin responded to predictable peasant resistance with hyperbolic constructions of threat, asserting that 'if the kulaks were to gain the upper hand they would ruthlessly slaughter hundreds of thousands of workers'.[52] In the regime's eyes, this eliminated all alternatives to outright mass violence: 'either the kulaks massacre vast numbers of workers, or the workers ruthlessly suppress the revolt of the predatory kulak minority ... there can be no middle course'.[53] The assassination of the senior Chekist Moise Uritskii and an attempt on Lenin's life in the summer of 1918 radicalized violence further. The 'official' Red Terror was initiated over the following months, targeting a broad array of 'counter-revolutionaries'.[54] Only as the Bolsheviks prevailed in the civil war was state violence relaxed.

The Bolshevik party and Soviet state that Stalin would come to control was a product of this radicalizing trajectory. Circumstances of autocracy, revolution, and war were key here, as rationalists might suggest. Leading Bolsheviks grew to adulthood in a repressive Tsarist political system that limited opportunities for peaceful opposition.[55] After their revolution, the Bolsheviks faced the task of governing a disintegrating Russian Empire in the face of hostile forces.[56]

---

[49] Cited in ibid. 88.
[50] Cited in Weitz 2003, 61.
[51] Cited in Ryan 2012, 84.
[52] Cited in ibid. 105.
[53] Cited in ibid.
[54] Holquist 2002, ch.6; Ryan 2012, 127-8.
[55] Ryan 2012, 7.
[56] Priestland 2007, 35.

Nevertheless, these circumstances alone do not explain the violent evolution of Marxism-Leninism in this period. As Ryan observes:

> The Bolsheviks in power were not merely reacting in a rational and calculated, though often brutal fashion, to circumstances and events ... Bolshevik perceptions, reactions and solutions were fundamentally informed by [their] eschatological and politically Manichean vision, and indeed contributed to the creation of hostile circumstances ... The Bolsheviks, acting on the basis of theoretical appraisals, made certain choices that ... resulted in the large-scale application of violence, whereas less abrasive choices were often possible.[57]

The Bolsheviks' utopianism did matter here. It underpinned the utilitarian revolutionary calculation that extreme violence would prove worthwhile, while the failure of revolutionary predictions encouraged the Bolsheviks to resort to violence to 'correct' the situation.[58] Nevertheless, Leninist violence primarily rested on the interconnection of ideology and security politics, in particular through Lenin's rejection of a less violent strategy of socialist coalition-building, and his predisposition to see all challenges the Bolsheviks faced as the result of pervasive counter-revolutionary enemies.

Some moderating influences on state violence remained under Lenin. His was a hardline rather than ultrahardline elite, and this shaped the scope of the Red Terror, which remained linked to ongoing armed struggle.[59] Even Lenin and Dzerzhinskii expressed concerns with the 'excessive' use of violence,[60] and the regime ultimately, in 1921, launched a major ideological retreat from the policies of the civil war in the form of the 'New Economic Policy'. Overall, though, moderating forces were not able to prevent significant radicalization in the Lenin era. Figes suggests that this reflected the 'psychological weakness of the moderates',[61] and that 'few dared question Lenin's judgement'.[62] This seems exaggerated, since there was some diversity of views and occasional opposition within the party.[63] Yet Lenin's authority was crucial, making his personal ideological inclinations essential in explaining the early character of the Soviet state and the Bolshevik party that Stalin inherited. Outside the party, moderates were also suppressed violently. The Bolsheviks enjoyed more support from soldiers and sailors in the first weeks of the revolution than their opponents, and were more ready to employ military force. Once the civil war was underway, the most important forces opposing the Bolsheviks' Red Army were not moderates, but right-wing White militarist forces, and

---

[57] Ryan 2012, 8.
[58] Ibid. 93.
[59] See ibid. 113, 117, & 120–1.
[60] Ibid. 170.
[61] Figes 1996, 511.
[62] Ibid. 504.
[63] As Figes observes; see ibid. 506–7. See also Swain 1996, ch.2; Priestland 2007, 60–1; Ryan 2012, 117 & 120.

they remained divided. The result by 1921 was an outcome that seemed remarkably unlikely from the perspective of 1917: a Bolshevik Soviet Union in control of most of the former Tsarist empire.

Stalinist Radicalization

This quite contingent pathway of Bolshevik ideological evolution under Lenin is critical in explaining Stalinist violence since, as Lynne Viola observes, 'Stalin both grew out of and contributed to the radicalization of this [Leninist] political culture'.[64] But Stalinist violence in the 1930s, much more expansive than anything seen under Lenin, reflected further radicalization of the Soviet regime—to a large extent through the gradual ascension of Stalin and his key followers following Lenin's death in January 1924. Stalin had long been an ultrahardliner on issues of revolutionary security and party unity. He adhered to a unitary image of the revolution, describing the revolutionary party in 1905 as a 'proletarian army' in which 'once the unity of views collapses, the party collapses'.[65] After the civil war, he was closely involved with the development of the political police, favouring expanded police powers and enforced unity of party thought, and often issuing dire warnings of internal subversion.[66] In contrast to Lenin, who accepted the need to employ 'bourgeois' experts, Stalin decried the infiltration of the state by 'alien, bureaucratic, half-bourgeois' elements.[67] By the late 1920s, the Stalinists increasingly saw alternative voices within the Bolshevik party as, in the words of Stalin's ally Vyacheslav Molotov, 'saboteurs of the party line'.[68]

These ultrahardline tendencies were strengthened by political developments over the latter 1920s. Many Bolsheviks became frustrated with the ideological compromises of Lenin's New Economic Policy, and were increasingly convinced that war, most likely with Britain, was imminent.[69] Partly in consequence, the regime pressed ahead with agricultural 'collectivization'—shifting peasants into state-run collective farms—as a strategy for economic modernization. By provoking major peasant unrest, however, this only intensified a sense of growing crisis. Stalin and senior secret police officials were, moreover, ideologically preconvinced that internal enemies existed. As James Harris observes: 'Dzerzhinskii and his successor Viacheslav Mezhinskii passionately believed that they were protecting the regime from very real dangers, but the general lack of rigorous scepticism of the evidence of that threat contributed significantly to the belief.'[70] Institutional interests probably played a role here too, since alarmism helped Dzerzhinskii

---

[64] Viola 2013, 17.
[65] Cited in Priestland 2007, 75–6. See also van Ree 1993.
[66] Shearer and Khaustov 2015, ch.1.
[67] Priestland 2007, 175.
[68] Doc.40, Viola et al. 2005, 159.
[69] Priestland 2007, 166–7.
[70] Harris 2016, 51.

and Mezhinskii maintain secret police power and financing.[71] Either way, they frequently diagnosed mundane international developments as evidence of imminent invasion,[72] while party and police reports continued to analyse economic problems and rural unrest as evidence of counter-revolutionary sabotage ('wrecking'). This was exacerbated by ideologically induced organizational pathologies. Soviet leaders received relentlessly positive feedback from their subordinates on state policies, since such policies were framed as *the scientifically correct Marxist-Leninist course*. Consequently, when problems became apparent, it was assumed that this could only reflect intentional sabotage.

In the late-1920s, we thus 'begin to see a kind of [secret police] "master narrative" evolving, one that would intensify Stalin's extremism by presenting the spectre of war, domestic economic crisis, and peasant rebellion as imminent threats to the regime'.[73] Guided by this ideological narrative, the Stalinists concluded that a major increase in state repression was necessary. In responding to a question concerning the GPU (the new name for the secret police) at a meeting with foreign socialist organizations in 1927, Stalin gave an illustrative account of his justifications for such repression:

> I refuse to understand some workers' delegates, who come to the USSR, and ask with concern: whether there have been many counterrevolutionaries punished by the GPU, whether the various kinds of terrorists and conspirators against proletarian power will still be punished, whether it is not time to end the existence of the GPU? Where do some workers' delegates get this concern for enemies of the proletarian revolution? ... They preach for maximum softness, and advise to eliminate the GPU ... But, is it possible to guarantee that, after eliminating the GPU, capitalists of all countries will stop organising and financing counterrevolutionary groups, conspirators, terrorists, instigators, bombers! Well, is this not foolishness, is it not a crime against the working class to disarm the revolution, without having any guarantees that enemies of the revolution would be disarmed! ... internal enemies are not isolated units. The fact is that they are connected by a thousand threads to the capitalists of all countries, who support them with all their force, with all their means.[74]

Our best evidence suggests that these claims were not mere rhetoric. Internal party documents are replete with similar assertions of counter-revolutionary conspiracy in the late 1920s and early 1930s.[75] In the 'Shakhty Affair' of 1928, the regime accused a set of 'bourgeois' engineers at the Shakhty coal mine in the North Caucasus

---

[71] Shearer and Khaustov 2015, ch.1. Ideological divergences within the Bolshevik party were often influenced by institutional positions; see Priestland 2007, 49 & 138.
[72] Shearer and Khaustov 2015, ch.2.
[73] Viola et al. 2005, 4. See also Shearer and Khaustov 2015, 59.
[74] Doc.38, Shearer and Khaustov 2015, 69–70.
[75] Viola et al. 2005; Shearer and Khaustov 2015, chs.1–3.

of deliberate wrecking. Stalin pressed for a show trial and, while the charges were baseless, Harris reports how:

> the correspondence of most other Politburo members ... suggests that they genuinely believed that economic sabotage had become a key weapon of hostile capitalist powers in their struggle against the Soviet Union and that that sabotage extended well beyond Shakhty and the Donets Basin.[76]

For most of the 1920s, however, repression did not reach its later heights. A relatively 'moderate' faction on the 'right' of the party continued to support the New Economic Policy, while a broader range of figures, such as the Commissar for Foreign Affairs Georgii Chicherin and Justice Commissar Nikolai Krylenko, and even some of Stalin's circle, such as Sergo Ordzhonikidze and Maksim Litvinov, were less inclined towards Stalin's exaggerated perceptions of internal and external enemies.[77] As Stalin and his allies extended their political and ideological dominance, however, radicalization accelerated from 1929 onwards.

5.3 Stalinism and Repressive Mass Killing

The Stalinist Elite

Stalinist mass killing in the 1930s was concentrated in two waves: the mass collectivization of Soviet agriculture in 1930–33, and the Great Terror of 1937–38. Both involved significant degrees of popular participation and 'bottom-up' escalatory dynamics, but they were fundamentally elite-led. Again, these campaigns are often interpreted through a traditional-ideological account: in which Soviet leaders identified certain groups as incompatible with their utopian ideological project to create a communist society, and therefore exterminated them. But this portrayal is misleading. The Bolsheviks did not have a clear long-term plan to purge certain groups from society. Again, Lenin's justification of repressive killings had been tightly linked to revolutionary struggle against capitalist resistance, and once the Russian civil war ended in 1921, mass killing ceased until the 1930s. Stalinist violence was likewise implemented, not as a long-accepted project of social reengineering, but as a response to mounting regime panic surrounding specific crises. This is especially clear in the Great Terror, which, as David Shearer points out, occurred as 'the regime moved *away* from a revolutionary program of social restructuring to one of protecting state interests and establishing public order'.[78]

---

[76] Harris 2016, 53–4. For examples see Priestland 2007, 202; Shearer and Khaustov 2015, 74.
[77] Viola et al. 2005, 120 & 177; Shearer and Khaustov 2015, 34–43, 58, & 62. See also references to internal disagreements in Docs.39 & 40, Viola et al. 2005, 154–9.
[78] Shearer 2009, 287 (*my emphasis*). See also Hagenloh 2009, 12.

Both campaigns are more effectively understood as forms of ideologically radicalized security politics. Far from reflecting self-evident strategic incentives that all regimes would have similarly reacted to, Stalinist mass killing depended on a fantastical ideological narrative of domestic insecurity, involving hyperbolic assertions of threat and criminal conspiracies, completely at odds with the reality of the regime's hegemonic dominance.[79] As John Arch Getty and Oleg Naumov observe, 'the Stalinists always believed themselves figuratively surrounded, constantly at war with powerful and conniving opponents'.[80] They retained the Bolshevik preconviction, developed under Lenin, that economic dysfunctions, crime, and unrest were not simply predictable consequences of regime policies, but the result of counter-revolutionary activities: 'wrecking'. Similarly, the elite continued to see disunity as intolerable, with political opposition within the party denounced, in Stalin's words, as 'the most vile form of wrecking [that] deserves the *harshest* punishment'.[81]

Throughout the 1930s, the regime therefore analysed domestic security through mutating notions of the internal enemy organized around vaguely defined ideological categories of political blocs ('Right-Oppositionists', 'Trotskyist-Zinovievite' conspiracies, 'Kadet-Monarchist' organizations), notional class categories (the 'kulak'), national minorities, and broader 'socially harmful elements'.[82] So ideologically embedded were such concepts that the language of Soviet repression can be bewildering to an outsider. A typical order was framed as 'striking a crushing blow at rightist-Trotskyist, military-fascist, White Guardist-insurgent, SR-ish, Menshevik, and nationalist organizations, in the secret service of Japan and other foreign countries'.[83] But such categories were made meaningful to Soviet officials and citizens through their relentless propagation in political discourse. At the same time, they were elastic, encouraging expansive reinterpretation by Soviet officials.[84] They also became more essentialist over the Stalinist era, as individuals in these categories were increasingly seen as incorrigible enemies of the people who were 'deemed threatening in essence rather than in action'.[85]

Such ideologically induced perceptions of extreme threat were apparent in the collectivization campaign of 1930–33. Massively ambitious, poorly administered, and rushed through at breakneck pace, the regime's efforts to transfer peasant farming into state-run collective farms did not initially involve plans for mass killing, but provoked chaos and bitter peasant resistance.[86] In response, the regime relentlessly asserted that the only explanation for the crisis

---

[79] Whitewood 2015, 13.
[80] Getty and Naumov 1999, 16.
[81] Cited in Priestland 2007, 200 (*emphasis in original*).
[82] See, in general, Getty and Naumov 1999; Viola et al. 2005; Shearer and Khaustov 2015.
[83] Cited in Starkov 1993, 31.
[84] Getty and Naumov 1999, 21. See also ibid. xiii.
[85] Hagenloh 2009, 269.
[86] Viola et al. 2005; Harris 2016, 86–8.

was counter-revolutionary sabotage, linked to foreign governments.[87] When, for example, large numbers of Poles in the Western Soviet Union predictably emigrated to Poland to avoid collectivization, the government paranoidly concluded: 'There is no doubt that this campaign has the goal of preparing popular opinion to justify an armed attack [by Poland] on the Soviet Union.'[88] Popular unrest associated with food price rises was blamed by both national and regional elites on 'counter-revolutionaries', 'Trotskyite elements', and 'class aliens'. As Harris emphasizes: 'This was not a code, or a smokescreen for covering up failures. They genuinely believed that their enemies were exploiting "temporary difficulties" in an effort to bring down the regime.'[89]

The regime consequently employed brutal repression against those sectors of society identified with such 'enemies'. The principal target was the 'kulak'—a folk term originally denoting 'tightfisted' peasants which the Bolsheviks ideologically reconstructed into a class category 'defined oxymoronically', as Viola points out, 'as a "capitalist peasant"'.[90] The category bore little resemblance to any meaningful demographic block, but kulaks were relentlessly demonized by Stalinist propaganda.[91] The Bolshevik Party Central Committee declared the kulaks a hostile threat, guilty of grain hoarding,[92] and blamed problems of collectivization on a 'kulak underground', and (as under Lenin) explained rural protests as the result of kulaks 'spreading various kinds of provocative rumours'.[93] The regime therefore adopted, in 1930, policies for the 'liquidation of the kulaks as a class' or 'dekulakization': the requisitioning of kulak property, the forced resettlement of two million kulaks in 'special settlements', and the direct execution of around 30,000 kulaks.[94] Soviet leaders anticipated that this would break peasant resistance, secure food, and allow the exploitation of natural resources in the USSR's less liveable territories. In practice, dekulakization produced massive rural famine in 1932–33 and never generated a meaningful 'profit' for the state.[95]

Dekulakization clearly comes closest to matching the 'projects of social transformation' focused on by traditional-ideological perspectives, given its roots in mass collectivization. Yet even collectivization policies were tied to security diagnoses and not simply a product of utopian ambitions. From Stalin's perspective, the Soviet Union was in mortal danger from foreign enemies, needed to radically expand heavy industry to strengthen the military, and therefore had to increase and control food supplies needed for industrial expansion.[96] As in Leninist violence,

[87] E.g. Doc.32, Viola et al. 2005, 142–4.
[88] Cited in Martin 1998, 839.
[89] Harris 2016, 108.
[90] Viola 2017, 13.
[91] Viola 2007, 6.
[92] See, for example, Doc.52, Viola et al. 2005, 202.
[93] Local OGPU (secret police) report, Doc.67, ibid. 248.
[94] Viola 2007, 6.
[95] Ibid. 7.
[96] Viola et al. 2005, 124. See also Priestland 2007, 174–200.

pragmatic concerns were therefore tightly interwoven with ideology: ultimately catastrophic policies were adopted because the Stalinists' ideological worldview suggested that they would improve agricultural productivity and enhance national security.[97]

The actual *violence* associated with collectivization, moreover, was not understood by the Stalinists as directly required due to long-term transformational goals, but as a defensive reaction to threat. Stalin argued that the peasants were 'willing to leave the workers and the Red Army without bread', and were thus engaged in a '"silent" war against Soviet power'.[98] As the new head of the secret police, Genrikh Iagoda, explained in a January 1930 memo to senior subordinates:

> [The Kulak] renders more and more brutal and fierce resistance, as we see already ... from insurrectionary plots and counterrevolutionary kulak organizations to arson and terror. [The kulak] will and is already burning grain, murdering activists and government officials. If we do not strike quickly and decisively ... we will face a whole series of uprisings.[99]

Stalin similarly diagnosed nationalist groups—especially in the Ukraine—as conspiratorial criminals representing deadly threats, which left the state with no alternatives to repressive killing.[100] Even as collectivization brought about massive famine in 1932, such security threats remained Stalin's focus. He wrote to Kaganovich on 11 August 1932, warning:

> If we don't take measures now to correct the situation in Ukraine, we may lose Ukraine. Keep in mind that [Polish leader] Piłsudski is not sleeping, and his agents in Ukraine are many times stronger than [Ukrainian secret police head] Redens and [Ukrainian First Party Secretary] Kosior think. Keep in mind that in the Ukrainian Communist Party (500,000 members, ha-ha) there are not a few (yes, not a few) rotten elements ... [and] direct agents of Piłsudski. As soon as things worsen, these elements will not hesitate to open fire within (and without) the Party, against the Party.[101]

As this last line illustrates, the elite was becoming increasingly suspicious that counter-revolutionary plots had spread to the Bolshevik party itself.[102]

Mass killing associated with collectivization diminished, however—not because the kulaks had been eliminated (they had not), but because of the regime's shifting strategic diagnoses. The leadership became both increasingly concerned about the

---

[97] Priestland 2007, 192–5.
[98] Cited in ibid. 274.
[99] Doc.54, Viola et al. 2005, 219.
[100] Priestland 2007, 287–8.
[101] Cited in Martin 1998, 844–5.
[102] Doc.66, Shearer and Khaustov 2015, 111.

counterproductive consequences of mass violence and satisfied that the strength of 'kulak resistance' had at least been broken, securing food supplies and protecting industrial needs. The result was a few years of *comparatively* relaxed repression from 1934 to 1936, in which the regime entered a more conservative phase focused on maintaining social order and consolidating the Soviet state.[103] The leadership nevertheless sensed, probably correctly, that the chaos of collectivization had produced significant if muted unease within the party. At the party congress in January 1934, Stalin ominously stated: 'We have smashed the enemies of the party, the opportunists of all shades, the nationalist deviators of all kinds. But remnants of their ideology still live in the minds of individual members of the party, and not infrequently they find expression.'[104]

The biggest puzzle is why extreme fears about internal subversion and threat resurfaced in 1936 and led to the start of the Great Terror over the winter of 1936–37, despite the regime's diminishing social radicalism and its secure domestic hegemony. A key trigger was the shocking murder, on 1 December 1934, of the popular Politburo member Sergei Kirov. Western historians during the Cold War tended to assume that Kirov's death had been orchestrated by Stalin, but this view has largely been abandoned.[105] The Stalinist leadership immediately initiated an investigation into the assassination, seemingly under the genuine suspicion that it may have been orchestrated by counter-revolutionaries.[106] Yet mass killing did not emerge immediately, but two years later. Moreover, while commonly associated with the regime's internal purge of Stalin's ruling Bolshevik party, the Great Terror incorporated broader campaigns of violence known as 'mass operations' against various categories of 'socially harmful elements' in Soviet society such as 'kulaks' and criminals,[107] as well as suspect ethnic minorities.[108] Eventually, the Stalinists even turned the terror against those secret police operatives who had been its principal perpetrators. This diverse and expansive character of the violence also stands in need of explanation.

Historians generally agree that Stalin's personal agency is a critical part of that explanation. But Stalin's behaviour in the Great Terror is, on the surface, mysterious.[109] His private correspondence offers strong evidence that he sincerely believed—and was sometimes shocked by—growing reports of counter-revolutionary conspiracies.[110] Yet he simultaneously instructed officials on what the likely conspiracies were and what evidence they should turn up in their investigations, which might appear to suggest more cynical motives. Occasionally

---

[103] But on continued repression see Priestland 2007, 282–4; Shearer 2009, chs.6–7.
[104] Doc.27, Getty and Naumov 1999, 130–1.
[105] Priestland 2007, 309–10, fn.10.
[106] Harris 2016, 145.
[107] Hagenloh 2000; Hagenloh 2009; Shearer 2009.
[108] Martin 1998.
[109] Priestland 2007, 220–6.
[110] Shearer and Khaustov 2015, 87. See also, in general, Lih, Naumov, and Khlevniuk 1995; Davies, Khlevniuk, and Rees 2003; Sebag Montefiore 2004.

Stalin also cast doubt on the security services' accusations or criticized subordinates for their excesses.

This curious behaviour is explicable as part of a key dynamic of mutual radicalization within the ideological infrastructure of the political elite and its security institutions. As already noted, Stalin had strong ideological prejudices as to what heresies and oppositional forces the regime faced—what Harris terms his 'revolutionary instincts'.[111] As his political dominance increased, these prejudices created institutional pressures for self-reinforcing feedback from security and intelligence chiefs and, in turn, from mid-ranking police, security, and party officials.[112] Such feedback depended primarily on rumours, escalating denunciations, and torture, which produced an endless stream of fictitious conspiracies. But, like advocates of 'enhanced interrogation techniques' all over the world, the Stalinist elite tended to assume that such methods could uncover valuable 'intelligence'. When confessions stated, for example, that 'I am a participant in the anti-Soviet Trotskyist-fascist military conspiracy ... The main objectives of the conspiracy were: overthrow of the central leadership of the Party and country by armed force',[113] Stalinists credulously accepted this as evidence of a 'successful' interrogation, rather than dismissing it as a dubious product of torture.[114] The reality of the regime's increasing domestic hegemony made no difference, since Stalin expressed his ideological conviction that enemies would inevitably become more numerous and dangerous as the victory of socialism neared.[115]

Thus, after the Kirov assassination, Stalin issued orders to 'look for the murderers among the Zinovievites' (associates of Stalin's former rival Grigory Zinoviev), despite scepticism from the secret police chief, Iagoda.[116] The secret police duly responded, from quite absurd starting points. Attempts to find an anti-Soviet plot within the Kremlin managed to turn up only a conversation between two cleaning women and a telephone operator containing complaints about life under Soviet rule. But interpreted as 'anti-Soviet' and 'counter-revolutionary' conversations, the secret police used this as a basis for 'discovering' a much broader counter-revolutionary terrorist network with links to party members, in particular Zinoviev and Lev Kamenev.[117] Wendy Goldman explains how, in January 1935:

> Stalin reviewed the political situation in a secret letter that was circulated to all party organizations for discussion. Summarizing the leadership's current thinking on the Kirov murder, the letter claimed that Nikolaev, Kirov's assassin, had

[111] Harris 2016, 94.
[112] Hagenloh thus highlights how 'police officers ... played the primary role in generating the criminological knowledge that supported the Stalinist repression of the 1930s'; see Hagenloh 2009, 330.
[113] Doc.103, Shearer and Khaustov 2015, 190.
[114] Priestland 2007, 222–3; Hagenloh 2009, 269–70; Harris 2016, 145.
[115] Viola 2017, 11. See also Doc.20, Getty and Naumov 1999, 116.
[116] Cited in Priestland 2007, 329.
[117] Shearer and Khaustov 2015, 174–81.

been a member of a 'Zinovievite group,' based in Leningrad, that was responsible for the crime. This 'Leningrad center' in turn reported to a 'Moscow center,' which had been unaware of the actual assassination plan, but was fully cognizant of the 'terrorist moods' of the Leningrad Zinovievites ... While outwardly professing loyalty to the Party's policies and leaders, they were really two-faced 'double dealers' (*dvurushniki*), 'Judas betrayers with party cards in their pockets,' who 'masked' their true intentions.[118]

A relatively limited series of arrests followed—and Kamanev and Zinoviev were tried and imprisoned, not for having organized the assassination (which, interestingly, the NKVD concluded could not be proved), but for 'moral complicity' for inspiring the terrorists.[119]

Matters could have stopped there, since Stalin's secret letter declared that 'the nest of villainy—the Zinoviev anti-Soviet group—has been completely destroyed'.[120] But, consistent with the ideological tendencies of the elite, interpretations of the Kirov murder continued to radicalize. In response to fresh secret police reports of counter-revolutionary plots, the investigation was reopened by Stalin and assigned to Nikolai Ezhov—a high-ranking party insider and 'arch-conspiracy theorist'[121]—who criticized the NKVD for its failure to prevent the assassination, and sought to develop a much more expansive account of its origins.[122] By spring 1936, Ezhov claimed that Zinoviev and Kamenev, under instruction from Trotsky, *had* in fact plotted the assassination of Kirov, and also planned to kill Stalin and other Politburo members.[123] Zinoviev and Kamenev were reinterrogated and eventually reprosecuted in the first of the major Moscow Show Trials in August 1936, found guilty, and executed.

This conspiratorial narrative might be seen as merely a rationalization for power-political manoeuvring among the elites, but the evidence leans against such an interpretation. Again, the elite was already secure and had little further to gain from a new wave of repression (although careerist interests may have influenced Ezhov—who gained much, initially, from stoking Stalin's fears). But the Stalinists appeared to sincerely believe the more expansive conspiracy theory Ezhov formulated. In response to the 'confessions' Ezhov's interrogations produced, Kaganovich wrote privately to Stalin in the summer of 1936:

> I read the testimony of the scoundrels Dreitser and Pikel'. Although it was clear enough before, in this one they reveal in fine detail the true criminal face of

---

[118] Goldman 2011, 33–4. This letter is reprinted as Doc.28, Getty and Naumov 1999, 147–50.
[119] Getty and Naumov 1999, 146.
[120] Doc 28, ibid. 147.
[121] Harris 2016, 187.
[122] Starkov 1993, 25; Harris 2016, 148–52.
[123] Getty and Naumov 1999, 248.

the killers and provocateurs Trotsky, Zinoviev, Kamenev and Smirnov. It's now absolutely clear that the mercenary whore Trotsky was the gang leader.[124]

Stalin likewise wrote to the Defence Commissar, Klimint Voroshilov:

> Did you read the testimony of Dreitser and Pikel'? What do you think of the bourgeois dogs in the camp of Trotsky-Marchkovskii-Zinoviev-Kamenev? They want to 'remove' all Politburo members, these, to put it mildly, shitheads! It's ridiculous, isn't it? The depths to which people can sink.[125]

Accordingly, a new secret letter of July 1936 explained to the regional and local party committees the new view of the leadership:

> On the basis of new materials gathered by the NKVD in 1936, it can be considered an established fact that Zinoviev and Kamenev were not only the fomenters of terrorist activity against the leaders of our party and government but also the authors of direct instructions regarding the murder of S.M. Kirov as well as preparations for the attempts on the lives of other leaders ... [I]t has been proven that the Trotskyist-Zinovievist monsters unite in their struggle against Soviet power all of the most embittered and sworn enemies of the workers of our country—spies, provocateurs, saboteurs, White Guards, kulaks, and so on ...[126]

This represented a significant shift. Stalin and his inner circle had moved from seeing their political opponents as bearers of dangerous ideas to the conclusion that 'these gentlemen have slid even deeper ... They must therefore now be considered foreign agents, spies, subversives, and wreckers representing the fascist bourgeoisie of Europe ...'.[127] Stalin replaced NKVD chief Iagoda with Ezhov, who immediately extended internal investigations to new circles of 'suspicious persons' within the party.[128] Other former political opponents were persecuted in brutal inquisitorial sessions of the Party Central Committee in December 1936.[129] Ezhov promised to 'pull up this Trotskyist-Zinovievist slime by the roots and physically annihilate them'.[130]

---

[124] Cited in Harris 2016, 152.
[125] Cited in ibid.
[126] Doc.73, Getty and Naumov 1999, 251 & 255.
[127] Doc.80, ibid. 273.
[128] Ibid. 274–84.
[129] Ibid. ch.8.
[130] Doc.94, ibid. 308.

Such actions escalated into the Great Terror over the course of 1937. It was formally initiated at the following Party Central Committee plenum in February–March 1937, which was dominated by discussion of the 'conspiracy'. Stalin's close ally Kaganovich told the plenum:

> We never imagined before 1936 to what depths Zinoviev and Kamenev ... could have sunk ... This is why we must no longer, in my opinion, continue this magnanimous [policy] of ours. Our party must be purged of these people ... We must do away with these people in order to keep them from harming us. (Applause.)[131]

As assertions of threatening, criminal enemies poured out from the central leadership, regional NKVD officials and party committees across the Soviet Union duly began 'identifying' local elements of the conspiracy. Large-scale arrests and executions followed, and Ezhov collected these reports and used them to further expand his diagnosis of immense plots against the state.[132] Stalin and Ezhov concluded that mass repression had to broaden beyond party circles to include all those deemed hostile to the regime, as Paul Hagenloh summarizes: 'former kulaks and ex-convicts ... religious figures, foreigners, individuals formerly disenfranchised (*lishentsy*), and eventually a vast number of individuals who supposedly belonged to active anti-Soviet insurgent organizations'.[133]

These 'mass operations' built on policing trends and institutional ideologies developed over the preceding years, which increasingly construed such socially marginal groups as intrinsically harmful.[134] Now the regime, encouraged by reports from regional party and NKVD officials, came to see them as a potential basis for anti-Soviet insurgencies in the event of future war.[135] As Ezhov declared in the infamous Operational Order 00447, which officially launched the mass operations:

> Significant cadres of anti-Soviet political parties ... as well as cadres of former active members of bandit uprisings, Whites, members of punitive expeditions, repatriates, and so on remain nearly untouched in the countryside ... all of these anti-Soviet elements constitute the chief instigators of every kind of anti-Soviet crimes and sabotage ... The organs of state security are faced with the task of mercilessly crushing this entire gang of anti-Soviet elements, of defending the working Soviet people from their counterrevolutionary machinations, and, finally, of putting an end, once and for all, to their base undermining of

---

[131] Doc.134, ibid. 389.
[132] Hagenloh 2009, 244; Harris 2016, 167.
[133] Hagenloh 2009, 243.
[134] Ibid. 8–12.
[135] Docs.107–8, Shearer and Khaustov 2015, 197–9.

the foundations of the Soviet state. Accordingly, I therefore order ... a campaign of punitive measures against former Kulaks, active anti-Soviet elements, and criminals ...[136]

Separate orders were issued targeting those national minorities also suspected of disloyalty and collaboration with the enemy in the event of war.

The Stalinist elite's rationale for mass killing thus lay, first and foremost, in an ideologically rooted perception of a vast array of civilian groups as deeply threatening to the Soviet state and as guilty of counter-revolutionary crimes. The importance of this extreme ideological portrayal of *certain* civilian groups is thrown into relief, moreover, by the regime's contrasting policies towards those groups diagnosed as suspect, but less actively threatening, such as Poles, Finns, Koreans, Kalmyks, and Chechens. Nothing 'objectively' made such groups less threatening than internal party enemies, kulaks, or socially harmful elements, and the regime did worry that they could collaborate with wartime enemies. But, seeing the enemy primarily in terms of criminal conspirators linked to capitalist classes and states, the Stalinists tended to assume that non-bourgeois elements of such ethnic groups were largely innocent and not yet engaged in active conspiracy against the Soviet state (more capitalistic elements, by contrast, were frequently killed).[137] The regime consequently adopted policies of mass internal (or sometimes external) deportation for such groups, which inflicted immense misery but generally fell short of mass killing. In many instances, the regime emphasized that ethnic deportees 'preserve all civil rights and are free to move within the regions in which they are settled' and that deportations were 'not a repression but a preventive measure'.[138] Distinctive ideological constructions of civilian groups therefore underpinned distinct, although still brutal, policies of state violence.

While ideological notions of victims' threat and guilt provided the central impetus for Stalinist mass killing, its vast scale reflected the regime's broader ideological conception of the legitimacy and necessity of state violence. Had Stalinism's value-system—however utopian its dreams—accepted internal debate and been meaningfully committed to individual rights, it is unlikely that even the perception of severe threats would produce blanket mass executions. Needless to say, the Stalinists' normative orientations were quite different. They inherited the Leninist assumption that massive revolutionary violence was essentially unavoidable, given the certainty of capitalist aggression and subversion. As already noted, Stalin consistently argued that the growth of internal and external enemies was an inevitable consequence of the Soviet Union nearing the achievement of socialism. Even as

---

[136] Doc.170, Getty and Naumov 1999, 474.
[137] Statiev 2009, 244.
[138] Ibid. 245.

the collectivization campaigns of the 1930s were drawing to a close, for example, the Central Committee ominously warned that:

> The class struggle in the countryside will inevitably become more acute. It will become more acute because the class enemy sees that the [collective farms] have triumphed, that the days of his existence are numbered, and he cannot but grasp ... at the harshest forms of struggle against Soviet power. For this reason there can be no question at all of relaxing our struggle against the class enemy. On the contrary, our struggle must be intensified with all the means at our disposal and our vigilance must be sharpened to the utmost.[139]

While the elite wavered over some elements of the state violence, there is consequently little evidence that alternatives were given significant consideration. As Kaganovich later stated: 'The situation was such in the country and in the Central Committee, the mood of the masses was such, that there was no thought of acting in any other way.'[140]

Intense valorization of ideological unity and loyalty likewise played a central role in the justification of Stalinist violence. As Sheila Fitzpatrick observes: 'Discipline and unity were high on the list of party values. They were spoken about in almost mystical terms even in the 1920s ... every Communist was bound to obey unswervingly any decision of the party's highest organs.'[141] In the early stages of dekulakization, members of the Party Control Commission, tasked with maintaining internal discipline, argued: 'that which confronts us in 1931 demands solid unity between the top echelons of the Soviet and party leadership. Not the slightest cleft should be permitted.'[142] Likewise in the Great Terror, Ezhov told an audience of students at the Soviet police academy: 'We must train chekists now, so that this becomes a close-welded, closed caste which will unconditionally fulfil my orders and be faithful to me, just as I am faithful to Comrade Stalin.'[143]

This valorization of absolute obedience also fed back into the perception of threats and guilty traitors. Stalinism aspired to be ideologically monopolistic: it 'tolerated no competing discourses, branding them "enemy propaganda" and equating their creation or distribution with treasonous acts.'[144] Again, Stalinists in the Party Control Commission emphasized this internally, persecuting voices of opposition by asserting:

> The party discipline of a true Bolshevik ... lies in this: that once a decision is taken by the party, one must defend it everywhere and at all times—not only at

---

[139] Doc.20, Getty and Naumov 1999, 115–16.
[140] Cited in Goldman 2011, 312.
[141] Fitzpatrick 1999, 19.
[142] Cited in Getty and Naumov 1999, 33.
[143] Starkov 1993, 33.
[144] Getty and Naumov 1999, 20. See also Arendt 1951/1976, 423–4.

meetings but also in one's personal conversations ... even the words which were uttered [by critics within the party] against the CC [Central Committee], against Stalin, constitute a piercing weapon ... [B]y their agitation against the CC, they placed a gun in the hands of the class enemies.[145]

The Stalinists simultaneously valorized militarist virtues of hardness and ruthlessness, and contemptuously denigrated moral reservations about violence as 'softness', 'sentimental humanitarianism', and 'rotten liberalism'. The elite sought to cultivate such hardness throughout the party and broader society, advising party members: '*Don't be afraid of taking extreme measures. The Party stands four-square behind you. Comrade Stalin expects it of you. It's a life and death struggle; better to do too much than not enough.*'[146] Security operatives were similarly told by their instructors: 'It is better to overstep the mark than to fall short ... Remember that we won't condemn you for an excess, but if you fall short—watch out!'[147] This was buttressed by the party's widespread utilization of military language, rituals, dress styles, and symbolism—communists were, as Stalin announced in 1924, 'the army of Comrade Lenin'.[148]

This valorization of obedience and harshness fed, in turn, into the pervasive Stalinist valorization of 'vigilance'—which in practical terms meant implementing terror, denouncing others to the security services, and supporting repressive violence. This concept saturates the archival documents from the period. Central Committee directives demanded that party organizations: 'focus their attention on raising Bolshevik revolutionary vigilance with every means at their disposal.'[149] Stalin's conception of proper vigilance was clear, as he told the military high command in June 1937: 'Every Party member has not only the right but also the duty to inform ... so that everyone will be looking, noting every shortcoming, every breakdown, seeking the enemy out.'[150] Regional party committees reciprocated, intoning: 'it is not complacency that we need but vigilance, a genuine revolutionary, Bolshevik vigilance'.[151] The absence of vigilance, in turn, provided the central explanation for many crises. The revelation of 'enemies' was consistently attributed to a 'totally extraordinary blunting of Bolshevik vigilance',[152] 'political myopia and loss of class vigilance'[153], situations where 'a proper Bolshevik vigilance is still lacking'.[154] This discourse of vigilance helped generate a climate of watchful fear

---

[145] Doc.10, Getty and Naumov 1999, 86–7.
[146] Cited in Glover 1999, 259.
[147] Cited in Figes 2002, 84.
[148] Cited in Hoffman 2003, 68.
[149] Doc.73, Getty and Naumov 1999, 254–5.
[150] Cited in Halfin 2009, 20.
[151] Doc.58, Getty and Naumov 1999, 208.
[152] Doc.51, ibid. 194.
[153] Doc.37, ibid. 166.
[154] Doc.78, ibid. 269.

throughout Soviet society, creating normative expectations for denunciation that then fed back into elite perceptions of threat.

The Stalinists obviously also retained Lenin's broad assumption that the eventual benefits of the Soviet Union's communist project were so immense that they overwhelmed the suffering 'required' by the defence of the revolution in the present. As Kaganovich later told his son: 'These are unpleasant acts, granted, but we do not find any of this immoral. You see, all acts that further history and socialism are moral acts.'[155] This eschatological utopianism is, however, less explicit in Stalinist violence—although this may reflect the fact that it was such as a basic assumption of the Bolsheviks that it did not need to be restated. Futurized justifications of Stalinist violence did, however, encourage one of the critical underlying logics of Stalinist repression—namely, what I called in Chapter 3, an *overkill* logic of violence. Stalinists were not blind to the fact that innocents were killed in the Terror. But they deemed this tolerable given the vast future stakes of the fight against counter-revolutionaries. As Figes summarizes:

> As Stalin said in June 1937, if just 5 per cent of the people who had been arrested turned out to be actual enemies, 'that would be a good result' ... According to Nikita Khrushchev ... Stalin 'used to say that if a report was ten per cent true, we should regard the entire report as fact.' Everybody in the NKVD knew that Stalin was prepared to arrest thousands to catch one spy.[156]

Similarly, Ezhov told his subordinates: 'There will be some innocent victims in this fight against fascist agents. We are launching a major attack on the enemy; let there be no resentment if we bump someone with an elbow. Better that ten innocent people should suffer than one spy get away.'[157]

The Stalinists' overarching justificatory narrative for mass killing did not only matter for the central leadership. It also influenced those *intermediary elites*—regional party, state, and NKVD bosses and their district-level subordinates—who oversaw the implementation of the Terror within their areas of command. Of course, such figures were subject to immense coercive pressure from the centre to implement repression.[158] But this offers only a partial explanation for their participation in mass killing, for two reasons.

First, intermediary elites often actively escalated the violence. Many expressed frustrations at constraints suggested by the central leadership,[159] and when the leadership formulated its initial 'quotas' for arrests in the Terror, many intermediary elites lobbied to *increase* their quotas or simply exceeded them.[160] Towards

---

[155] Cited in Glover 1999, 256.
[156] Figes 2002, 239.
[157] Bellamy 2012b, 110.
[158] Vatlin 2016, 39.
[159] Docs.23–5, Viola et al. 2005, 130–2.
[160] Getty 2002, 131–5; Hagenloh 2009, 246; Viola 2017, 14.

the end of the Terror, moreover, the central leadership handed over sentencing of many cases to special NKVD courts known as 'troikas' comprised of regional NKVD and party and procuracy (judicial) heads. Such troikas were freed from centrally issued quotas and could have restrained violence at this point. Yet there was no decline in executions—indeed, the special troikas considered 108,000 individuals in two months, sentencing 72,254 to death.[161] So intermediary elites do not generally seem to have been reluctant participants in mass killing. As Hagenloh reports:

> all the available evidence suggests that local and central officials were in agreement about the nature of the cohorts to be repressed and the procedures to be followed ... they agreed that their removal, 'once and for all,' was a necessary part of the drive to secure social and political order in the Soviet Union.[162]

Second, however, intermediaries' behaviour—and apparent enthusiasm for the Terror—did *vary*, despite shared exposure to top-down coercive pressure. We lack a detailed overarching picture of such variation, but Alexander Vatlin notes how at the local level: 'Some district heads undertook only the minimum of cases and received reprimands for their laxness. Others attempted to overfulfill their quotas.'[163] Examples can be found of local authorities and commanders expressing scepticism towards justificatory accusations, seeking to limit violence, or condemning atrocities as violations of 'revolutionary legality'.[164] Intermediary elites therefore evidently possessed meaningful agency: there was, as Viola notes, a 'key role [for] the regional NKVD officials in establishing the [local] parameters of the terror'.[165] Since the central leadership—as it often complained—was unable to perfectly micromanage intermediary elites, intermediaries' level of ideological support (or the strength of local ideological structures) appears to have been consequential. Energetic mobilization of hardliners was an important component of the regime's ability to propagate mass killing, and the local acceleration of state violence was often closely connected to the arrival of especially committed and energetic local leaders.[166]

As with all perpetrators, intermediary elites were not solely guided by ideology. But more self-interested or bureaucratic motives were often interwoven with ideology, since willing endorsement of ideological justifications of violence allowed intermediaries, practically and psychologically, to pursue such interests through violence. As Getty observes: 'On one level, they jealously protected their position as the elite ... On another level, though, there is no reason to believe that they

---

[161] Hagenloh 2009, 279–81.
[162] Ibid. 264–5.
[163] Vatlin 2016, 42–3. See also Shearer 2009, 324–6.
[164] E.g. Statiev 2009, 249, 254, & 256.
[165] Viola 2017, 98.
[166] Ibid. 93–5.

were not also true believers in communism. In fact, there was little contradiction between the two ...'[167] Had most intermediary elites been reluctant participants, the terror would have looked quite different, the centre's coercive power notwithstanding. The vast bulk, however, were longstanding Bolsheviks, and both their self-interest and worldview were attached to the Soviet project, and to Stalin's leadership.

Nevertheless, the fact that there appears to have been behavioural variation amongst intermediaries is a reminder that, even in Stalinist Terror, the elite was not homogenous. This is true even for the central leadership, where there were some divergences—in particular, between Ezhov and a *somewhat* 'milder' approach favoured by the head of the Procuracy, Andrey Vyshinsky.[168] Competing institutional positions could generate disagreements, though these sometimes had little to do with ideology.[169] Even Stalin's own worldview contained divergent tendencies, although distinguishing these from his tactical manoeuvring is challenging. He was convinced of the need for mass violence, but also appeared to hope for the gradual consolidation of a rule-governed institutional order. He drove forward vast purges of the party, while on occasion counselling moderation and holding the door open to redeemable wrongdoers.[170] He imposed huge pressure on the regions to meet execution quotas, yet was suspicious of regional elites and liked to position himself as a defender of the 'little person' against regional excesses.[171] We might hesitate to call such elements 'restraints' in a regime of such vastly murderous capacities, but they nevertheless influenced patterns of violence in the Terror.

Outside the distinctive Stalinist worldview, Soviet mass killings are largely inexplicable. That the Stalinist leadership would engage in *some sort* of repression is no real puzzle. This was an authoritarian regime, originally placed in power by a narrow coup and resulting civil war, facing tremendous domestic and international challenges in governing the largest contiguous political territory on Earth. Yet these facts apply to the entire history of the Soviet Union, and cannot explain why the Stalinist regime went so far beyond limited repression to launch vast campaigns of mass killing against largely imaginary enemies in years of relative security. But the violence becomes intelligible when the broad challenges facing the Stalinist leadership are analysed in combination with the leadership's distinctive ideology—in particular, with its preconceptions of likely enemies and the legitimacy of revolutionary violence, and the particular narratives of threat and crisis that those preconceptions generated over the course of the 1930s.

---

[167] Getty and Naumov 1999, 11.
[168] Hagenloh 2009, 239–41 & 284.
[169] Priestland 2007, 356–60.
[170] See, for example, Getty and Naumov 1999, 409–19.
[171] See, in general, ibid. chs.4–6.

The Rank-and-File of Stalinist Violence

Until recently, the actual rank-and-file perpetrators of Stalinist violence were relatively understudied.[172] But new research, in particular two key studies by Lynne Viola and Alexander Vatlin, has reversed this trend, in part by examining data generated by the 'purge of the purgers' at the end of the Great Terror, when Stalin blamed Ezhov and lower-level officials for committing excesses. Unlike the Terror itself, investigations in the purge of the purgers tended to be genuinely 'investigative', and generated significant usable data for historians.[173] The findings of these studies allow us to construct a much more detailed picture of rank-and-file perpetration than would have been possible even a decade ago.

This picture does not match a traditional-ideological portrayal of mass killing. Revolutionary zeal and transformational ideals do not generally appear to have been primary motivations for most Soviet police operatives. Instead, the most obvious drivers of rank-and-file participation were coercive pressures and careerist self-interest operating within institutionalized bureaucracies.[174] Local NKVD branches in the Great Terror received 'quotas' of arrests which they were expected to fulfil, often translated into individual quotas for each local NKVD operative. Failing to meet quotas carried the high probability of being arrested oneself for counter-revolutionary or anti-Soviet activity.[175] When rank-and-file perpetrators were later interrogated, almost all identified the fear of being labelled an enemy of Soviet power as a central motive for their participation in torture and killing.[176]

Such prosaic motives became more central, moreover, as the Great Terror escalated. For roughly the first half of the Terror, local NKVD units primarily targeted individuals who had already been recorded as 'suspicious elements', and for whom some compromising material or identity, however trivial, could be found. After a few months, however, the NKVD exhausted such categories, and shifted to arbitrary arrests in which the falsehood of extracted confessions became increasingly blatant. In such circumstances, many NKVD operatives recognized that the accusations they were producing were bogus.[177] One operative later explained how: 'During the day we usually made-up fabricated interrogations for the accused and at night we made them sign under compulsion'.[178] Victims who survived also suggested that their interrogators recognized the falsehood of the charges. As one

---

[172] Viola 2013; Viola 2018; Whitewood 2018.
[173] Obviously such data cannot simply be taken at face value; see Viola 2017, 5–8.
[174] A subset of NKVD officers used the terror for personal material gain through the requisitioning of victims' possessions; see Vatlin 2016, 58–68; Viola 2017, 91–2, 97, 100–1, & 134. A small number seem sadistic; see Shearer 2009, 348; Viola 2017, ch.3.
[175] Hagenloh 2009, 255. Vatlin 2016, 29 & 42;
[176] Vatlin 2016, 75.
[177] Shearer 2009, 351–2.
[178] Cited in Vatlin 2016, 31.

said of her interrogator: 'He knew I was not guilty but [said] I had to sign the confession like a conscientious Soviet young woman.'[179] Another reported: 'The investigator gave me a prewritten confession ... When I said it was all lies, the investigator answered: In the NKVD we know that and have nothing against you but it is necessary.'[180]

Bureaucratic interests and expediency also played a significant role in facilitating and escalating rank-and-file perpetration. One NKVD officer explained how his boss 'demanded that we only get [confessions] from arrestees about espionage because in spy cases it was easier to complete the investigation and to indict the arrestee'.[181] Bureaucratic demands allowed dynamics to enter the terror at the rank-and-file level that truly had nothing to do with the regime's ideology. The use of what contemporaries sometimes called the 'biological approach'—arresting individuals based on familial ties to other victims—had no basis in Marxism-Leninism and was repeatedly repudiated by party leaders.[182] Yet it was typically tolerated as an easy way to impute criminal leanings to individuals and fulfil arrest quotas.

Nevertheless, the evidence does not support a portrayal of rank-and-file perpetrators of Stalin's Terror as largely free from ideological influence. For a start, ideological structures were clearly central in guiding implementation of violence. Coercive micromanagement of the rank-and-file by the central leadership was not possible in Stalin's Soviet Union—the state was hugely overburdened and the NKVD desperately understaffed for the demands placed upon it.[183] The Terror depended on local operatives understanding and willingly adhering to the elite's justificatory narrative and resulting repressive logics.[184] As Vatlin puts it, rank-and-file perpetrators 'agreed to perform the leading roles in a play written in the Kremlin and at the Lubianka [the NKVD Headquarters]'.[185] They targeted specific categories of the population according to shifting elite discourses and directives, sought out denunciations and confessions that aligned with the elite's conspiratorial theories, and relied on Stalinist ideology in the framing, processing, and justification of actions against individual targets. They sometimes also acted on initiative in response to leaders' speeches and press reports even in the absence of direct orders.[186] The bureaucratic processes of the terror were, in consequence, saturated in esoteric Stalinist ideological language. One typical charge against 39 supposed Ukrainian counter-revolutionaries, for example, read as follows:

---

[179] Cited in ibid. 32.
[180] Cited in ibid.
[181] Cited in ibid. 42.
[182] Priestland 2007, 326 & 397–8; Viola 2017, 25.
[183] Overy 2004, 210–11; Viola 2017, 19–20.
[184] See also Hagenloh 2009, 246 & 269–71; Vatlin 2016, 106.
[185] Vatlin 2016, 143.
[186] Thurston 1986, 228, fn.84.

162   IDEOLOGY AND MASS KILLING

> On the assignment of the Ukrainian counterrevolutionary bourgeois nationalist centre ... Vasilenko (arrested) set up a Ukrainian nationalist insurrectionary organization led by the former secretary of the Skvirskii District party committee, Kokhanenko (arrested), and subsequently by the boss of the district land department, Zalevskii (arrested) with the aim of overthrowing the socialist system and government of the USSR and resurrecting capitalism and breaking off Ukraine from the USSR and establishing a fascist dictatorship. The organisation united anti-Soviet Petliurists[187] and repressed kulak elements with tasks to prepare insurrectionary-diversionary cadres in the event of war against the USSR to raise an uprising in the rear ... breaking Ukraine from the USSR and orienting Ukraine to the fascist countries of Poland and Germany.[188]

Needless to say, such accusations are almost incomprehensible, let alone implausible, outside the Stalinist ideology of the time.

This ideological framework was not mere window dressing. Again, it drove the targeting of specific categories of the population: 'kulaks' throughout the collectivization campaign and the Great Terror, those identified as 'social harmful elements' in the mid-1930s and the mid-to-latter stages of the Terror, and particular national minorities in the latter half of the Terror. Moreover, while the capacity of rank-and-file actors to invent plots and conspiracies to rationalize arrests was *almost* unlimited, they did face ideologically imposed constraints at the outer edges. Cases had to fit the justificatory narrative, and were occasionally dismissed by the courts or higher authorities when they did not do so.[189] Anticipating this, cases were sometimes abandoned—not, of course, for lack of *actual* evidence (since this was almost always absent) but because the right kinds of 'witnesses' couldn't be found, and the ideologically appropriate conspiratorial linkages couldn't be made.[190]

Indeed, the very fact that the NKVD worked so arduously to generate forced confessions and testimonies 'raises the question', as Kotkin points out, 'of why people were not just rounded up and shot or deported without the labor-intensive formalities ... [especially for] people whose arrests were not reported publicly and who would never be heard from again'.[191] If the terror had purely been a repressive power-grab, such efforts would be otiose. But they make sense within a form of ideological security politics, in which the NKVD's leaders and rank-and-file operatives held on to an image of their organization as a professional security service engaged in the patriotic defence of the Soviet Union. Consequently, the

---

[187] Symon Petliura had led the Ukrainian nationalist forces and the short-lived Ukrainian republic (1918–21) in the Russian civil war against Soviet forces.
[188] Cited in Viola 2017, 74.
[189] Vatlin 2016, 47 & 130.
[190] Ibid. 127.
[191] Kotkin 1995, 335.

NKVD intensely promoted an internal ethos and set of organizational norms rooted in Stalinist ideology. As with the elite, these norms exalted harshness and communist duty, glorified participation in violence, and sought to eliminate moral reservations and compassion for victims.[192]

The pressure of such norms was important, because NKVD operatives were by no means uniformly eager to implement the Terror, and a minority queried directives and expressed a reluctance to participate.[193] As part of the ideological mobilization of the NKVD's workforce, however, such scepticism was aggressively denigrated as shameful weakness and 'liberalism'.[194] Many operatives testified that it was 'considered shameful when someone under arrest was released', since organizational norms held that 'once you arrest someone, it means he is guilty'.[195] An especially brutal perpetrator later reported how local NKVD leaders 'poured shame on the workers ... who did not use physical methods of influence [the euphemism for torture], called them *intelligenty* [intellectuals], unable to work with the arrested, and they set me up as an example to others. And, I, fool, was proud of this.'[196]

Valorization at the rank-and-file level also depended heavily on the 'doctrine of superior orders'[197]—casting unquestioning and energetic obedience as essential, and invoking higher authorities to render violence legitimate and honourable. As Vatlin reports:

When [a] new operative asked for normative instructions about the spy operation, [his superior] answered, 'You're just a young operative and don't understand what's going on here. After all, we're carrying out an operation to remove Germans, Poles, and other alien nationalities by order of the Politburo of the Central Committee of the Communist Party.' The mention of this authoritative body was enough to remove doubt from the operatives' minds about the legitimacy of their actions.[198]

When themselves later arrested and tried for their participation in the Terror, almost all operatives invoked this obligation to unquestioningly obey communist authorities as part of the explanation and justification of their actions.[199]

Irrespective of the extent to which rank-and-file perpetrators of Stalinist violence genuinely internalized the regime's justificatory narrative, then, they were clearly subjected to strong ideological pressures that structured the violence

[192] Viola 2013, 15.
[193] Viola 2017, 58–9, 99, & 102–3.
[194] Ibid. 58, 67, 98, 108, & 121.
[195] Cited in ibid. 41.
[196] Cited in ibid. 68.
[197] Hilberg 1985, 288.
[198] Vatlin 2016, 43.
[199] Viola 2017, 106–7 & 178.

according to the prevailing justificatory narrative. But the vast majority also appeared to have sincerely internalized Stalinist justifications of the violence *to some meaningful extent*. It should be remembered, for a start, that most of these operatives were not ideologically disengaged policemen. The NKVD recruited operatives with sound revolutionary credentials, and almost all the rank-and-file perpetrators in Vatlin's and Viola's studies were longstanding members of the Communist party, with a majority having served in the Red Army or fulfilled party roles as far back as the civil war.[200] Such operatives sometimes harboured doubts or disbelief about the *specific cases or actions* they were involved in, but there is little evidence that they were sceptical about the legitimacy of Stalinist mass killing as a whole. Like intermediary elites, they often went well beyond the directives of the central leadership in enthusiastically repressing the regime's 'enemies'.[201] Many later emphasized their sincere belief, asserting: 'I consider that those charged in these cases were tried correctly';[202] 'We blindly believed and thought that we were doing the right thing';[203] 'I did not understand then that what happened was criminal. I did not understand its enemy essence. I considered that all these measures must be directed only against the enemy'.[204]

Such later testimony, most of which comes from the 'purge of the purgers' at the end of the Terror, cannot be taken at face value, since there were self-serving reasons for such pleas. But nor should it be dismissed out of hand. That NKVD operatives would believe in the broad justification of repressive policies authorized by the Soviet leadership is unsurprising given their backgrounds and the broader ideological environment and institutional culture within which they worked. Aside from the general power of Soviet propaganda in the 1930s, operatives had spent years socialized by Communist party discourses and NKVD training. Even if they did not feel deep personal commitment to those organizations' values (though many seemed to), they appeared to at least adopt much of the justificatory narrative offered for NKVD activities. As one operative subsequently emphasized: 'I was trained to consider that if the NKVD organs arrested a person, then he was an enemy and he must confess his enemy activity'.[205] Senior NKVD officers, he elaborated, 'taught us that it is necessary to relate brutally to the enemy, that it is necessary to force them to surrender to the party and Soviet power'.[206] Another investigator testified:

> Ezhov's authority in the organs of the NKVD was so high that I, like the other employees, did not doubt the guilt of individuals who were arrested on his direct

[200] Ibid. 21.
[201] See, for example, Docs.53, 56, & 66, Viola et al. 2005, 216–18, 220–1, & 245–6.
[202] Cited in Viola 2017, 43.
[203] Cited in ibid. 124.
[204] Cited in ibid. 109.
[205] Cited in ibid. 55.
[206] Cited in ibid. 70.

orders even when the investigator did not have any materials which compromised the given individual. I was convinced of the guilt of such an individual even before the interrogation and then, during the interrogation, tried to obtain a confession from that individual using all possible means.[207]

Moreover, while self-interested motives existed for such statements, perpetrators did not consistently stick to the most self-serving rationalizations of their actions. Many did not appeal to coercion, even when this would seem a logical defence, some did acknowledge doubts and disbelief, and some specifically delineated their belief in certain arrests from disbelief in others.[208] Even when it was clear they would be convicted, and were pressured to repent for their crimes, many did not. One argued: 'I never engaged in any kind of criminally harmful work. At that time, I considered the methods used in the operational-investigative work absolutely correct and believed that the party required it.'[209] Another likewise stressed: 'I do not recognize my guilt because I could not not [sic] believe the directives of Ezhov and [the regional NKVD chief] Uspenskii, and everyone believed them.'[210] Shearer summarizes how:

> Depositions of officers arrested after the purges reveal no remorse, only regret, at most, that 'some' procedural rules were not followed. One officer ... later recorded that he and his colleagues received their commander's [operational briefing] with 'boisterous approval'. To these men, the danger was real and the enemy everywhere.[211]

This image is also supported by testimony from witnesses of the Terror who became émigrés. Two former inmates of Stalinist prisons wrote that, from their experience: 'at the end of the [Ezhov] era, the great majority of the young examining magistrates were more or less genuinely convinced of the guilt of the 'enemies of the people' and of the reality of their confessions'.[212] Another émigré recalled an NKVD acquaintance telling him: 'if you fell into our hands, it certainly wouldn't be without reason ... Purges are absolutely necessary. The Politburo is more than right about that.'[213] Mikhail Gorokhov, an émigré who was actually trained as an NKVD officer, stated: 'at that time ... no doubts assailed my mind, and I accepted the accusations as genuine'.[214] Of other NKVD recruits, he said:

---

[207] Cited in Starkov 1993, 33.
[208] Viola 2017, 138 & 143.
[209] Cited in ibid. 139.
[210] Cited in ibid. 141.
[211] Shearer 2009, 337.
[212] Cited in Thurston 1986, 228.
[213] Cited in ibid. For similar examples see ibid. 228, fn.79.
[214] Cited in ibid. 226.

they are, of course, Party members, simple boys, who have been told that 'enemies of the Socialist society' try to wreck our Soviet system and kill our leaders, and that these wreckers must be exterminated. The fellows do what they are told and quietly accomplish their job.[215]

The available evidence thus strongly suggests that NKVD perpetrators, as Viola concludes, 'mainly, at least at that time, thought that they were acting correctly'.[216]

Of course, the NKVD was not homogenous.[217] As should be expected, there was diversity in how rank-and-file members of Soviet organizations responded to the Terror, and alongside those who internalized the Stalinist justificatory narrative were a significant number of more reluctant perpetrators.[218] A very limited number displayed the courage and disposition to obstruct the Terror at a local level.[219] More common was variation in the eagerness and intensity with which local operatives participated. A report from 1938, as the Terror was beginning to reverse course, criticized the Moscow provincial leadership for its 'competitive' efforts to raise arrest numbers, but also pointed to recalcitrance elsewhere in the repressive apparatus, observing how '[the provincial leadership] viewed some operative departments (XI, VI departments) as lagging behind in this competition because they did not produce large numbers of confessions. Investigators [in these departments] tried to question arrestees conscientiously, more or less.'[220] Moreover, many NKVD operatives found their work psychologically straining and sought to evade participation by taking leaves of absence, and there were several suicides.[221] So the extent to which rank-and-file operatives willingly accepted the ideological justification of the Terror varied in ways that were consequential at the micro-level.

As the Terror escalated, many NKVD operatives increasingly recognized their methods as fraudulent.[222] But in later testimony, most espoused a common standpoint: that errors occurred, but within practices of state repression that were fundamentally legitimate.[223] Like members of most security services, while they did not privately approve of all the actions their work involved, they did buy into their organization's official purposes and professional values. They were not motivated in the main by Stalinist 'high doctrine': most NKVD had not even completed secondary education, and often weakly comprehended the regime's more abstruse theories.[224] But they accepted the Stalinists' justificatory narrative for violence. As Vatlin concludes:

---

[215] Cited in ibid. 228.
[216] Viola 2017, 142. See also ibid. 28, 71, 87, 109, & 113–14.
[217] Ibid. 21.
[218] Vatlin 2016, 23.
[219] Shearer 2009, 360–3; Viola 2017, 151–3.
[220] Vatlin 2016, 43.
[221] Shearer 2009, 352; Vatlin 2016, 44, 76, & 112; Viola 2017, 21–2 & 101.
[222] Shearer 2009, 351; Vatlin 2016, 74; Viola 2017, 40, 49, & 80.
[223] Shearer 2009, 351–2.
[224] Vatlin 2016, 73.

[NKVD operatives] were not standard cogs in a machine of repression. They were living people who had to answer to their own consciences. It is important to exercise care in generalizing about them. The majority of NKVD workers accepted directives from above as the latest development in the war against the counter-revolution. Others thought their orders were the expression of the great leader's genius, incomprehensible to mere mortals ... [but] some workers also revelled in their power over helpless people and manipulated the tragedy for their own benefit ...[225]

Viola likewise concludes that 'these operatives ... were not simply automatons fulfilling orders from above, cogs in the machine of terror. Each had a distinct personality and voice, each was an individual with a role to play and a degree of agency, however constrained, in helping to enact the Great Terror.'[226]

Publicly Justifying Stalinist Repression

Stalinist mass killing did not require enthusiastic mass support. While crude portrayals of Soviet rule as built purely on coercion have been abandoned in contemporary scholarship, there is no question that the Stalinist state was highly repressive, and the capacity for opposition to state policies limited. Some campaigns of Stalinist killing were also conducted by police and security organizations with limited public participation.[227] While dekulakization and the Great Terror were exacerbated by social tensions within the Soviet population, there is no evidence that they primarily reflected popular antagonisms.

But the state did depend, in various ways, on public collaboration and acquiescence. Relying solely on mass coercion of a seethingly discontented population would have presented unacceptable risks of popular uprisings in any impending war, and been unsatisfactory given the elite's ideological self-conception of the Soviet Union as a revolutionary state that truly served the interests of the masses. More importantly, while the Stalinist state could mobilize immense force, its capacity for coercive micromanagement of the population was limited, given the hugely overstretched nature of the security and broader state apparatus. How Soviet policies played out 'on the ground' therefore depended not only on rank-and-file agents, but also on the degree to which Soviet citizens willingly orientated themselves to official policies. As Wendy Goldman summarizes:

---
[225] Ibid. 61.
[226] Viola 2017, 29.
[227] Hagenloh 2009, 12, 260, & 284–5.

mass participation played a critical role in the terror's acceleration and growth ... The Terror did not consist only of targeted attacks from above, but also a self-devouring process from below that took place in every institution, organization, and workplace.[228]

The Stalinist regime remained highly concerned about the state of public opinion. Rising discontent may even have been a factor in the elite's moves to bring the Great Terror to a close in 1938, as 'criticism from the Procuracy ... and countless citizen letter writers began to wear down the machinery of terror'.[229]

Consequently, the regime expended considerable effort publicly legitimating its repressive activities, through a justificatory narrative that was largely consistent with the elites' internal ideological rationale for state violence. The state monopolized most public political discourse, seeking to control all major epistemic authorities in Soviet society and deny the population alternatives to the regime's construal of events.[230] In this the Soviet judicial apparatus, and the major show trials of 1928, 1930, 1936, 1937, and 1938, played a central role, providing an especially visible platform for advancing the narrative of conspiratorial threats and a valiant state campaign against nefarious criminals. As Priestland observes: 'The trial allowed the leadership to define and present the beliefs, character, and behaviour of the internal enemy ... It was particularly good at demonstrating ideas of class infection, by describing and 'proving' the existence of the conspiracy'.[231] The show trials convinced many, in the West as well as in the Soviet Union, that the Stalinists had uncovered very real conspiracies.[232]

At the first of the large Moscow trials in 1936, the Soviet Chief Procurator Andrey Vyshinsky portrayed the defendants as 'underground terrorist groups' intending to 'seize power at all costs', who 'made use of the most detestable method of fighting, namely, terrorism'.[233] He detailed how 'while preparing the assassination of Comrade Stalin and other leaders of the [party], the United Center simultaneously strove by all means in its power to give assurances of its loyalty and even devotion to the Party'.[234] At the second Moscow show trial in 1937, he likewise framed the defendants as 'a brigand gang', 'murderers', and 'Not a political party ... merely a gang of criminals ... hardly to be distinguished from gangsters who use blackjacks and daggers on the high-road on a dark night ...'.[235] Vyshinsky also

---

[228] Coleman et al. 2019, 232.
[229] Viola 2017, 22. See also ibid. 168–70.
[230] Brandenberger 2011, 137.
[231] Priestland 2007, 229.
[232] Goldman 2011, 47.
[233] Cited in ibid. 40.
[234] Cited in ibid. 41.
[235] Cited in Fitzpatrick 1999, 194–5.

made extensive use of dehumanizing rhetoric, variously labelling the defendants 'alien elements', 'dogs', 'reptiles', 'predatory beasts', 'vermin', and 'lice'.[236]

For most citizens, access to the Terror's justificatory narrative was mediated through the press. Newspapers proliferated in the Soviet Union—alongside the party daily *Pravda* and the state daily *Izvestia*, there were a vast array of local newspapers, and major factories would often have their own newspaper (and sometimes several). These were crucial in circulating the ideological messages of the show trials, which received 'saturation coverage ... as well as being broadcast on radio and filmed'.[237] In such press output, the security-centric justificatory narrative was dominant. 'The war motif', notes Fitzpatrick, 'was constantly elaborated in the press'.[238] *Pravda* warned readers that: 'An enemy with a party card in his pocket is more dangerous than an open counterrevolutionary'.[239] Factory newspapers explained how: 'The parallel center has been unmasked by the NKVD. They organized spying, wrecking, and diversionary acts to weaken defence'.[240] This newspaper went on to advise: 'We must not be calm. Every honest citizen should think about what these despicable traitors are preparing to do to us and our beloved country'.[241] Stalinist valorization of discipline, hardness, and vigilance was similarly promoted amongst the mass population. Party organizers encouraged workers to reject the 'petty-bourgeois servility' and 'passivity' which restrained them from denouncing possible wreckers.[242] In the factories, 'newspapers advised employees to resist the inclination to be kindhearted or soft in dealing with suspicious coworkers or comrades'.[243] Editors explained how 'the task of every honest Soviet citizen is to know how to unmask enemies in any mask, to discern and to prevent their insidious, traitorous activities'.[244]

The regime also played a dominant role in broader cultural output, which was consistently mobilized to support Stalinist justificatory narratives of repression.[245] The 1936 play *Party Card*, for example, portrayed the dangers of factories and party organizations being infiltrated by hostile elements,[246] while the popular *The Confrontation* (1937) depicted an evil conspiracy against the Soviet Union which was successfully defeated by the courageous efforts of Soviet citizens.[247] *The Great Citizen*, drafted between 1935 and 1937 but released to cinemas in the midst of the Terror in February 1938, pitted 'the progressive, patriotic Shakhov and his dream of building socialism in one country against concealed enemies within the party

---

[236] Weitz 2003, 73.
[237] Fitzpatrick 1999, 203.
[238] Ibid. 10.
[239] Cited in Kotkin 1995, 299.
[240] Cited in Goldman 2011, 98.
[241] Cited in ibid. 98–9. For further examples see ibid. 2 & 59.
[242] Ibid. 47.
[243] Ibid. 60.
[244] Cited in ibid. 59.
[245] Brooks 1994.
[246] Fitzpatrick 2000, 31–2.
[247] Fitzpatrick 1999, 203.

who conspire to undermine the USSR and turn the revolutionary society over to foreign capitalists abroad'.[248] Again, the press promoted such films and their correct interpretation. The publication *Sputnik agitatora* explained to its readers how: '*[The Great Citizen]* teaches us to recognize the methods and tactics of the Trotskyite-Zinovievite-Bukharinite scum and provokes in us a feeling of anger and hatred for this human waste. The film teaches us the necessity of keeping vigilant and being steadfast, firm, and observant.'[249]

It is difficult to precisely establish how the Soviet public responded to such efforts at ideological mobilization and legitimation, but there is some agreement among historians on the overall picture.[250] A significant subset of Soviet society rejected the regime's propaganda, resented its policies, and never accepted its basic legitimacy.[251] While Soviet policies benefitted certain groups, they imposed immense suffering on others, especially the peasantry.[252] There was a constant subculture of complaints, anti-regime jokes, and occasional expressions of hope at the regime's downfall throughout the Stalin era, visible in the mountains of (more credible) reports on popular dissent in the NKVD archives.[253]

Even amongst those who truly believed in Stalinism, many did not match the stereotypical image of ideological fanaticism. Of the roughly 2.4 million members of the Communist party in the 1930s,[254] a significant proportion were deeply committed to the party's mission but struggled to understand its more intellectual doctrines. Many 'had only an elementary education and were ignorant of political theory and history, decisive areas of knowledge for a Communist'.[255] Party members were often acutely aware of their ignorance, although in ways that illustrate their credulity for the regime's ideological narratives. As one admitted in 1935:

> I am politically illiterate and know how difficult this makes things for me ... if [non-party members] were to ask me about the former Zinovievite opposition, I would not be able to answer and would begin to hem and haw. This is why we must increase our political literacy along with our class vigilance.[256]

Another stated: 'I am ashamed to admit that although I've been a party member since 1924, I didn't know about this whole struggle that Zinoviev and Kamenev were waging against the party.'[257]

---

[248] Brandenberger 2011, 170.
[249] Cited in ibid.
[250] On addressing this methodological challenge see ibid. 121 & 303, fn.3.
[251] For examples see Thurston 1986, 223.
[252] For examples see Doc.27, Viola et al. 2005, 133–6.
[253] Overy 2004, 309–10 & 346; Brandenberger 2011, 60; Vatlin 2016, 93–100. See also Thurston 1986, 219, fn.26.
[254] This is an average, calculated from figures in Overy 2004, 138–9.
[255] Kotkin 1995, 300. See also Brandenberger 2011, 48 & 176.
[256] Cited in Brandenberger 2011, 43.
[257] Cited in ibid.

This all re-emphasizes the problems with the 'true believer' model of ideology. If, in looking for the influence of ideology and propaganda, one expects to find ideological sophistication, wholesale indoctrination, and mass enthusiasm, one will inevitably be disappointed by the more complex reality of Soviet public opinion. But once we drop such inflated standards, it becomes clear that Stalinist propaganda was hugely consequential. Internalization was uneven, but remarkably extensive given the vast disconnect between most regime claims and reality. 'Campaign after campaign in the press against Trotskyists, spies, and saboteurs', Vatlin notes, 'drew a response from the public.'[258] As Viola observes: 'Dissidents from the 1970s often recalled, with a sense of retrospective bewilderment if not shame, that they fully believed in Stalin and the Soviet system in the 1930s.'[259]

Large numbers, indeed, strongly identified with the Stalinist project. Lev Kopelev, initially a passionate Bolshevik who later turned dissident against the Stalinist regime, recalled: 'I convinced myself and others that ... all our ills, malefactions and falsehoods were inevitable but temporary afflictions in our overall healthy society. In freeing ourselves from barbarity, we were forced to resort to barbaric methods, and in repulsing cruel and crafty foes, we could not do without cruelty and craftiness.'[260] Utopian ideals did matter here. The novelist Boris Pasternak likewise wrote in 1935:

> The fact is, the longer I live the more firmly I believe in what is being done, despite everything. Much of it strikes one as being savage [yet] the people have never before looked so far ahead, and with such a sense of self-esteem, and with such fine motives, and for such vital and clear-headed reasons.[261]

Citizens later testified that they had been 'convinced that we were creating a Communist society, that it would be achieved by the Five-Year Plans, and we were ready for any sacrifice',[262] and accepted that 'when a new world is being built, injustice and victims are unavoidable.'[263] Such strong supporters of the regime also accepted the moral reorientation of violence propagated in Stalinist ideology. Figes notes how, later in life, Kopelev:

> recall[ed] his efforts to subordinate his moral judgement (what he called 'subjective truth') to the higher moral goals ('objective truth') of the Party. Kopelev and his comrades were horrified by what they were doing to the peasantry [during the collectivization campaigns], but they deferred to the Party: the prospect of retreating from this position, on the basis of what they had been brought up to

---

[258] Vatlin 2016, 79.
[259] Viola 2013, 13
[260] Cited in Thurston 1986, 226.
[261] Cited in Figes 2002, 190. See also Fitzpatrick 1999, 37; Brandenberger 2011, 139–40.
[262] Cited in Figes 2002, 91.
[263] Cited in Thurston 1986, 224.

dismiss as the 'bourgeois' ideals of 'conscience, honour [and] humanitarianism,' filled them with dread.[264]

Indeed, a significant subset of those targeted by the Terror so fully accepted the justificatory narratives of the regime that even when they or a relative was arrested, they would conclude, as Robert Thurston puts it: 'that everyone else detained by the NKVD was indeed guilty but that the case in question was a mistake'.[265] Even after his own father's arrest and execution, the writer Anton Antonov-Ovseenko continued to accept the purges as legitimate until 1939, as he explained:

> A year had not passed since the death of [my father], and his son was glorifying his murderer. For me, a youth of nineteen, Stalin's name was sacred. As for the executions of enemies of the people, what could you say? The state had the right to defend itself. Errors were possible in such matters, but Stalin had nothing to do with it. He had been and remained the Great Leader.[266]

Some continued to believe this for years.[267] In numerous cases, 'even as firing squads raised their rifles', Halfin reports, 'their victims cried, "Long live Comrade Stalin!"'[268]

Ideological justifications of Stalinist violence thereby supported various forms of indirect participation by Soviet citizens, most noticeably through denunciations of individuals to authorities, or agitation for repressive measures at party meetings. While coercion and social pressure certainly encouraged such activities, it operated alongside belief. As Goldman explains:

> Many people were originally persuaded that real enemies threatened Soviet security. A potent mixture of truth, half-truth, and lies, widely disseminated by newspapers, discussion circles, and party leaders, had considerable popular support. People's fears were thus first created and then shaped by their faith in the Party and its message. They wrote numerous *zaiavleniia* [denunciations] ... Even at moments when everyone might have remained silent at no cost, a voice from the floor would invariably shout, 'The punishment is too lenient!' or 'Exclude!'[269]

Again, such sincere internalization did not require blind fanaticism. Many clearly developed doubts, and some, whilst committed to the Soviet project, protested against aspects of the purges.[270] But in general the more ideologically

---

[264] Figes 2002, 191.
[265] Thurston 1986, 226. See also Kotkin 1995, 591, fn.339.
[266] Cited in Thurston 1986, 227.
[267] See, for example, Vatlin 2016, 105.
[268] Halfin 2009, 3.
[269] Goldman 2011, 196. See also Hoffman 2003, 84; Goldman 2011, 230 & 233.
[270] Hoffman 2003, 84–5.

committed portions of the Soviet citizenry readily adopted justifications that would assuage uncertainties. As one citizen later explained: 'I had my doubts about the Five-Year Plan ... But I justified it by the conviction that we were building something great ... a new society that could not have been built by voluntary means ... Today, I understand that it was very harsh ... but I still believe that it was justified.'[271] A mid-ranking Moscow communist likewise confessed to his diary: 'How can I judge, a rank-and-file party man? Of course sometimes doubts sneak in. But I cannot fail to believe the party leadership, the Central Committee, Stalin. Not to believe the party would be blasphemy.'[272]

Such strong ideological commitment was still probably a minority position largely confined to the party faithful. Most Soviet citizens had limited interest in politics and were dissatisfied with elements of Soviet rule. The émigré Viktor Kravchenko observed that in the early phases of the Great Terror, 'the population at large ... were pretty indifferent to what seemed to them a family quarrel among their new masters'.[273] For this larger subset of the population, revolutionary goals had limited appeal, and Stalin himself recognized that 'the people do not like Marxist analysis, big phrases, and generalized statements'.[274] But the regime proved skilful at mobilizing more conventional patriotic sentiments and concerns with self-defence to cultivate broader ideological acceptance of the Terror.[275] Its film and media output appears to have been popular. A rural activist's diary records a mass visit to one patriotic film:

February 23 [1935]. I took a hundred members of the [village] youth to Tirsopol. There they saw the film *Chapaev*. The impression was indescribable. Our audience saw in Chapaev their own peasant fellow, who conquers everyone with his fearlessness, bravery, ability to fight, and his ability to organise Red Army troops and lead them into battle ... Vasily Ivanovich, Commissar Furmanov, Petka, and Annushka have become heroes for our youth, who greeted an array of moments in the film with loud, approving applause.[276]

Coercion also cannot explain the 'mountain of fan mail' received by F.M. Ermler, the director of several other Stalin-era propaganda blockbusters,[277] nor, indeed, the welter of supportive letters that Soviet citizens spontaneously sent to the political leadership or the press.[278]

---

[271] Cited in Figes 2002, 111.
[272] Cited in Fitzpatrick 1999, 215.
[273] Cited in Thurston 1986, 217.
[274] Cited in Brandenberger 2011, 97.
[275] See also Priestland 2007, 284–92.
[276] Cited in Brandenberger 2011, 127.
[277] Ibid. 171.
[278] Overy 2004, 344.

In addition, Stalinist ideological lexicons were disseminated downwards through the dense institutional apparatuses of the Soviet state, filtering into local political and economic conflicts which otherwise might have had a quite minimal ideological quality. Citizens had to master the complex activity that Kotkin terms 'speaking Bolshevik',[279] as the regime's ideological language 'entered the daily parlance of factory employees, lending import, purpose, and direction to daily resentments and workplace quarrels'.[280] As already noted, this promoted the escalatory linkage of economic problems and mechanical or operational failures with conspiratorial crimes. As one factory organizer reported: 'The [Second Moscow show] trial was a great lesson for us. We should now approach every accident, every breakage in a different way. It's no secret that we sometimes overlook the facts after enemy hands have acted in the factory and the shop.'[281] Party members likewise wrote to local newspapers to assert: 'The class enemy rules here with great energy. Not long ago we had repeated instances of oil-fuelled explosions in the electric stove. Doesn't this show that class enemies are doing their evil work?'[282] Factory meetings to discuss breakdowns and accidents involved directors and party organizers telling the workforce that:

> There are facts indicating wrecking ... We have a group of wreckers. This was a defence order that burned: magnets wrapped in rubber and then in linen. Someone unwrapped the linen, cut the rubber, and then rewrapped the magnets. We have wheels with shavings in the bearings ... Now we need to look at certain people.[283]

The resulting conclusions of this investigation were predictable: 'The facts of the fire show that it was undoubtedly the act of class enemies ... The fire started on the eve of the trial of the Trotskyists. It was unquestionably a demonstration by enemies in response to the trial.'[284]

It thus seems that many ordinary workers did come to internalize the Stalinist justificatory narrative to some degree, and the regime's diatribes often meshed with workers' own frustrations with local communist bosses, generating grassroots enthusiasm for repression.[285] Sometimes, indeed, bottom-up support exceeded or ran counter to the intentions of the leadership. As Harris reports: 'When the regime had called for vigilance against the wrecking activities of the "bourgeois specialists" after the Shakhty Affair, and whipped up suspicion of "class aliens"

---

[279] Kotkin 1995, ch.5.
[280] Goldman 2011, 78–9.
[281] Cited in ibid. 47.
[282] Cited in ibid. 48–9.
[283] Cited in ibid. 85.
[284] Cited in ibid. 86.
[285] Ibid. 300–1. See also Doc.67, Viola et al. 2005, 246–8.

generally, [workers] responded with almost worrying enthusiasm, attacking all "bosses" instead of just the "bourgeois" ones.'[286]

So the Soviet public's relationship to ideology in the Great Terror was complex. Contemporaneous observers and modern historians tend to conclude that the majority 'seriously believed in the reasonableness and justice of what had happened'.[287] But as Kotkin elaborates:

> Elements of 'belief' and 'disbelief' appear to have coexisted within everyone, along with a certain residual resentment ... even in the category of 'true believers' it is necessary to think in terms of a shifting compromise, of rigidity and the search for slack, of daily negotiation and compromise within certain well-defined but not inviolable limits. Those limits were defined by recognition of the basic righteousness of socialism—always as contrasted with capitalism—a proposition that few people did or could have rejected ...'[288]

Critical, here, was the regime's domination of public discourse, and the epistemic dependence of most citizens on such discourse in the domain of security politics. Unless one was strongly predisposed to reject regime propaganda, then its claims, even if doubted, often provided the only available way of plausibly comprehending events. As one of Figes' interviewees later stated: 'it was impossible to doubt the existence of a dreadful conspiracy [against the Soviet State]. Any doubts on that score were inconceivable—there was no alternative. I am talking of the spirit of those times: either [those killed] were guilty or it was impossible to understand.'[289] The Soviet war poet Konstantin Simonov recalled his belief in the regime's assertions of a terrible conspiracy in the military, centred around the Soviet war hero Mikhail Tukhachevsky:

> Probably like the majority of people—at least the majority of the young people of my generation—I thought that the criminal case against Tukhachevsky and other military officers was probably correct ... Doubts simply never came to my mind because there was no alternative: *either they were guilty or what was happening was utterly incomprehensible.*[290]

Many recalled this feeling of being cognitively unable to formulate opposition— as a youth activist put it, 'I was not yet capable of appraising the situation' and assumed that what the 'government did was always for the best'.[291]

---
[286] Harris 2016, 88. See also Priestland 2007, 350–2.
[287] Cited in Thurston 1986, 227. See also ibid. 216; Davies 1997, 113.
[288] Kotkin 1995, 228.
[289] Cited in Figes 2002, 278.
[290] Cited in Brandenberger 2011, 181 (*my emphasis*).
[291] Cited in Thurston 1986, 224.

Again, then, the tendency to associate ideology with deep convictions, and assume that without such convictions ideology is rendered largely unimportant, profoundly obscures the popular dynamics of Stalinist repression. Stalinist ideology generated real, though often rather vernacularized, political support and legitimacy from significant sections of the population. For a much broader proportion, it provided a justificatory narrative for the violence that was sufficiently plausible and/or inescapable that outright opposition felt impossible. For both groups, and even for those more hostile to the regime, Stalinist ideology was also manifest in pervasive political practices, norms, and institutions, so that ordinary citizens faced immense pressures and incentives to actively participate in it. In all these roles, Stalinist ideology was vital in enabling and shaping the implementation of mass killing.

5.4 Conclusion

By comparison with the other cases discussed in this book, the conclusion that ideology mattered in Stalin's Soviet Union may be unsurprising. Quite how it mattered, however, has remained intensely debated. The political elite under Stalin appear to have been convinced by ideological justifications of mass killing. Decades later, Kaganovich, Molotov, and many participants in the Great Terror continued to justify the violence, with Molotov claiming:

> 1937 was necessary. Bear in mind that after the Revolution we slashed right and left; we scored victories, but tattered enemies of various stripes survived, and as we were faced by the growing danger of fascist aggression, they might have united. Thanks to 1937 there was no fifth column in our country during the war.[292]

The evidence suggests that this was how the Stalinist elite genuinely saw the violence at the time. Such subjective beliefs were crucial, since there were few 'objective' incentives for state violence on such a scale, which did little to advance Stalin's already hegemonic personal authority and undermined a range of regime interests.

In this, the totalitarian and utopian nature of Stalinism did matter. The regime's revolutionary ideals shaped perceptions of who the enemies were, while dogmatic beliefs that its ideology reflected scientific truth led to a murderous intolerance towards opposition.[293] But even in this case, a focus on revolutionary projects of social transformation or processes of radical moral disengagement mischaracterizes the central ideological drivers of the regime's repressive campaigns.

---

[292] Cited in Viola 2017, 10. See also Shearer 2009, 337.
[293] Kotkin 1995, 337.

Stalinist violence was not generally an attempt to directly implement a blueprint for communist utopia.[294] Even the relatively transformational collectivization campaign was in part a security strategy intended to fuel industrialization and military expansion. It involved mass killings of kulaks, moreover, not because of longstanding ambitions to create a classless society, but because the Stalinists became convinced that the kulak opposition represented a desperate security threat. There is also little evidence to support the thesis that Stalinist perpetrators 'knew' that the mass killing was wrong but were morally disengaged from their actions—although selective forms of moral disengagement probably played a role. Ultimately, Stalin's mass killings were a form of ultrahardline security politics: a 'prophylactic' strategy, as many historians have described it, aimed at protecting the Soviet political order against those who, within Stalinist ideology, were easily viewed as dangerous threats and criminals.[295]

That ideology was neither homogenous nor static, and its narratives of threat waxed and waned. For example, having practically waged war against the peasantry in the early 1930s,[296] the regime declared in the mid-1930s that former kulaks on collective farms were now, by definition, loyal Soviet citizens, before in the Great Terror again asserting that vast kulak conspiracies riddled the countryside.[297] Some scholars might fear that ideology appears so unstable, here, that it cannot really explain anything.[298] But while Stalinist conceptions of the peasantry were fluid, this reflected, not the primacy of non-ideological factors, but the way tensions in Stalinist ideology *interacted* with changing circumstances. In the early 1930s, actual peasant unrest was a product of the regime's ideological-strategic prioritization of collectivization. By the mid-1930s, this resistance had been defeated, there were reasons within Marxist-Leninist theory to conclude that peasants would now be loyal, and the Stalinists hoped to consolidate a more stable form of state governance consistent with the 'socialist' stage of Soviet history. At the same time, Stalinist elites still saw the world as intensely threatening, and tended to assume that threats were tied to internal conspiratorial networks. As the regime's ideologically informed diagnosis of the Kirov assassination and a worsening international situation radicalized, ideological suspicion of peasants again gained the upper hand.[299]

Stalin's personal ideological worldview bears a huge weight in all this, given the enormous preponderance of power concentrated in his hands. Yet the Great Terror took the form it did because Stalinist narratives and values extended beyond the

---

[294] Getty and Naumov 1999, xiii; Getty 2002, 135.
[295] McLoughlin and McDermott 2003, 2; Baberowski and Doering-Manteuffel 2009, 213; Viola 2017, 11, 172–3, & 176.
[296] See, in general, Viola et al. 2005.
[297] Shearer 2009, 331–2.
[298] See also Fearon and Laitin 2000, 863–4.
[299] Similar points could be made about the regime's paradoxical attitudes towards ethnic minorities; see Martin 1998.

leader to shape both private understandings and political norms and institutions across the Soviet Union. As Harris concludes:

> Stalin's actions and reactions were by no means unique. His inner circle shared his reaction to incoming intelligence. Many of those who opposed his rule also believed in the existence of foreign and domestic counter-revolutionary conspiracies, as did the population at large, fed as they were on a steady diet of campaigns of vigilance, news reports of conspiracies, and novels, plays, and films about wreckers and spies ... It made sense to people at all levels that there were wreckers and saboteurs at work ...[300]

Why had the Soviet Union become dominated by such an ideology? Certain broad societal factors were crucial—especially the legacy of autocracy and the experience of revolution and civil war. Yet it was not especially likely that such an ultrahardline ideology would dominate the territory of the Soviet Union in the 1930s. It emerged out of specific patterns of ideological activism by Lenin, Stalin, and other elites, shaped by their own personalities and the pre-existing Marxist frameworks they worked under, as well as by recurring circumstances of crisis. An extremely dangerous international environment also mattered, but explains relatively little about either dekulakization or the Terror, since they were guided by fictitious diagnoses of threat and worsened the real capacity of the USSR to protect itself.[301]

Especially notable, here, was the profound epistemic breakdown of the elite's security politics. Central aspects of security politics, such as the gathering and assessment of intelligence, became radically divorced from reality, not just in Stalin's mind, but throughout the party and state apparatus. That epistemic breakdown was itself rooted in the ideological preconvictions of the elites, as well as various institutional and private interests that became intertwined with ideology.[302] As I suggest is true in all mass killings, therefore, Stalinist repression was neither simply 'rational' nor inexplicable 'madness'.[303] It *appeared* like a rational security strategy from inside a certain radically hardline ideological worldview—one that had become significantly divorced from empirical realities, and untethered from meaningful constraints on violence.

---

[300] Harris 2016, 187.
[301] Kotkin 1995, 351.
[302] Getty and Naumov 1999, 259–61.
[303] Cf. Thurston 1986, 232.

6
Allied Area Bombing in World War II

6.1 Overview

By contrast with Stalinist repression, the bombing of German and Japanese civilians by Allied air forces in World War II is rarely portrayed as an 'ideological' case of violence against civilians. To some, it may seem contentious to even classify Allied area bombing as mass killing—it was certainly different, in several respects, from many more infamous cases. Area bombing killed civilians from vast distances, rather than in face-to-face massacres. It was part of a genuinely desperate war against murderous and hugely threatening Nazi and Japanese regimes. It was sustained only so long as those regimes refused to surrender. I nevertheless include Allied area bombing in my comparative examination of mass killings for two key reasons.

First, this remains one of the twentieth century's largest and well-evidenced instances of systematic violence against civilians. Many wrongly believe that Allied air forces in World War II restricted their bombing to military and industrial targets, only killing civilians as accidental 'collateral damage'. This view, rooted in Allied wartime propaganda, is wholly rejected by historians.[1] Allied bombers did conduct many crucial operations *besides* bombing civilians: destroying transport barges intended for an invasion of Britain, targeting oil production, directly attacking industrial sites, and engaging in tactical missions in support of ground or naval forces. But they also engaged in a campaign of 'area bombing', which, as officials explicitly declared, intentionally targeted civilians in broad urban areas as a strategy to undermine enemy morale and kill workers. Indeed, Arthur Harris, head of Britain's Bomber Command from February 1942, was contemptuous of 'sentimental' opposition to bombing civilians and frustrated by government efforts to mislead the public. He demanded that the real aims of bombing be made clear, writing:

> The aim of the Combined Bomber Offensive ... is the destruction of German cities, the killing of German workers and the disruption of civilized community life throughout Germany. It should be emphasized that the destruction of houses,

---

[1] Schaffer 1988, 69–70; Edgerton 1991/2013, 93–108; Connelly 2002; Davis Biddle 2002, 255–60; Overy 2005b; Bellamy 2008, 55–8 & 62, fns.34 & 40; Schaffer 2009; Bellamy 2012b, 141–5.

public utilities, transport and lives ... on an unprecedented scale ... are accepted and intended aims of bombing policy. They are not by-products of attempts to hit factories.[2]

By comparison with the Royal Air Force (RAF), the US Army Air Force (USAAF) showed some initial reluctance for such a campaign.[3] But from 10 October 1943, the USAAF joined the British in targeting civilian populations.[4] As historian Thomas Searle observes: 'the leaders of the USAAF knew exactly what they were doing, and civilian casualties were one of the explicit objectives of area incendiary bombing approved by both the USAAF and the Joint Chiefs of Staff'.[5] The bombing campaigns consequently killed 300,000–600,000 German civilians, 268,000–900,000 Japanese, and tens of thousands of civilians of other states.[6]

The second reason I examine Allied area bombing is that it nevertheless represents a hard case for many of the arguments of this book. To most observers, the leaders of the Allied powers in World War II will not look like radical ideologues. They are prime candidates for relatively 'rational' actors, who target civilians due to objective strategic incentives in a desperate war. Unsurprisingly, many prominent explanations of the bombing campaign are broadly rationalist. The leading comparative study of mass killing to include Allied area bombing, Alexander Downes' *Targeting Civilians in War*, presents the bombing of Japanese civilians as explained by desperate strategic circumstances: the need to win the war without catastrophic losses, and the infeasibility of more discriminate "precision" bombing for delivering such a victory.[7] Christopher Harmon endorses a similar perspective on the early RAF bombing of Germany, writing that: 'Precision bombing, the strategy which leaders had hoped would be the most moral (being discriminate) and the most militarily effective, was flatly unworkable. The Royal Air Force was compelled to choose another ... "area bombing"'.[8]

Such explanations rightly stress the roots of area bombing in the strategic goals and circumstances of the Allied war effort. But this chapter shows that hardline ideas about security and war, embedded in the broader liberal/conservative[9] ideological infrastructure of Britain and the United States, were equally crucial.

---

[2] Cited in Overy 2005b, 290.
[3] Overy 2012, 20–1. See also Davis Biddle 2002, 239–40 & 245.
[4] Schaffer 1988, 66–7. While other states were involved—Canada, Australia, and New Zealand in particular—I focus here on the United States and United Kingdom, given their dominant role.
[5] Searle 2002, 114:
[6] Schaffer 1988, 148; Hastings 2010, xi & 353.
[7] Downes 2008. See also Downes 2006.
[8] Harmon 1991, 17. Harmon nevertheless acknowledges significant flaws in this area bombing strategy.
[9] While it is typical in international relations and security studies to characterize Britain and the United States as 'liberal democracies', British and American elites in the 1930s and 1940s were at least as influenced by conservatism as liberalism. On liberal ideology and war see Cromartie 2015; Howard 1978; Owen 1997; Dillon and Reid 2009. On conservatism see Welsh 2003; Scanlon 2013.

Although there is much room for debate on the exact impacts of bombing and its contribution to Allied victory, the strategy of bombing civilian areas was at best strategically indeterminate. Even in the desperate circumstances of the war, its 'effectiveness' was highly questionable, and alternative strategies did exist. Nor, moreover, did policymakers come to see mass killing as justified only in response to desperate wartime circumstances. In truth, as Bellamy observes: 'A preference for area bombing was pervasive in the RAF and strategic circles well before the Second World War.'[10] Prior ideological doctrines were thus critical, and with different preconceptions, Allied policymakers could have pursued alternative strategies. Simultaneously, situationist explanations have a role to play in this case, in explaining lower-level involvement in the bombing by ordinary aircrews. But as in other cases, situationism tells us less about why the area bombing campaigns were launched in the first place. Only in an ideological context dominated by hardline ideas about modern war, one rooted in specific paths of ideological development in Britain and the United States, did strategic calculations and situational pressures come to promote a policy of directly bombing civilians.

6.2 The Roots of Allied Bombing Doctrines

As just suggested, the Allied justificatory narrative for bombing civilians did not emerge suddenly in response to crisis, but through a long-term process of doctrinal development over the decades prior to World War II. The new technology of air power was interpreted in particular hardline ways by British and American liberal and conservative elites, which were not simply a rational response to available evidence or strategic realities. Nor were they shared by similarly placed policymakers in other leading world powers.

Hardline ideas about bombing civilians were never hegemonic even in Britain and the United States. Indeed, the 1918–39 period represented the relative apogee of pacifist and peace movements in both countries,[11] and public opinion surveys before World War II generally displayed significant opposition to the bombing of civilians.[12] This period also saw efforts by the world's leading statesmen to restrict the future use of air power.[13] While the 1907 Hague Convention had already prohibited the bombardment of 'towns, villages, dwellings or buildings which are undefended',[14] legal experts attempted to go further in 1922–23 to formulate a code

---

[10] Bellamy 2008, 50. See also Overy 2014, 3.
[11] See also Ceadel 1980; McCarthy 2010; Overy 2013.
[12] Overy 2014, 32 & 34.
[13] Legro 1995, 94–100; Overy 2014, 29–31.
[14] Article 25, *Convention (IV) respecting the Laws and Customs of War on Land and its annex: Regulations concerning the Laws and Customs of War on Land*, The Hague, 18 October 1907, https://ihl-databases.icrc.org/applic/ihl/ihl.nsf/article.xsp?action=opendocument&documentid= d1c251b17210ce8dc12563cd0051678f (accessed: 26 August 2021).

on legitimate aerial warfare, while the World Disarmament Conference of 1932 established a commission to regulate aggressive bombing (though both efforts failed to produce agreement).[15] Roosevelt issued a prominent appeal in September 1939 to the states engaged in the war in Europe to avoid bombarding civilian populations, and even Churchill was, initially, sceptical that bombing civilians would deliver results.[16]

Yet notions that the area bombing of civilians was effective and legitimate had, by 1939, attained an ascendency in the circles that most mattered—the elites of the RAF and the USAAF—and had significant backing elsewhere in the political establishment. An influential hardline bombing faction had emerged, in other words, which would come to dominate policymaking during the war. While this is a matter of scholarly consensus for the RAF, several scholars suggest that the USAAF was, by contrast, hostile to area bombing, preferring 'precision' bombing of industrial and military targets.[17] I show that this is misleading. Leading US military figures generally assumed that *both* area bombing and precision bombing were legitimate and would play an important role in the future war effort.

Early Ideas of Air Power

The single biggest influence on Allied ideological conceptions of bombing in World War II was the attempt to learn from the experiences of air warfare in World War I. But that early learning experience did not, itself, occur off a blank slate free from ideological preconceptions. On the contrary, the advent of air power had been anticipated for decades, most clearly in an expanding lineage of futurist literature in the nineteenth and early twentieth centuries, including Jules Verne's *Clipper in the Clouds* (1886), William Delisle Hay's *Three Hundred Years Hence* (1881), S.W. Odell's *The Last War; Or, Triumph of the English Tongue* (1889), Roy Norton's *The Vanishing Fleets* (1907), and H.G. Wells' *The War in the Air* (1908).[18] Such stories portrayed air technology as vastly powerful, envisaging a future in which huge air fleets would appear over enemy cities during war and rain death upon their citizens, to devastating effect.[19] Parallel speculation in military and political circles emphasized air power's tremendous likely impact on enemy morale given the presumed weakness of civilian populations. It was assumed that when air power was fully realized, as a British officer predicted in 1893, 'the arrival of the aerial fleet over the enemy capital will probably conclude the campaign'.[20] So awesome

---

[15] Indeed, the scale and impacts of concrete pacifist or peace activism on military policy were generally limited; see Edgerton 1991/2013, 60–9.
[16] Harmon 1991, 4; Davis Biddle 2002, 79; Hastings 2010, 46–7.
[17] See, in particular, Downes 2008, 140–1.
[18] Davis Biddle 2002, 12–13; Hastings 2010, 32.
[19] Sherry 1987, 7.
[20] Ibid. 4.

was the anticipated power of the aircraft that many optimistically predicted that it would end war itself.[21] No less an authority than Orville Wright argued that 'the aeroplane has made war so terrible that I do not believe any country will again care to start a war'.[22] One American newspaper more poetically anticipated that 'behind the images of carnage shines the light of universal peace'.[23]

With the outbreak of World War I, more systematic policies were formulated under the influence of these high expectations for air power. Germany, France, Italy, Britain, and the United States all mobilized air forces and tried to engage in attacks on the presumed-to-be-vulnerable civilian 'rear' of their enemies. British policies were heavily influenced by the Smuts Report of 1917, which echoed the earlier futurist literature by claiming that: 'the day may not be far off when aerial operations with their devastation of enemy lands and destruction of industrial and populous centres on a vast scale may become the principle operations of war'.[24] Assertions that air power could end the war quickly and thereby save thousands of lives were increasingly articulated, and were supported by vociferous public demands for reprisal attacks on German civilians following the Zeppelin airship raids on Britain.[25]

By and large, however, air attacks on civilians in World War I were highly ineffectual. Reports on the German bombing of England noted public responses of anger rather than panic, while the British, American, and French bombing of German industry was assessed to have caused damage equivalent to less than a tenth of 1% of German war expenditure.[26] Yet this did not discourage confident claims that air power had been critical as a means of shattering enemy morale. Official British bombing surveys conducted between 1918 and 1920 dismissed available evidence that the effect of bombing was minimal, concluding that: 'Had the war continued a few months longer, a more or less total breakdown of labour at several [factories] might have been confidently expected.'[27] An equivalent American survey was more cautious, but also concluded that: 'It is certain that air raids had a tremendous effect on the morale of the entire people.'[28] A British parliamentary paper from 1919 likewise asserted: 'The effect, both morally and materially, of the raids on German territory carried out during the summer of 1918 can hardly be overestimated.'[29] Such surveys encouraged an ideological assumption of bombing's effectiveness largely disconnected from hard evidence, and this assumption proceeded to guide 20 years of interwar intellectual developments in military,

---

[21] Ibid. 5–8; Davis Biddle 2002, 12–13.
[22] Sherry 1987, 10.
[23] Ibid. 5.
[24] Hastings 2010, 33.
[25] Davis Biddle 2002, 30–1.
[26] Ibid. 23 & 44.
[27] Cited in ibid. 59.
[28] Cited in ibid. 66.
[29] Cited in ibid. 61.

political, and public circles. Area bombing was expected to be capable of winning wars without—and this was perhaps its most important attraction—the need for the catastrophic slaughter of trench warfare.[30]

The Italian theorist Giulio Douhet is often accorded pre-eminent influence in the development of such ideas. His major work, *The Command of the Air*, was widely read in both Britain and the United States in the 1930s and offers the most extensive elaboration of the interwar picture of future air power. It contained claims that would become central elements of the Allied justificatory narrative for bombing civilians, three of which deserve emphasis. First, Douhet engaged in the threat construction of civilians as no less dangerous than soldiers, since in modern war 'the soldier carrying his gun, the woman loading shells in a factory, the farmer growing wheat, the scientist experimenting in his laboratory' all played an equally important part in the war effort, and thus 'any distinction between belligerents is no longer admissible today either in fact or theory'.[31] Second, Douhet offered one of the first futurized defences of area bombing, arguing that civilian casualties would be easily outweighed by the ultimate benefits he was sure would accrue: the shortening of war and avoidance of bloody land battles. In future wars, Douhet predicted, bombers 'could spread terror through the nation and quickly break down [its] material and moral resistance'[32] and under such conditions 'how could a country go on living and working under this constant threat, oppressed by the nightmare of imminent destruction and death?'[33] For this reason, Douhet also presented the bombing of civilians as more virtuous than alternative strategies. 'The more rapid and terrifying the arms are', he wrote, 'the faster they will reach vital centres and the more deeply they will affect moral resistance. Hence the more civilized war will become.'[34] Finally, Douhet depicted the bombing of civilians as simply an inevitable consequence of abstract technological forces and strategic realities rather than human choices. 'Owing to unavoidable necessity …', he argued, bombing campaigns 'would be characterised by hideous atrocities … it is as sure as fate that, as long as such a direct method of attack exists, it will be used.'[35] As Ronald Schaffer summarizes, Douhet:

> presented himself not as one who consciously willed the change, but as one who had come to understand that the military and technological evolution, particularly the evolution of strategic air warfare, had made the barrier [between civilian and combatant] obsolete. Douhet urged his readers to confront the brutal facts of the future war, to view them 'without false delicacy and sentimentalism.'[36]

---

[30] Sherry 1987, 24 & 27; Davis Biddle 2002, 7; Shaw 2003, 172; Bellamy 2008, 49; Bellamy 2012b, 133. See also Shaw 2002, 348–9.
[31] Douhet 1921/1983, 196
[32] Ibid. 57.
[33] Ibid. 22.
[34] Ibid. 196.
[35] Ibid. 282.
[36] Schaffer 1988, 23.

But Douhet's arguments were mediated (or sometimes arrived at independently) by thinkers in the emerging British and American air forces, which would come to dominate Allied uses of air power in World War II.

The Development of Air Doctrine in Interwar Britain

Hugh Trenchard had been the head of Britain's 'Independent Force' of front-line bombers during World War I, and became the country's first post-war Chief of the Air Staff. No other figure played such a central role in establishing bombing campaigns against civilians as part of British security doctrine. 'His passionate belief in the potential of a bomber offensive against an enemy nation', Max Hastings writes, 'was to dominate the Royal Air Force for more than twenty years.'[37] In December 1918, three years prior to Douhet's *Command of the Air*, Trenchard composed his final dispatch on his World War I command, which was reprinted in the *London Gazette* in January 1919. In it, he famously asserted that 'at present the moral[38] effect of bombing stands undoubtedly to the material effect in a proportion of 20 to 1', a formula that was to become axiomatic in political and military circles despite having, as Tami Davis Biddle notes, 'no scientific or mathematical basis.'[39] This was combined with a conviction that defensive air forces would never be able to stop a determined bomber offensive. Bombing civilians was thus the only way to achieve a quick victory, Trenchard claimed, and would represent a net saving of lives in the long run. As he suggested in a 1923 address: 'if we could bomb the enemy more intensively and more continuously than he could bomb us, the result might be an early offer of peace.'[40]

Like the post-war bombing studies on which they drew, such claims were in large part the product of ideologically biased analyses, and a resistance to acknowledge contrary evidence in favour of prevailing ideological assumptions.[41] Trenchard and his followers failed to appreciate the potential power of organized fighter squadrons to inflict prohibitive losses on bombers,[42] while hugely exaggerating the impact of bombing on civilian morale—often thanks to the influence of classist, sexist, and racist attitudes which encouraged assumptions that the ordinary urban civilian population would immediately panic when bombed.[43] Evidence of the actual steadfastness and resistance of civilian populations under attack, or their limited ability to pressure their own governments to surrender during wartime, was largely ignored.[44]

---

[37] Hastings 2010, 34.
[38] In the documents of the period, the concept of 'morale' was commonly spelt 'moral'. See also Ussishkin 2017.
[39] Davis Biddle 2002, 48.
[40] Ibid. 85.
[41] Ibid. 69–81.
[42] Sherry 1987, 27; Davis Biddle 2002, 27, 89–90, & 95; Hastings 2010, 35, 38, 40, & 51.
[43] Sherry 1987, 26; Davis Biddle 2002, 15–16, 64, & 109; Slessor 2009, 65; Overy 2014, 26.
[44] Davis Biddle 2002, 31 & 78–9.

While Trenchardian views were not universally endorsed, they saturated formal training, RAF public statements, and internal memoranda in the 1920s and 1930s.[45] The 1935 RAF War Manual was explicit that bombing civilian targets to undermine enemy morale was the principal offensive function of the air force:

> the most important and far-reaching effect of air bombardment is its moral effect ... The moral effect of bombing is always severe and usually cumulative, proportionately greater effect being obtained by continuous bombing especially of the enemy's vital centres.[46]

Again, little evidence existed to support these assertions. Instead, 'RAF leaders', as Richard Overy observes, 'continued to rely on unverifiable assumptions about the social fragility of the enemy'.[47] But such expectations were endorsed by a substantial number of leading British strategists, including the influential Basil Liddell Hart and J.F.C. Fuller, sustained by deep-seated ideological convictions about the weak character of civilian populations.[48] Envisaging a full-scale aerial attack on Britain, Fuller argued:

> [P]icture, if you can, what the result will be: London for several days will be one vast raving Bedlam ... the city will be in pandemonium ... Then will the enemy dictate his terms, which will be grasped at like a straw by a drowning man. Thus may a war be won in forty-eight hours and the losses of the winning side may actually be nil![49]

The limited available data that could be garnered on aerial attacks—from air power's use in the British colonies, the Spanish Civil War, and the Italian invasion of Abyssinia—did not tend to support this view, but were again interpreted in ways that supported received wisdom.[50] Bomber advocates, as Davis Biddle summarizes, 'found a range of ways to see what they wanted to see, and to avoid seeing what contradicted their views'.[51]

Such thinking was not limited to the military, being propagated by a powerful air lobby with roots in conservative sections of the press and public.[52] A pamphlet by the 'Hands Off Britain Air Defence League' distributed in 1933 called on the government to 'create a new winged army of long range British bombers to

---

[45] Ibid. 79. See also ibid. 92–3.
[46] Cited in Overy 2014, 48.
[47] Ibid.
[48] Sherry 1987, 23–9; Davis Biddle 2002, 24–5 & 104–5; Hastings 2010, 37.
[49] Cited in Sherry 1987, 24.
[50] Davis Biddle 2002, 82–3, 90–2, & 115–20; Tanaka 2009b. On RAF uses of air power within the British Empire see: Overy 2014, 48–9.
[51] Davis Biddle 2002, 91.
[52] Ibid. 82–4.

smash the foreign hornets in their nests!'[53] In a famous statement to the House of Commons on 10 November 1932, Conservative Prime Minister Stanley Baldwin espoused a classic statement of hardline Trenchardian theory:

> I think it is well for the man in the street to realize that there is no power on earth that can protect him from being bombed. Whatever people may tell him, the bomber will always get through. The only defence is in offence, which means that you have to kill more women and children more quickly than the enemy if you want to save yourselves.[54]

Many, Baldwin included, were hardly enthusiastic about this vision of future war. As one member of Britain's House of Lords observed: 'It is a poor consolation that the only answer we can find to the destruction of half of civilization is that we should be able to destroy the other half.'[55] Yet British politicians and civil servants appeared to accept that this vision was accurate even if regrettable. 'For years before the outbreak of war', Overy concludes, 'the common assumption among those who planned civil defence … was the unavoidability of war waged against civilians.'[56]

British hardline views of air power do not appear, then, to have simply been a rational response to the real security environment Britain faced. Not only did such beliefs often fail to reflect the available evidence, but they also actively undermined British security in certain key respects. By intensifying fears of war and encouraging exaggerated perceptions of German air power, they increased pressures for appeasement.[57] They also had tragic consequences at the outbreak of World War II. Underinvestment in British fighter forces, justified by beliefs in the supreme power of the bomber, made defeat at the hands of the Luftwaffe a real possibility in 1940, while British bombers proved unable to inflict meaningful damage on Germany.[58] As Davis Biddle summarizes: 'Bomber Command had only very limited effectiveness in any of the roles it attempted to undertake … the gap between rhetoric and reality proved to be nothing less than an abyss.'[59] Yet even in 1939, the retired Trenchard criticized efforts to build up British fighter defences—which ultimately prevented the German invasion of Britain—as a needless diversion from bombers.[60] Writing to congratulate Sir Charles Portal on his appointment as Commander-in-Chief of Bomber Command in April 1940, the retired Trenchard noted that he was 'sorry that you could not [yet use British

---

[53] Hastings 2010, 39.
[54] HC Deb (10 November 1932) vol.270, col. 632. https://hansard.parliament.uk/commons/1932-11-10/debates/4bccc63e-763c-4b01-ac54-9708217b3c57/internationalaffairs (accessed 10 August 2021).
[55] Cited in Overy 2014, 28.
[56] Ibid. 35.
[57] Sherry 1987, 76-7; Davis Biddle 2002, 2 & 126; Johnson 2004, 92; Overy 2014, 29.
[58] Davis Biddle 2002, 88-94; Hastings 2010, 41-3, 50-3, & 58-9.
[59] Davis Biddle 2002, 183.
[60] Hastings 2010, 51.

air power] where I and others think it probably would have ended the war by now'.[61] Such an extraordinary detachment from reality—as though a few robust bombing runs on German cities in 1940 would have halted the Nazi war machine—was not, by this stage, widely shared. But Trenchard's faith in the area bombing of civilians had been sedimented into British security doctrine, and would continue to dominate during World War II.

The Development of Air Doctrine in the Interwar United States

The development of ideas about air power in the United States during the interwar years followed a less linear course towards bombing civilians than in Britain. There were a variety of reasons for this, but in particular the lack of US air force independence (the USAAF remained a section of the US Army) and notable resistance from influential policymakers presented obstacles to ideological radicalization towards the killing of civilians.[62] Indeed, American politicians, military officers, and ordinary citizens maintained ambivalent attitudes towards area bombing up to the final years of World War II. Early statements of US air power doctrines did endorse Trenchardian assumptions about the importance of the morale effect and reproduced his 20:1 ratio with the same absence of evidence.[63] But at the same time they expressed caution. A 1936 memorandum by the War Plans division of the War Department (still dominated by the army and navy) noted that: 'The effectiveness of aviation to break the will of a well-organized nation is claimed by some, but this has never been demonstrated.'[64] Throughout the 1920s and 1930s, many who did discuss air power were sceptical of its use against civilian populations.[65]

Nevertheless, as in Britain, core ideas on the power and legitimacy of bombing civilians were increasingly widespread from the mid-1920s onwards, especially amongst Air Corps leaders. In this respect, and especially in the eyes of the American public, the flamboyant Billy Mitchell was a major ideological influence as a campaigner for an independent Air Service and a general publicist for new visions of air power. Mitchell 'assaulted the public with speeches, articles, books, and endless appearances before congressional committees'[66] and organized a famous demonstration of air attacks on (stationary) battleships. He echoed Douhet's faith in the power of aerial attacks on civilian morale to bring wars to a swift and more humane conclusion, and also, like Douhet, saw civilians as contributing to war as directly as soldiers. In a 1921 book he argued that:

[61] Ibid. 59.
[62] Davis Biddle 2002, 128–9 & 163.
[63] Ibid. 132 & 134.
[64] Ibid. 128.
[65] Sherry 1987, 62–75; Davis Biddle 2002, 150–3 & 171.
[66] Davis Biddle 2002, 136.

Warfare today between first-class powers includes all of the people of the nations so engaged—men, women and children ... We must expect, therefore, in case of war, to have the enemy attempt to destroy any or all of our combatant or industrial forces—his attacks being entirely controlled by the dictates of strategy, and the means of bringing the war to a quick conclusion ... The personnel of entire cities—men, women and children—can be destroyed by gas attacks from the air.[67]

As with the statements of Douhet and Trenchard, this passage largely matches the eventual justificatory narrative of World War II area bombing. Attacks on civilians were presented as an *inevitable* consequence of 'the dictates of strategy' rather than policymakers' choices.[68] Their deaths were justified by promises that this would shorten war in the future, and by threat construction to convert civilians into a target that seemed actively engaged in 'warfare'. Mitchell would go on to argue in 1930 that bombing would be 'really much more humane than the present methods of blowing up people to bits by cannon projectiles or butchering them with bayonets'.[69]

Mitchell's deviation from the official stance of the Air Corps was unacceptable to senior army officers, and he ultimately left the military. But his ideas reflected growing ambitions amongst US air theorists. In 1924, the Harvard chemistry professor Norris Hall defended the use of gas and air strikes against civilian populations, on the grounds that 'however vast the destruction, however "inhuman" the methods used, however appalling the sacrifice of life, such a solution might well be preferable to its alternative, a long war ... of starvation, exhaustion, lying, brutalization, and madness'.[70] A 1926 Air Corps training text advocated air attack 'as a method of imposing will by terrorizing the whole population', though also spoke of 'conserving life and property to the greatest possible extent'.[71] Air officers placed increasing emphasis on 'morale effect' and the targeting of enemy 'will', and confidently predicted the power of the bomber to end wars quickly.[72] As one air force doctrinal statement explained:

Where is that will to resist centred? How is it expressed? It is centered in the mass of the people. It is expressed through political government. The will to resist, the will to fight, the will to progress, are all ultimately centered in the mass of the people—the civil mass—the people in the street ... Hence, the ultimate aim of all military operations is to destroy the will of those people at home ... The Air Force can strike at once at its ultimate objective; the national will to resist.[73]

---

[67] Ibid. 345, fns.30 & 32.
[68] See also ibid. 147.
[69] Sherry 1987, 30. See also Davis Biddle 2002, 136–7.
[70] Hall, Jr., and Hudson 1925, 37.
[71] Cited in Schaffer 1988, 27.
[72] Davis Biddle 2002, 139–42 & 157–9.
[73] Cited in Overy 2014, 53.

Like their British counterparts, such arguments often involved a radical underestimation of the power of organized fighter defences. In an oft-repeated statement, the Air Corps Training School instructor Kenneth Walker stated: 'a well-organized, well-planned and well-flown air force attack will constitute an offensive that cannot be stopped'.[74] An umpire of Air Corps training exercises in 1931 concluded simply that 'it is impossible for fighters to intercept bombers'.[75]

In addition to these strategic assessments, air power was frequently presented with a glamorous sheen that connected it to broader ideological visions and values of American society.[76] The Air Corps played a leading role in propagating its doctrines through organized efforts to create an 'airminded' US culture, with leading officers producing books for popular consumption, and several children's adventure series.[77] Michael Sherry observes that, whereas traditional army and naval forces were associated with militarism and the attritional slaughter of World War I, 'Americans saw [military aviation] as a way to uphold New Era virtues of economy, efficiency and technological innovation'.[78] The dramatic and brilliantly publicized strides made by civil aviation (especially the achievements of Charles Lindbergh and Amelia Earhart) further encouraged positive attitudes towards air power.[79]

Interwar air power doctrines were not, as such, simply military theories unrelated to broader ideological worldviews. Allied understandings of air warfare were rooted in distinctive ideological conceptions of strategy, drawing on characteristically democratic assumptions that, since governments ultimately depended on their civilian population for authority, pain inflicted on civilians would translate into pressure on governments. They also reflected broader liberal/conservative conceptions of war, technology, and militarism. As Davis Biddle argues, Britons and Americans 'reinforced national self-identities that celebrated mastery of science and technology',[80] and bombing was attractive in large part because it promised 'efficiency' and 'precision' and a war conducted at the frontiers of modernity.[81] Sherry has influentially described this form of 'technological fanaticism' as definitive of American attitudes towards air power since the late nineteenth century. As he writes:

> People did not look at the airplane and then deduce its impact on human affairs. Rather they took general propositions about the benefits of technology and applied them as confidently to the imagined airplane as they did to other weapons and inventions—or more confidently, since the airplane was endowed with more virtues.[82]

---

[74] Cited in Davis Biddle 2002, 142.
[75] Cited in ibid. 168.
[76] Young 2009, 156.
[77] Davis Biddle 2002, 149.
[78] Sherry 1987, 35.
[79] See also Edgerton 1991/2013, xxi–xxii.
[80] Davis Biddle 2002, 4.
[81] Ibid. 161.
[82] Sherry 1987, 5.

As David Edgerton has argued, Allied enthusiasm for bombing was therefore rooted in a kind of 'liberal militarism', in which bombing and air warfare appeared to offer attractive, efficient, and even heroic alternatives to the futile, barbarous kind of militarism associated with large-scale land warfare and the trenches of World War I.[83] The Allies' vision of area bombing had little appeal, by contrast, to fascist regimes which saw in large-scale land warfare a heroic enterprise necessary for keeping the nation socially and biologically healthy, as I show in section 6.3.[84] Liberal/conservative political ideology was neither necessary nor sufficient for justifications of area bombing to flourish, but there was a strong elective affinity between them.

6.3 Ideology and Area Bombing in World War II

Policymakers and the Ideology of Area Bombing

At the outbreak of war in Europe, Germany and Britain defied enthusiastic predictions about air power by showing initial restraint in bombing civilians, at the insistence of civilian political leaders.[85] But with the defeat of France, the contraction of British military options, and the growing realization that bombing specific military targets was costly and having little impact, area bombing took centre stage.[86] After the Luftwaffe accidentally bombed London on the night of 24 August 1940, Churchill ordered a reprisal attack on Berlin, and both sides thereafter engaged in increasing attacks on 'civilian morale' and industry.[87] When the United States entered the war it initially attempted daylight 'precision' bombing, rather than night-time area bombing.[88] But as the challenges of such an approach became clear, it gradually joined the RAF in night-time area attacks on civilian populations.[89]

Wartime desperation and the failure of 'precision' bombing were, as such, key factors in the choice by Allied policymakers to shift to area bombing. Yet there are four problems with a purely rationalist explanation of Allied area bombing, which depicts it as simply a natural response to such strategic circumstances.

---

[83] Edgerton 1991/2013, xxxi–xxxiii & 66–92. Note that Edgerton associates this ideological support for air power principally with the political right—as is generally true in the British context, 'liberal' does not simply denote 'left-wing'.

[84] See, however, the links between British fascism and air power in ibid. 75–9 & 91–2.

[85] Two principal factors explain such restraint: first, the continued presence of opponents to area bombing within senior military circles, and second, (misplaced) concerns that targeting civilians would alienate neutral countries and provoke a backlash from the British public. See Harmon 1991, 13; Legro 1995; Bellamy 2008, 46 & 50–1; Schaffer 2009, 31–4; Hastings 2010, 3–17, 54–7, 73, 95–6, & 216; Farr 2012, 136.

[86] Davis Biddle 2002, 176–7; Hastings 2010, 147–70.

[87] Davis Biddle 2002, 188.

[88] Ibid. 209.

[89] Ibid. 228–9 & 243–5.

First, as shown in section 6.2, many British and American policymakers were already convinced, before the start of the war, that area bombing of civilians was morally and strategically justifiable, and not merely as a response to desperate circumstances, but as the best way to fight modern wars *in general*. This was true even in the USAAF. Downes, indeed, cites Thomas Searle's observation that: 'The orders for the 9 March 1945 [firebombing of Tokyo] reflected *the longstanding interest of the Air Staff* ... in using urban incendiary raids to cut Japanese industrial production by (among other things) killing Japanese civilians.'[90] While Downes suggests that this reflects strategic decisions made in 1943 and 1944 and was therefore still a response to desperation, Searle's actual point is that such an interest *predated* the failure of precision bombing, dating back to at least 1941 and being rooted in ideas from the interwar period.[91] Official USAAF doctrine at the start of World War II may have placed more explicit emphasis on 'precision' bombing, in part for political reasons. But plenty of influential USAAF leaders assumed that area bombing was an entirely justified and, indeed, unavoidable tool of warfare.[92]

Second, while the ineffectiveness of precision bombing increased the attraction of area bombing,[93] it did not render it a necessary or even particularly logical alternative.[94] Allied options were certainly constrained. But with the possible exception of 1940–41, various options existed between precision bombing on the one hand and the massive area bombing of civilian populations on the other. These included the effective aerial mining and bombing campaigns against German and Japanese shipping, the redeployment of air forces to the Atlantic, Middle Eastern, and Asian theatres (where they were desperately needed), the use of bombers to attack Nazi Germany's vulnerable oil industry or to support the blockade of Japan, the reassignment of more bombers to tactical roles,[95] or merely a tighter focus on bombing specific high-priority industrial zones and armaments factories rather than seeking to obliterate whole cities or residential areas.[96] Indeed, Hitler's Minister of Armaments, Albert Speer, later declared that Germany may have been forced to surrender as early as 1943 if the bombing campaign had been more focused on German armaments.[97] Such options would still have involved significant loss of civilian life, but on nothing like the scale of intentional area bombing. Moreover, alternatives expanded over time as the Allies' strategic position and technological capacities improved.[98] Yet, contra rationalist explanations,

---

[90] Downes 2008, 126 (*my emphasis*).
[91] Searle 2002, 115–28.
[92] See also Schaffer 1988, ch.2.
[93] Downes 2008, 122–9.
[94] Christopher Harmon similarly suggests that 'the bomber offensive was all the prime minister had with which to hit Germany back' (see Harmon 1991, 15), but while this may be true of *some sort of bomber offensive*, it is not true of area bombing civilians specifically.
[95] While strategic bombing advocates insisted that their bombers could not be used to support armies in a tactical role, on occasions when they were overruled—such as during the Allied landings in France—they found that their crews often performed effectively in such operations; see Davis Biddle 2002, 234–7.
[96] Hastings 2010, 82, 149–50, & 434.
[97] Smith 1977, 182.
[98] See also Harmon 1991, 33, fn.58.

the reduced 'need' to target civilians did not produce a shift away from area bombing. Targeting civilians remained a dominant strategy in Europe and Asia right up to the end of the war, with some of the heaviest operations occurring in its last few months.[99]

Third, by diverting resources away from other options, the area bombing of German civilians risked being outright counterproductive, since its strategic gains were limited and its costs in materiel and aircrews' lives severe. Both the official British and American bombing surveys after the war were, indeed, highly critical on these grounds—especially for the European theatre.[100] As the post-war British Bombing Survey Unit reported: 'In so far as the offensive against German towns was designed to break the morale of the German civilian population, it clearly failed. Far from lowering essential war production, it also failed to stem a remarkable increase in the output of armaments.'[101] Overy likewise concludes that 'the claims of Bomber Command were clearly false'.[102] In Japan the line between the bombing of industry and the targeting of civilians was more blurred, since firebombing strategies aimed to wipe out diffuse small industrial capacity across urban areas.[103] Such strategies probably did contribute to the Japanese government's surrender, but even here, several US studies suggest that the Japanese economy was largely crippled by naval blockade and attacks on Japanese shipping *prior* to the onset of heavy area bombing.[104] Area bombing was strategically indeterminate, and different ideological preconceptions could easily have encouraged Allied warmakers to pursue alternative strategies.

Most notably, despite the persistent post-war myth to the contrary,[105] the atomic bombs did not play a decisive role in ending the pacific war, nor save the hundreds of thousands of lives their proponents constantly invoked to justify them. The US government's own Strategic Bombing Survey concluded:

> The Hiroshima and Nagasaki atomic bombs did not defeat Japan, nor, by the testimony of enemy leaders who ended the war, did they persuade Japan to accept unconditional surrender ... certainly prior to 31 December 1945 and in all probability prior to November 1945 Japan would have surrendered, even if the atomic bombs had not been dropped, even if Russia had not entered the war, and even if no invasion had been planned or contemplated.[106]

---

[99] Harmon affirms this, although dates the declining justifiability of area bombing later (towards the end of 1944) than I would; see ibid. 23–4.
[100] Although the survey findings are not uncontested, see Gentile 1997.
[101] Davis Biddle 2002, 281. See also Hastings 2010, 288.
[102] Overy 2005a, 112. See also Hastings 2010, 324.
[103] Selden 2009, 83.
[104] Gentile 1997, 68–72 & 77–8; Davis Biddle 2002, 278–9.
[105] Messer 2005, 308–9.
[106] Cited in Messer 1995, 14.

America's most senior military officer in World War II, Admiral William Leahy, likewise later affirmed: 'the use of this barbarous weapon at Hiroshima and Nagasaki was of no material assistance in our war against Japan. The Japanese were already defeated and ready to surrender.'[107] Historians generally support this evaluation.[108] As J. Samuel Walker summarizes: 'The consensus among scholars is that the bomb was not needed to avoid an invasion of Japan and to end the war within a relatively short time. It is clear that alternatives to the bomb existed and that Truman and his advisers knew it.'[109]

Finally, the role of ideology in shaping interpretations of strategic circumstances is suggested by the fact that 'in other air forces', as Overy observes, 'different cultures prevailed and produced contrasting strategic choices'.[110] Japan, despite a similar maritime geostrategic orientation to Britain and the United States, never placed area bombing at the centre of its conception of warfare (although weaknesses in Japan's industrial base also played a role here). Contrasting ideological interpretations of modern war are especially clear in the German case. If clear strategic incentives for area bombing existed in World War II, they should have been felt by Germany in its confrontation with Britain. Germany came to recognize the war as a total and existential confrontation, and, like the Allies in the case of Japan, faced in Britain an island nation that presented formidable and costly obstacles to a land invasion. Yet the German government was never enamoured by the idea of area bombing civilians, being guided by different ideological assumptions that were institutionalized within German strategic doctrine, procurement choices, and military training. Whereas British and American theorists reasoned that war involved a clash of wills ultimately rooted in the civilian population, the German military concluded the exact opposite. As its 1935 air doctrine statement emphasized: 'the will of the nation finds its greatest embodiment in its armed forces. Thus, the enemy armed forces are therefore the primary objectives in war.'[111] Where Germany did conduct air operations against civilians—such as in London, Rotterdam, and Warsaw—these were generally ad hoc reprisals or tactical operations against urban defences. Only in certain operations during the Blitz did Germany resort, comparatively briefly, to area bombing.

Of course, the Nazi military was extraordinarily abusive to civilian populations more generally, and this highlights how the key ideological contrast here was not

---

[107] Cited in Goldhagen 2010, 201.
[108] Although see Gentile 1997, 55–8.
[109] Walker 1990, 110. For summaries of the evidence see Alperovitz, Messer, and Bernstein 1991/1992, 205–13; Messer 1995. See also Barton Bernstein's divergence from Alperovitz and Messer in the same correspondence article: Alperovitz, Messer, and Bernstein 1991/1992, 214–21. Even if Bernstein's view was accepted entirely, however, it remains largely consistent with my argument here—since Bernstein mutually emphasizes 'the evolution of attitudes, the growth of national hatreds, and the practice of virtually total war' (ibid. 220).
[110] Overy 2014, 54.
[111] Cited in ibid. 43.

about radical moral values or ideological goals. The crucial difference lay in the specifics of British, American, and German security doctrines. German policymakers, including Hitler, Göring (as head of the Luftwaffe), and other senior air force officers, never presented the area bombing of civilians as an important element of their vision of modern war. This led to strategic conclusions that contrasted radically with those of British and American policymakers:

> A July 1939 *Luftwaffe* Intelligence report pointed out that the British were known for their toughness and it was not certain that an air campaign alone would bring their defeat. Already in May of 1939, Hitler had told his service chiefs that Britain could not be defeated by air attack alone. In July 1940, Reich Marshall Hermann Göring ... challenged the assumption that an independent *Luftwaffe* campaign would destroy the British will to fight when no one believed that the Germans would stop fighting if Berlin were bombed ... By the beginning of the war ... the *Luftwaffe* had become increasingly dominated by a philosophy of airpower attached to land operations ... strategic bombing was rarely associated with the type of morale bombing thinking that emerged in the Royal Air Force.[112]

This partly reflected the brand of militarism that dominated Nazi ideology.[113] The Nazis lacked the Allies' aspirations for cost-efficient warfare at a distance, and Hitler was an 'army man' who glorified land warfare and had little interest in air power theories.[114] The ideological conception of modern war and air power that was so central to Allied area bombing thus never gained a serious foothold amongst German elites.

Allied area bombing was, then, not a straightforward consequence of wartime realities and clear strategic needs, but resulted from the *interaction* between strategic context and policymakers' distinctive ideological conceptions of war and security. Allied elites were not ideologically homogenous, but the most important figures embraced a vision of warfare in which bombing civilians was effective, practically unavoidable, and morally appropriate. As Churchill stated, in a famous memorandum of 8 July 1940 to the Minister for War Production, Lord Beaverbrook:

> When I look around to see how we can win the war I see that there is only one sure path. We have no Continental Army which can defeat the German military power. The blockade is broken and Hitler has Asia and probably Africa to draw from. Should he be repulsed here or not try an invasion he will recoil eastward and we have nothing to stop him. But there is one thing that will bring him back and bring him down, and that is an absolutely devastating, exterminating attack by

---
[112] Cited in Legro 1995, 109–10.
[113] Ibid. 110–18.
[114] Ibid. 115–18.

very heavy bombers from this country upon the Nazi homeland. We must be able to overwhelm him by this means, without which I do not see a way through.[115]

This was the essential ideological understanding of area bombing that guided Allied policy, revolving around two key elements.

First, Allied policymakers assumed that the targeting of civilians was a *strategically necessary* and *valorous* course which was highly likely to deliver *huge future benefits*. The known destruction of civilian life in the present was justified because bombing would either decisively end the war on its own, or at least massively shorten its duration.[116] Such claims flowed directly from interwar air power theories, and were expressed in hugely exaggerated expectations about the effects of area bombing. As head of Bomber Command in January 1941, Sir Richard Peirse justified focusing bombers against urban areas on the grounds that 'we are likely to precipitate a crisis in Germany's war economy *this year*'.[117] Sir Charles Portal wrote to Churchill in September 1941 suggesting that if a 4,000 bomber force could be mobilized 'it could break Germany in six months'.[118] In March 1942, Churchill's principal scientific advisor, Professor F.A. Lindemann, authored an infamous memorandum to the Prime Minister arguing for the 'de-housing' of millions of German workers by massive bombing, on the grounds that 'there seems little doubt that this would break the spirit of the [German] people', specifically by the middle of 1943.[119] In November 1943 the Assistant Chief of the Air Staff (Intelligence), F.F. Inglis, claimed to Churchill that 'we are convinced that Bomber Command's attacks are doing more towards shortening the war than any other offensive including the Russians'.[120]

But Harris was the most ardent and erroneous predictor. In February 1943 he asserted that German surrender due to bombing was imminent,[121] in August 1943 he told Portal that 'we can push Germany over by bombing this year',[122] and in November, declaring himself 'certain that Germany must collapse',[123] he told Churchill: 'We can wreck Berlin from end to end ... It will cost between us 400–500 aircraft. It will cost Germany the war.'[124] By December 1943, Harris was specifying even more precisely that victory could be achieved *through area bombing alone* by 1 April 1944, through the 'destruction of between 40% and 50% of

---

[115] Cited in Bellamy 2012b, 138–9. See also Slim 2007, 154.
[116] Churchill's view, somewhat less optimistic than the main bomber advocates, tended towards the latter expectation; see Harmon 1991, 16.
[117] Cited in Davis Biddle 2002, 192 (*my emphasis*).
[118] Cited in Berrington 1989, 30.
[119] Cited in Hastings 2010, 154. Notably, Lindemann's memorandum supported this claim by citing a study of the effects of bombing on British cities, but effectively reversed the study's findings, since it had actually found little evidence of serious damage to civilian morale; see Berrington 1989, 21–2.
[120] Cited in Hastings 2010, 330.
[121] Tanaka 2009a, 3.
[122] Cited in Hastings 2010, 329.
[123] Cited in Overy 2005a, 112.
[124] Cited in Davis Biddle 2002, 230. See also Hastings 2010, 232.

the principal German towns'.[125] The repeated failure of these predictions did not appear to dent such confidence. As late as March 1945, Harris wrote to Norman Bottomley, his eventual successor as head of Bomber Command, asking: 'Japan remains. Are we going to bomb their cities flat—as in Germany—and give the armies a walk over—as in France and Germany—or are we going to bomb only their outlying factories and subsequently invade at a cost of 3 to 6 million casualties?'[126]

Similar claims were made by American policymakers. Roosevelt gave support to a 1942 plan proposed by Claire Lee Chennault, commander of the principal USAAF forces in China, that purported to be able to bomb Japan to defeat within a year.[127] US Senator Sheridan Downey declared in October 1943 that area bombing was going to 'bring the Nazis to their knees in four months', whilst General Eaker expressed his faith that 'the German people cannot take that kind of terror much longer' in November 1943.[128] Like the British, American air force reports emphasized that 'all evidence points to the fact that conditions in Germany are resolving themselves into ... a marked deterioration in morale', sidestepping the need to identify such evidence by arguing that bombing was 'likely to have produced ... far in excess of the sum of the visible damage'.[129] American advocates of area bombing also, like their British counterparts, assured their colleagues that such a strategy was essentially *saving* lives. After the Allied Casablanca conference in 1943, General Henry 'Hap' Arnold, a longstanding acolyte of Billy Mitchell and head of the USAAF, sent an aide to the Air Staff to explain that 'this is a brutal war and ... the way to stop the killing of civilians is to cause so much damage and destruction and death that the civilians will demand their government cease fighting'.[130] Colonel Ewell, a US expert on incendiary weapons, similarly wanted leaders to recognize that firebombing was 'the key to accelerating the defeat of Japan, and if as successful as seems probable ... might shorten the war by some months and save many thousands of American lives'.[131]

As several of these statements suggest, justifications of area bombing also involved the destruction of alternatives. As in interwar theories of area bombing, targeting civilians was comparatively evaluated only against massive and bloody land warfare. Harris thus felt able to claim in his post-war memoirs that 'bombing proved a comparatively humane method ... it saved the flower of the youth of this country and of our Allies from being mowed down by the military in the field, as it was in Flanders in the war of 1914–18'.[132] Similarly, LeMay wrote in his memoirs:

---

[125] Cited in Davis Biddle 2002, 230.
[126] Cited in ibid. 261.
[127] Sherry 1987, 124.
[128] Cited in Davis Biddle 2002, 224–5.
[129] Cited in ibid. 222.
[130] Cited in Bellamy 2012b, 148.
[131] Cited in Schaffer 1988, 120. See also ibid. 93.
[132] Cited in Hastings 2010, xii.

> No matter how you slice it, you're going to kill an awful lot of civilians. Thousands and thousands. But, if you don't destroy the Japanese industry, we're going to have to invade Japan. And how many Americans will be killed in an invasion of Japan? Five hundred thousand seems to be the lowest estimate. Some say a million ... We're at war with Japan. We were attacked by Japan. Do you want to kill Japanese, or would you rather have Americans killed?[133]

Such arguments depended upon a dramatic and artificial narrowing of the field of possibilities. One had to *either* bomb Japanese civilians in their hundreds of thousands *or* kill hundreds of thousands of Americans—other options simply did not exist. This sustained the claim that area bombing civilians was therefore 'saving lives', visible most clearly in Truman's defence of the dropping of the atomic bombs:

> I ordered the Atomic Bomb to be dropped on Hiroshima and Nagasaki. It was a terrible decision. But I made it. And I made it to save 250,000 boys from the United States and I'd make it again under similar circumstances. It stopped the Jap War.[134]

Again, none of these claims reflected clear strategic realities—indeed, they were frequently false. Yet whether as instrumental justification or tacitly internalized assumption, the presentation of bombing civilians as the *only way* to avoid a slaughter of American forces was pervasive. Cumulative radicalization was also a factor here: the more area bombing was employed, the more it appeared an irrevocable feature of 'modern war'.[135] As Robert Messer observes of Truman's choice to use the atomic weapons:

> The historical record shows that for him it was really a nondecision, the answer to which was implicit in the context of the question. Bombs had been used and were being used daily to win the war. The new atomic weapon was a bomb; an exponentially bigger bomb ... but still a bomb. Under the 'doctrine' of total war bombing civilians was acceptable, legitimate, a regrettable but unavoidable product of the fact that 'machines were ahead of morals' ... [Truman] always maintained publicly that he had no choice.[136]

In this sense, Allied justifications were 'deagentifying'—bombing appeared dictated by war, not a strategic choice for which leaders bore responsibility.[137]

Allied policymakers' presentations of area bombing were also strongly *valorized*. The advocates of bombing employed not only arguments about strategic

---

[133] Cited in Downes 2008, 135.
[134] Cited in Schaffer 1988, 172. See also Churchill's similar justification in Bellamy 2012b, 156.
[135] Overy 2005b, 295.
[136] Messer 2005, 312–13.
[137] See also Zanetti, cited in Schaffer 1988, 157.

effectiveness and necessity, but also a powerful normative rhetoric which lauded bombing as a hard-headed, manly, and clear-sighted strategy. This involved the intense denigration of moral opposition, which was framed as sentimentality and effeminacy, and a product of a naïve failure to understand modern war. Thus, the British Admiral of the Fleet, Lord Fisher, reportedly declared that 'moderation in war is imbecility',[138] and after the war LeMay expressed his view that: 'All war is immoral, and if you let it bother you, you're not a good soldier.'[139] He elaborated his view that:

> [T]here was no transgression ... no venturing into a field illicit and immoral ... Soldiers were ordered to do a job. They did it ... The military man carries out the orders of his political bosses ... so that didn't bother me at all.[140]

The American Chiefs of Staff further denied the possibility of *any* moral criticism of war-waging with the reasoning that it was 'folly to argue whether one weapon is more immoral than another. For in the larger sense, it is war itself which is immoral and the stigma of such immorality must rest upon the nation which initiates hostilities.'[141] Colonel Weicker, one of the leading planners of bombing campaigns in Germany, ridiculed critics with the retort that 'you cannot always use the Marquis of Queensberry's rules against a nation brought up on doctrines of unprecedented cruelty, brutality, and disregard of basic human decencies.'[142] Harris likewise urged his superiors to disregard 'the sentimental and humanitarian scruples of a negligible minority.'[143] After the war Truman described nuclear theoretician Robert Oppenheimer's moral concerns about the atomic bombs as typical of 'crybaby scientists.'[144]

In this manner, embracing area bombing came to look admirably 'hard' and 'tough'. On the eve of war, Army Chief of Staff George C. Marshall had warned that: 'if war with the Japanese does come, we'll fight mercilessly. Flying fortresses will be despatched immediately to set the paper cities of Japan on fire. There won't be any hesitation about bombing civilians—it will all be out.'[145] 'We must not get soft', demanded Arnold. 'War must be destructive and to a certain extent inhuman and ruthless.'[146] The US Assistant Secretary of War for Air, Robert Lovett, called for the use of new white phosphorous shells with the claim that 'if we are going to have a total war we might as well make it as horrible as possible.'[147] After the

---

[138] Cited in Hastings 2010, 147.
[139] Cited in Schaffer 1988, 150.
[140] Cited in ibid. 152.
[141] Cited in ibid. 200.
[142] Cited in ibid. 77.
[143] Cited in Overy 2005b, 291.
[144] Cited in Messer 1995, 15.
[145] Cited in Bellamy 2012b, 149–50.
[146] Cited in Schaffer 1988, 103.
[147] Cited in ibid. 93.

largest firebombing raids on Tokyo that killed as many as 80,000 civilians in one night, General Arnold wired LeMay to say: 'Congratulations. This mission shows your crews have got the guts for anything.'[148]

This ideological characterization of the area bombing of civilians as beneficial, unavoidable, and valorous provided its central justification amongst Allied policymakers. But a second element was the ideological representation of civilian victims. It is doubtful that area bombing would have enjoyed such strong backing had political and military elites seen the German and Japanese civilian populations as blameless, as victims of their regimes, or as part of the community for whom the Allies fought. Policymakers did express significant reservations, for example, about bombing plans that would result in far lower numbers of French civilian deaths.[149] Indeed, if policymakers had accorded the lives of German and Japanese civilians moral weight vaguely comparable to other human lives, it is hard to see how many of the later area bombing operations, such as the bombing of Dresden or the use of the atomic bombs, could have been justified when victory was already imminent.

But Allied policymakers did not accord German and Japanese civilian lives such a moral weight. Instead, the German and Japanese civilian population was consistently ideologically framed as simply part of 'the enemy'. As in other mass killings, this ideological representation revolved around three justificatory mechanisms: the assertion that civilians were fundamentally *constitutive of the major threats* to the Allies, the framing of civilians as sweepingly *complicit in the crimes* of their governments, and the taken-for-granted assumption that German and Japanese civilians *lacked relevant moral identity* with the Allies.

Conceptions of German and Japanese civilians as part of the enemy threat, and thus only questionably civilians at all, were built on the broader ideological understandings of 'modern' and 'total' war formulated in the interwar period. Because modern wars relied on industrial production, civilian workers and other civilians who sustained them were seen by bombing advocates as no less central to the enemy threat than military forces.[150] Harris stated: 'It is clear that any civilian who produces more than enough to maintain himself is making a positive contribution to the German war effort and is therefore a proper though not necessarily a worthwhile object of attack.'[151] Harris's predecessor as head of Bomber Command, Sir Richard Peirse, likewise asserted that 'there is no distinction between combatant and non-combatant',[152] while US General Ira Eaker likewise declared his view that: 'the man who builds the weapon is as responsible as the man who carries it into battle'.[153] Most victims of Allied bombing were not, in truth, workers building weapons. But the fact that targeting workers would

---

[148] Cited in ibid. 132.
[149] Berrington 1989, 23–4.
[150] Shaw 2003, 23–6. See also Overy 2005b, 284.
[151] Cited in Overy 2005b, 281–2.
[152] Cited in Farr 2012, 144.
[153] Cited in Schaffer 1988, 92. See also ibid. 151.

necessarily kill huge numbers of non-workers was not incorporated into the assessment of the permissibility of targeting workers. Instead, moral evaluation was compartmentalized: threat construction construed workers as legitimate targets without any consideration of non-workers, whose deaths were then deemed legitimate because they were unavoidable consequences of the 'legitimate' targeting of workers.

More generally, Allied policymakers often refused to differentiate the civilian population of Germany and Japan from a military conception of 'the enemy'. Portal affirmed that: 'I have for some time been expounding that the whole of an industrial city is in itself a military target.'[154] Meeting Stalin in the summer of 1942, Churchill told him: 'We looked upon [German] morale as a military target ... We sought no mercy and we would show no mercy.'[155] Guido Perrera, secretary of the American Committee of Operational Analysis—the principal body for formulating bombing war plans—explained in his post-war memoirs that 'if [the enemy's] cities were indeed honeycombed with small war making plants and were a vital source of his war making power ... there were logical grounds for attacking them'.[156] Again, there was nothing at all obvious or 'natural' about such classifications. As Downes, indeed, observes: 'the American blockade had already strangled the Japanese economy by the time the firebombing began ... In many cases, therefore, the B-29s simply wiped out idle production capacity and workers who were already out of a job.'[157] Indeed at times, as when Truman described the entire city of Hiroshima as a 'military base' to the American public, such claims were absurd misrepresentations.[158] But by slotting civilian populations into broad untextured categories—'military targets', 'industrial areas'—civilians could appear as merely part of the enemy threat.

Similarly, area bombing's supporters consistently generalized the crimes of Nazi and Japanese military and state organizations to the entire populations of Germany and Japan. While many German and Japanese civilians did bear some complicity in their state's crimes, large numbers did not, so justifications of bombing in terms of punitive justice amounted to demands for indiscriminate collective punishment.[159] John Weiss reminds us that 'even at the height of Hitler's popularity about half of all Germans ... rejected the racist violence of the Nazis, though they could not halt it'.[160] The Japanese military regime also faced dissent. Above all else, a vast proportion of civilian deaths from bombing were children, about whom no reasonable claim of guilt can be made.[161]

But this was not how bombing was construed in Allied ideology. For its organizers, executors, and supporting publics, a *logic of reprisal* was central to

---

[154] Cited in Bellamy 2012b, 141.
[155] Cited in Harmon 1991, 15.
[156] Cited in Schaffer 1988, 163–4.
[157] Downes 2008, 285, fn. 77.
[158] Messer 1995, 12.
[159] Slim 2007, 175.
[160] Weiss 1997, ix. See also Burleigh 2001, 144–5 & 153.
[161] Markusen and Kopf 1995, 13 & 16.

the perception of area bombing as justified. 'Certainly', Davis Biddle reports, the British Air Staff 'believed that their own actions had been justified by German attacks on such places as Warsaw, Rotterdam, and Coventry—as well as the V-weapon attacks on British soil'.[162] Churchill frequently embraced this defence of area bombing, talking approvingly of the general German populace 'tasting and gulping each month a sharper dose of the miseries they have showered upon mankind'.[163] At a lunch speech in July 1941 he likewise stated:

> If tonight the people of London were asked to cast their vote as to whether a convention should be entered into to stop the bombing of all cities the overwhelming majority would cry, 'No, we shall mete out to the Germans the measure and more than the measure they have meted out to us'.[164]

In one of the first British War Cabinet meetings to formally approve of wider bombing against Germany, such retaliatory logics appear to have been crucial, with a 'general accord as to the justice of retaliating for all that Germany had already done'.[165]

The assertion that guilt for German crimes justified bombing German civilians was equally endorsed by American policymakers. Roosevelt argued that 'The Nazis and the Fascists have asked for it—and they are going to get it'[166] and made the remarkably explicit biopolitical assertion that: 'We either have to castrate the German people or you have got to treat them in such a manner so they can't just go on reproducing people who want to continue the way they have in the past.'[167] General LeMay similarly reported himself unconcerned with civilian casualties during the firebombing of Japan since 'we knew how the Japanese had treated the Americans—both civilian and military—that they'd captured in places like the Philippines'.[168] Schaffer reports a telling interaction after the dropping of the atomic bomb:

> When General Leslie R. Groves, the director of the Manhattan Project, told [General] Arnold and General Marshall about the attack on Hiroshima, Marshall suggested that it would be a mistake to rejoice too much, since the explosion had undoubtedly caused a large number of Japanese casualties. Groves replied that he was not thinking as much about those casualties as about the men who had made

---

[162] Davis Biddle 2002, 253.
[163] Cited in Slim 2007, 155.
[164] Cited in Connelly 2002, 49.
[165] Harmon 1991, 8.
[166] Cited in Downes 2008, 138.
[167] Cited in ibid.
[168] Cited in Bellamy 2012b, 151.

the Bataan Death March [where Japanese soldiers marched US soldiers to death]. Afterwards ... Arnold slapped Groves on the back and exclaimed, 'I am glad you said that—it's just the way I feel.'[169]

As Schaffer concludes: 'even highly educated Americans sometimes lumped together all persons of Japanese ancestry with the particular organizations and individuals responsible for attacking Pearl Harbor and for later Japanese atrocities'.[170]

In addition, while the Allies did not generally claim that German and Japanese civilians lay entirely outside the universe of moral obligations, they engaged in various ideological manoeuvres that quelled moral identification with those civilians. Often, as noted above, this simply involved a radical devaluation of civilian deaths if *any* Allied lives could be asserted as saved by them. As Schaffer reports: '[Colonel] Lowell Weicker suggested that enemy lives and Allied lives had wholly different values, for he maintained that if terror attacks saved just a few British and Americans, the price paid by the enemy should not be an object of serious consideration.'[171] Harris, likewise, stated: 'I would not regard the whole of the remaining cities of Germany as worth the bones of one British grenadier.'[172] And a group of scientists lobbied for the use of the atomic bombs with the demand: 'If we can save even a handful of American lives, then let us use this weapon—now!'[173]

Such ideas sometimes descended into blatant dehumanization, especially towards the Japanese. General Haywood S. Hansell testified to a 'universal feeling' amongst the upper echelons of the US government that the Japanese were 'subhuman'.[174] The British government's chief diplomatic advisor, Baron Robert Vansittart, described the Germans as a 'degradation of the human species',[175] Harris 'regularly used the metaphor of insect extermination',[176] and operations analysts in both governments regularly used 'biological analogies' in discussing and conceptualizing the bombing campaigns against civilian areas.[177] More common, however, was deidentification via abstract euphemistic discourses, in which bombing was presented as attacking 'cities', the 'enemy's interior', the 'social body', or 'morale',[178] as if these entities could somehow be attacked without the human beings who actually constituted them being targeted in the process.[179] This was, as

---

[169] Schaffer 1988, 154.
[170] Ibid. 155. See also ibid. 63.
[171] Ibid. 106. See also Slim 2007, 167.
[172] Cited in Hastings 2010, 449.
[173] Cited in Schaffer 1988, 158.
[174] Cited in ibid. 153. See also Selden 2009, 87.
[175] Cited in Farr 2012, 138.
[176] Overy 2005b, 284.
[177] Sherry 1987, 198.
[178] As, for example, in the Area Bombing Directive of 14 February 1942 or the British and American Casablanca agreement on bombing; see Bellamy 2008, 54; Farr 2012, 144. See also Schaffer 1988, 67; Overy 2005b, 278 & 283–4; Bellamy 2008, 48; Young 2009, 158 & 163; Hastings 2010, 124 & 153–5.
[179] Bellamy 2012b, 136.

Overy characterizes, 'an anatomical language that created a deliberate abstraction in place of the real bodies that bombing would damage'.[180]

The policymakers who initiated and organized Allied policies of area bombing civilians were guided, then, by these two critical sets of ideas: an image of bombing as effective, unavoidable, and virtuous warfare, and a portrayal of the civilians it killed as dangerous, guilty, and morally dispensable. Key hardline advocates of area bombing appeared to internalize these ideas with high levels of conviction, with questionable strategic assessments framed by clauses that: 'all the evidence goes to prove',[181] 'these figures are conservative and can be absolutely relied upon';[182] 'the word "minimum" cannot be repeated too often',[183] 'there seems little doubt',[184] 'it is clear',[185] 'certain',[186] 'we are convinced',[187] 'that conclusion emerges from a deliberately conservative examination',[188] 'certain knowledge',[189] and so on. Some of this reflected the need to lobby more ambivalent policymakers in a context of stiff competition over resources between different departments. But many appeared to genuinely embrace such overconfidence out of ideological precommitment. As the British scientific advisor Solly Zuckerman reported, regarding his efforts to shift more bombing onto transport infrastructure targets in the face of resistance from area bombing advocates: 'Most of the people with whom I was now dealing seemed to prefer an *a priori* belief to disciplined observation.' Indeed, Harris set up an explicitly skewed epistemic framework for regulating the permissibility of bombing, arguing that 'to my mind, we have absolutely no right to give up [area bombardments] unless it is certain they will not have [their promised] effect'.[190] On this logic, even a 1% possibility that Harris's promises would obtain would be deemed sufficient to justify the certain deaths of hundreds of thousands of civilians.

Again, such hardline ideas were not a necessary consequence of the strategic circumstances the Allies found themselves in, and not all policymakers accepted them. Army and navy officers were often critical, as were some scientific advisors to the US and UK governments.[191] Harris, in particular, faced internal criticism for not maintaining a tighter focus on the destruction of industrial areas.[192] Churchill's own views fluctuated,[193] and he sometimes mused apprehensively on the measures the Allies were resorting to—'are we beasts?' he asked a war cabinet colleague

---

[180] Overy 2014, 48.
[181] Cited in Overy 2005a, 38–9.
[182] Cited in Hastings 2010, 231.
[183] Cited in ibid. 337.
[184] Cited in ibid. 154.
[185] Cited in Overy 2005b, 281–2.
[186] Cited in Overy 2005a, 112.
[187] Cited in Hastings 2010, 330.
[188] Cited in ibid. 216.
[189] Cited in Schaffer 1988, 63.
[190] Cited in Davis Biddle 2002, 257.
[191] Berrington 1989, 19–21 & 30.
[192] See, for example, Wilson 2007, 73–4.
[193] Berrington 1989, 27–8.

on watching footage of RAF bombings of the Ruhr in June 1943.[194] Eisenhower attempted to introduce some rigour into the planning process for bombing, warning subordinates that: 'the phrase "military necessity" is sometimes used where it would be more truthful to speak of military convenience or even of personal convenience.'[195]

But Sherry's comments about Generals Arnold and LeMay can be extended to most involved in the bombing campaigns: 'the limitations of these men were generally the product of the political, cultural, and intellectual environment in which they worked'.[196] Even beyond the ranks of the ideologically committed, the justificatory narrative that underpinned area bombing achieved sufficient predominance in crucial institutional settings in the interwar period to generate a prevailing coalition of support in policymaking circles. As Overy summarizes: 'because so many influential people said that bombing was important, the belief grew in political and military circles that it must be so'.[197] The justificatory narrative was also strongly institutionalized into the norms, prevailing policies, and operational frameworks of those political and military organizations responsible for bombing. Dissenting positions were never successfully mobilized into effective resistance to bombing, either because they did not hold critical influence over the policymakers who were in a position to shape the bombing campaigns, or—as in the case of Churchill—because doubts were not held with sufficient conviction to constrain policy.

The Rank-and-File Bombers

The rank-and-file aircrews of the Allied area bombing campaign played a very different role from rank-and-file agents in most other cases of mass killing. For a start, area bombing was a form of what Laia Balcells labels 'indirect violence'[198]: the RAF and USAAF bomber crews did not confront civilian victims face-to-face, but from a height of several kilometres.[199] As Charles Lindbergh, then serving as an informal advisor to General Arnold, observed across two entries in his wartime journals from 1944:

> You press a button and death flies down … It is like listening to a radio account of a battle on the other side of the earth. It is too far away, too separated to hold reality … In modern war one kills at a distance, and in doing so does not realize that he is killing.[200]

---

[194] Cited in Harmon 1991, 3.
[195] Cited in Sherry 1987, 140.
[196] Ibid. xi.
[197] Overy 2005a, 118.
[198] Balcells 2017, 21.
[199] Sherry 1987, 209–13; Selden 2009, 88; Tanaka 2009a, 1–2.
[200] Cited in Sherry 1987, 209–10.

Such 'indirect' violence is always likely to lessen the difficulty of ideological justification amongst rank-and-file agents. Most obviously, as Lindbergh suggests, killing-at-a-distance reduces many of the fundamental psychological obstacles to killing civilians that arise in proximate massacres—obstacles that ideological legitimation may help overcome. In addition, indirect violence increases the epistemic dependence of rank-and-file agents on their elite superiors: it is easier for elites to ideologically construct the meaning of violence when the rank-and-file's actual experience of the ultimate effects of that violence is so limited.

Second, by contrast, the bombing aircrews were *not distant from genuine warfare*. Unlike many perpetrators of mass killing, who kill civilians far from any real battlefront, the aircrews risked death daily in carrying out bombing raids, with a life expectancy measured in weeks, and only a quarter of airmen completing their assigned tour of duty.[201] As Kevin Wilson writes: 'it took courage of the kind it's difficult to comprehend today to fly into the apparently unbroken wall of bursting flak that was Berlin or the Ruhr, then—having survived by an apparent miracle—do it again the next night or the one after that.'[202] Bomber crews were full participants in a very real war, who spent much of their time engaged in direct operations against military and industrial targets, or evading deadly assaults by enemy fighter formations. The strains of aircrew life also made reflection on the specific moral and strategic character of bombing different targets neither especially likely nor psychologically helpful. 'The men who fared best', Hastings observed, 'were those that did not allow themselves to think at all.'[203]

Third, while *violent* coercion was minimal, Allied area bombing is as close as we get to a campaign of mass killing that could be micromanaged by central authorities. Reluctant participants could have sought transfer to other duties or have applied to different service arms on normative grounds,[204] but once in the bombing aircrews, only a dishonourable discharge provided an exit path—and this involved abandoning front-line service in the war effort entirely, which few airmen considered palatable. Aircrew were not required to have any detailed understanding of the strategic and moral arguments elites saw as justifying area bombing, only to have mastered the skills required to operate the bombers and carry out terse, abstract mission directives. The contents of those directives, which determined the scale, shape, and intensity of area bombing, were all set by elites.

For all three reasons, the strength of ideological justifications amongst the rank-and-file plays a much less central explanatory role in Allied area bombing than in other mass killings. The attitudes of ordinary Allied servicemen to the war varied, and many lacked strong convictions about its ultimate purpose and character. The average soldier, one of the most in-depth studies of the American military

---
[201] Ibid. 205.
[202] Wilson 2007, 70.
[203] Hastings 2010, 221.
[204] See also Terkel 1984, 194.

suggested, was 'typically without deep personal commitment to a war which he nevertheless accepted as unavoidable'.[205] Aircrews appear primarily motivated by patriotic duty, comradeship, and personal interest in military service.[206] While the capacity of the Allied governments to render area bombing consistent with such motives was important, this was not difficult when many bombing actions were of clear military significance, when some civilian casualties were a certain consequence of such actions, and when the line between such operations and area bombing was so abstract from several kilometres distance. In short, the nature of area bombing and its place within the overall context of World War II lessened the need for ideological mobilization of the bombing crews themselves, and there is much less observable variation in those crews' behaviour according to their ideological understandings of the campaign than in most other cases of mass killing. I consequently dwell less on ideology's role amongst the rank-and-file here than in my other case studies.

Still, acceptance of the ideological narrative that justified area bombing was not completely irrelevant. For all the physical and psychological distance created by bombing, no-one involved could be oblivious to the massive civilian death toll. Area bombing could have provoked disquiet from the rank-and-file of the RAF and USAAF, and their agency was not nil. Had bombing been widely perceived as a 'dirty' business by comparison to service with fighter squadrons or the army and navy, this could have had serious recruitment or operational implications. The aircrews, moreover, were not mindlessly credulous of superior authority. They expressed scepticism concerning many elements of the bombing campaign, including the equipment they had to work with, the operational plans of their missions, and Bomber Command's explanations of especially costly raids.[207]

Yet aircrews appear to have generally accepted the Allied justificatory narrative for area bombing, and to have used it, if not as a central motive for participation, then to legitimate the violence and eliminate moral or psychological anxieties. In most cases, this did not reflect raging hatred. As Sherry argues, 'the enemy was usually faceless ... crews had "no personal feeling against Germans" or "an almost impersonal attitude toward the enemy"'.[208] Instead, willing participation was facilitated by the assumption that area bombing was legitimate warfare that would end the war quicker, coupled with the sweeping guilt-attribution of the German and Japanese civilian population espoused in Allied wartime propaganda, and the logic of reprisal this implied.[209] A young British navigator, en-route to Berlin, wrote to his girlfriend:

---

[205] Stouffer et al. 1949, 149.
[206] See, in general, Lewis 1991; Wilson 2007. See also Stouffer et al. 1949, 149–72.
[207] Lewis 1991; Wilson 2007.
[208] Sherry 1987, 134.
[209] Goldhagen 2010, 201–2.

I hope we knock the blazes out the target (which incidentally is the post office in the centre of the city). Before, I have always felt sorry for the people down below, but the other night I came over Portsmouth on the way home and saw it afire. I saw an explosion about 2,000 feet high. So now I feel different about it ...[210]

A rear gunner who was captured and taken to Berlin after his bomber was downed likewise later recalled:

Berlin was in fact in a terrible mess ... You had a tiny sense of guilt when you thought, 'My God, I helped to do this,' but then you thought about Coventry and Plymouth and the other bombed British cities and that the Germans had started it all. The Germans had, of course, developed the V-weapons at that time and if we hadn't finished the war in time they would have come up with something bigger, perhaps the atom bomb.[211]

A study of aircrews' letters from June 1942 offered a similar characterization: 'Expressions of satisfaction that the Germans are having to undergo the punishment they have hitherto meted out to others are found in almost all letters, but there is an absence of vindictiveness or fanaticism in the phrases used.'[212] The logic of reprisals and collective punishment embedded in Allied guilt-attribution allowed many to see area bombing as justified even when endorsed with lesser degrees of conviction, consciousness, or emotional fervour.

Since aircrews were diverse, attitudes naturally varied. A small minority were reluctant participants. Airman Ted Hallock observed: 'When you're only one of the hired hands, who's being carried along to do the dirty work, to drop the bombs and do the killing, you don't feel so good about it.'[213] Others involved in the campaigns had mixed opinions. An aerial navigation instructor (and former professor of German) observed:

London was a mess. Though the Germans asked for it, the destruction of their cities was vastly greater. Dresden is the one thing I'm really ashamed of. I mean hellishly sorry. We were behaving like the Nazis. The war was as good as over. It was an open city, full of refugees coming back from the eastern front. Who can say that wrecking this beautiful, non-military city shortened the war?[214]

Others suggested more enthusiastic resonance with Allied justifications, especially sweeping guilt-attribution, although opinions often remained nuanced. The

---

[210] Cited in Hastings 2010, 122–3.
[211] Cited in Wilson 2007, 90–1.
[212] Cited in Hastings 2010, 178. See also Overy 2013, 229.
[213] Cited in Sherry 1987, 136.
[214] Cited in Terkel 1984, 353.

American poet John Ciardi, who served as a navigator in bombing missions in the Pacific theatre, commented:

> We were in the terrible business of burning out Japanese towns. That meant women and old people, children. One part of me—a surviving savage voice—says, I'm sorry we left any of them living. I wish we'd finished killing them all. Of course, as soon as rationality overcomes the first impulse, you say, Now, come on, this is the human race, let's try to be civilized. I had to condition myself to be a killer ... One measure of that is hatred. I did want every Japanese dead. Part of it was our own propaganda machine, but part of it was what we heard accurately. This was the enemy. We were there to eliminate them. That's the soldier's short-term bloody view.[215]

More unambiguously, American flying ace Robert Scott, Jr., though not specifically speaking of area bombing, wrote in his memoirs: 'Personally, every time I cut Japanese columns to pieces in Burma, strafed Japs swimming from boats we were sinking, or blew a Jap pilot to hell out of the sky, I just laughed in my heart and knew that I had stepped on another black-widow spider or scorpion.'[216] Even more explicit was an article from the Fifth Air Force's Weekly Intelligence Review, in July 1945, by Colonel Harry F. Cunningham:

> We military men do not pull punches or put on Sunday School picnics. We are making War and making it in the all-out fashion which saves American lives, shortens the agony which War is and seeks to bring about an enduring Peace. We intend to seek out and destroy the enemy wherever he or she is, in the greatest possible numbers, in the shortest possible time. For us, THERE ARE NO CIVILIANS IN JAPAN.[217]

Simultaneously, however, the memoirs and testimony of rank-and-file participants in the bombing campaigns suggest that the moral and strategic justifications of mass killing were of relatively low priority in their thinking, offering some support to situationist theories of violence. Aircrews were consumed with the operational and psychological challenges of the bombing missions themselves. Bruce Lewis's semi-autobiographical collection of stories from 'the men who flew the bombers' contains abundant stories of such operations, of relationships between aircrew, of especially harrowing near-death experiences, but almost no mention of attitudes towards civilian casualties. Lewis himself only disparagingly comments that: 'There were no arguments at that time about it being morally wrong to bomb Germany. Such academic "afterthoughts" only emerged when victory was in

---

[215] Cited in ibid. 196–7.
[216] Cited in Sherry 1987, 134.
[217] Cited in Schaffer 1988, 142 (emphasis in original).

sight.'[218] Some of the official denigration of moral critique can be discerned here, but concerns with enemy civilians were typically, as Sherry observes, 'dwarfed by more immediate emotions, anxiety and grief over loss of buddies'.[219]

Whether as institutionalized expectations, tacitly adopted legitimation, or a more enthusiastic basis for willing participation, the Allied justificatory narrative for area bombing does appear to have been accepted by most rank-and-file agents. Since the provision of willing participants in a campaign to area bomb civilians is not automatic, this mattered. The justificatory narrative injected an acceptable social meaning into area bombing, blocking framings of it—whether by accusers in broader society or amongst the bombers themselves—as mass atrocity. The narrative encouraged expressions of hatred and vindictiveness from some airmen, which were notably stronger against the more racially dehumanized Japanese.[220] That aircrews consistently referenced this justificatory narrative when thoughts of the suffering of German and Japanese civilians arose suggests that it facilitated willing involvement. But the necessary production and acceptance of such a narrative was fundamentally a low bar. It proved unsurprisingly easy for the organizers of the bombing campaign to meet that bar in a mass killing perpetrated at a distance, during a genuinely desperate war.

The Public Legitimation of Area Bombing

As in other cases of mass killing, area bombing in World War II was not primarily driven by public attitudes. As Sherry observes: 'What facilitated official reliance on air power was less irresistible public pressures than the previous acceptance and general attractiveness of strategic bombing.'[221] Yet there are good reasons to think that public attitudes mattered. Even in wartime conditions, the United States and Britain were open societies by comparison with most regimes that engage in mass killing, and the governments of both states expressed intense concern about public opinion towards the bombing campaigns, while engaging in extensive efforts to legitimate the strategy. Indeed, when such legitimation efforts have struggled in other conflicts (such as the War in Vietnam), democratic governments have often felt it necessary to show more restraint, at least in areas within the range of media scrutiny.

In World War II, by contrast, the British and American publics generally gave their governments a free hand in bombing civilians, pockets of opposition notwithstanding. The evidence suggests that they largely accepted their governments' justifications of area bombing (alongside outright disinformation which

---

[218] Lewis 1991, 12.
[219] Sherry 1987, 211.
[220] See also Stouffer et al. 1949, 158–61.
[221] Sherry 1987, 144. See also Downes 2008, 137.

obscured the extent to which the military was intentionally targeting civilians). As John Dower observes:

> [N]o sustained protest every materialized. The Allied air raids were widely accepted as just retribution as well as sound strategic policy, and the few critics who raised ethical and humanitarian questions about the heavy bombing of German cities were usually denounced as hopeless idealists, fools, or traitors. When Tokyo was incinerated, there was scarcely a murmur of protest on the home front ... Japan had merely reaped what it had sowed.[222]

Indeed, direct surveys of Allied soldiers and civilians show quite shocking levels of support for mass killing by the end of the war. A survey of US public opinion after the dropping of the atomic bombs on Japan found that 5% of respondents opposed any use of atomic weapons, 14% wanted a warning demonstration first, 53.5% endorsed the use of two bombs 'just as we did', whilst 23% expressed regret that *more* atomic bombs had not been dropped.[223] Twenty percent of respondents in a December 1944 study likewise wanted to see Japanese citizens tortured, exterminated, or otherwise harshly punished. Ordinary American infantry expressed even more extreme views when asked, in 1943 and 1944: 'What would you like to see happen to the Japanese after the war?' Of 4,000 infantrymen surveyed in the Pacific theatre, 43% responded 'punish leaders but not ordinary Japanese', 9% 'make Japanese people suffer plenty', and 42% '*wipe out whole Japanese nation*'. Amongst a thousand veteran infantrymen in the European theatre, the proportion of American soldiers supporting the genocide of the entire Japanese nation *rose* to 61%.[224]

Available data suggest that British attitudes towards bombing civilians were somewhat more mixed, but from early stages the bombing of German civilians occurred against the background of public calls for retaliation, which intensified as the war proceeded.[225] Asked solely about reprisal bombings against Germany, 46% of survey respondents in December 1940 said they approved, 48% disapproved, and 8% did not know. By May 1941, support for reprisal bombings had risen to 53%, while disapproval had fallen to 38%.[226] The British government's efforts to gauge public opinion likewise showed significant support for the Allied justificatory narrative surrounding area bombing. Home Intelligence reports suggested that sweeping attributions of guilt to the German public were central.[227] Newspapers' letter columns also offered many examples of such sentiments, endorsing the view that, as one *Daily Telegraph* reader put it: 'We ought, with utter

---

[222] Cited in Downes 2008, 136.
[223] Messer 2005, 308.
[224] Stouffer et al. 1949, 158. See also Chirot and McCauley 2006, 216.
[225] Bellamy 2012b, 138–9.
[226] The Lighter Side 1941, 4; Connelly 2002, 54–5; Overy 2013, 229.
[227] Davis Biddle 2002, 221.

impunity, to bomb Berlin and bomb it unmercifully.'[228] Most also appeared to internalize the government's presentation of area bombing as effective warfare. In March 1941, Home Intelligence reported to Churchill that 'people will want a lot of convincing that really heavy raids on civilian centres in Germany are not our most efficacious weapon.'[229] The American air attaché in London likewise noted in April 1942 how 'the British public have an erroneous belief, which has been fostered by effective RAF publicity, that the German war machine can be destroyed and the nation defeated by intensive bombing'.[230]

While providing the infrastructural support needed for policymakers to feel unconstrained in prosecuting the campaign, such attitudes were typically nuanced rather than fanatical, and varied rather than homogenous. A British censor's report on a sample of civilian letters after the first 1,000 bomber raid in May 1942 summarized that:

> There are those who are pleased, and those who regret that so much suffering should have to be inflicted. There are those who fear reprisals. Many of the letters contain two or more of these elements. Predominant is satisfaction, but many women express regret ...[231]

The leading survey organization *Mass Observation* published its similar 1944 assessment of British attitudes to the bombing campaign, as reported by Connelly:

> In London six out of ten people gave unqualified verbal approval to the raids. Two said they were necessary, but expressed major qualms about their effects on the civilian population of Germany. Only one in ten felt they were too awful to be approved in any way, 'though few go so far as wanting them stopped.' It was found that very few expressed gloating or vengeful sentiments. Only one in six felt that bombing would end the war, but considerably more believed that it would shorten it 'and this is the most usual reason for approval of our raids.'[232]

Indeed, justifications of bombing were often accepted selectively, or in ways compatible with moral reservations. Max Hastings quotes a letter to *The Times* by one Brigadier, which he deems typical of popular opinion:

> Britain and her Allies and well-wishers must all be devoutly thankful that the RAF is at last able to repay Germany in her own coin and to inflict upon her cities the same devastation that she has inflicted on ours. But it must offend the

---

[228] Connelly 2002, 48.
[229] Cited in ibid. 55.
[230] Cited in Hastings 2010, 130.
[231] Cited in ibid. 216–17.
[232] Connelly 2002, 55.

sensibilities of a large mass of the British population that our official broadcasts, when reporting these acts of just retribution, should exult at and gloat over the suffering which our raids necessitate ...[233]

There was also opposition to bombing. As noted above, between a third and just under half of the British public declared some form of opposition to the bombing of German civilians, and more strongly dissenting conscientious objectors existed in both Britain and the United States.[234] But intense and vocal opposition was the province of only a small minority. Overy estimates around 150,000–170,000 formal members of the main active British pacifist organizations—barely a third of 1% of the UK population—although these were accompanied by a larger number of sympathizers.[235] Sherry likewise notes how: 'Dissent from the course of bombing was confined to the fringes of American politics, and even there few mounted an effective case that bombing ran beyond the requirements of victory.'[236]

What explains this large-scale public endorsement of Allied ideological justifications of area bombing? While not absent, longstanding intergroup animosities do not seem to have played a major role. A rough legacy of hostility towards both states—mainly rooted in World War I in the German case and racism in the Japanese case—certainly existed and was exploited during the war, but was hardly remarkable compared to other historical legacies of conflict worldwide. Again, both the United States and Britain were also relative focal points in the growth of peace activism and opposition to bombing civilians in the interwar period. Two main factors, by contrast, appear more central to the public legitimation of area bombing.

The first was the wartime crisis itself. Some shifts in public attitudes appeared to occur fairly 'spontaneously' from exposure to enemy attacks, especially the Blitz and Pearl Harbour.[237] As early as autumn 1940, British intelligence reported 'increased hatred of Germany' and demands for reprisal among the British population,[238] while immediately after Pearl Harbour, 67% of Americans surveyed favoured aerial bombardment of Japanese cities.[239] Yet these endogenous effects of wartime crisis are insufficient to explain the extent and character of the public legitimation of mass killing. The concrete effects of wartime crisis depended heavily on how public political discourse was generated and funnelled by ideological authorities to construct the political meaning and apparent implications of the war. Indeed, it was not those who most directly experienced German and

---

[233] Cited in Hastings 2010, 217.
[234] Ceadel 1987, 137; Connelly 2002, 52–5; Overy 2013, 208–9.
[235] Overy 2013, 206–7.
[236] Sherry 1987, 140.
[237] Even many pacifists shifted their stance; see Overy 2013.
[238] Harmon 1991, 29, fn.29.
[239] Downes 2008, 136.

Japanese attacks that were most radicalized.[240] As in many other cases, those who relied purely on media coverage of events often seem to have been most accepting of hardline positions, with Sherry noting that 'hatred of the enemy was generally strongest among the civilians furthest removed from him.'[241]

What mattered, then, was the combination of the war context with a second factor: sustained ideological activism, through which key justifications for area bombing were disseminated to the public through the joint efforts of official government agents and a hugely supportive private media sector. A lengthy demand for more intense area bombing of Germany by one of the UK's largest newspapers, the *Daily Mirror*, was representative of much media discourse:

> This is the only policy. This is the only effective method available to us in self-defence. This is the offensive ... Bomb for bomb and the same all round! The only policy ... The air war is no time for lecturers, and gloved persons wishing to live up to a high standard of ancient chivalry. The invention of the bombing plane abolished chivalry for ever. It is now 'retaliate or go under'. We are not dedicated to passive and polite martyrdom. We *must* hit back ... Also the dislocation of German communications and nerve-centres is essentially a 'military objective'—if really it is reasonable to go on making this almost obsolete distinction. A distinction that wears very thin. People are killed, in the devilish war of today, everywhere, anyhow. People killed are, in tens of thousands, useful workers; mainly war workers.[242]

As in policymaker circles, then, public discourses presented bombing civilians as effective, unavoidable, and valorous warfare that would end or shorten the war in a hard-headed, tough, but ultimately humane manner. Bombing civilians 'would be a kindness—even to the enemy' because it would save lives in the long run.[243] US newspapers and magazines supported a reliance on air power as 'relatively cheap in men' by comparison with the 'nightmare' of land warfare.[244] *Harper's* magazine contended that: 'It seems brutal to be talking about burning homes. But we are engaged in a life and death struggle for national survival, and we are therefore justified in taking any action which will save the lives of American soldiers and sailors.'[245] Serious strategic and moral assessment of such claims was largely avoided or aggressively shut down. As Sherry summarizes, by the time the war had begun:

---

[240] Overy 2013, 229. See also Stouffer et al. 1949.
[241] Sherry 1987, 134.
[242] Cited in Connelly 2002, 47–8.
[243] Cited in Sherry 1987, 141.
[244] Cited in ibid. 137.
[245] Cited in Downes 2008, 136–7.

Where once alternatives (however arbitrary) had been offered in the clash between [bombing] prophets and skeptics, now public argument was largely confined to disagreement over the techniques ... Americans were given no alternative to substantial reliance on widespread destruction by air as one method to win the war. Strategic as well as moral debate had not ceased, but it ran in narrow channels.[246]

Bombing advocates' inaccurate presentations of bombing's effectiveness were also echoed uncritically in the media. Even one of the more sceptical leading US reporters, Allan Michie, suggested to readers that US bombing could bring Germany 'to her knees before the end of the summer of 1943' so that the Allies could simply 'walk across Europe'.[247] The air correspondent of *The Times* likewise assured readers in June 1943: 'If the present bomber offensive could be multiplied by four, [Germany's] war production would be completely halted. That conclusion emerges from a deliberately conservative examination of the damage done by the big bombers.'[248] The interwar valorization of bombing was also continued during the war. Newspapers were consumed by fervent excitement on the announcement of new super-large bombs and vast thousand-bomber raids.[249] American films and books 'cultivated popular expectations for a virtuous campaign of annihilation against Japan'.[250] Even one prominent Christian magazine endorsed the abandonment of moral restraints, asserting that it was 'idle to try to put a check upon the way in which weapons are used. If we fight at all, we fight all out.'[251]

Simultaneously, public discourse intensely promoted the threat construction, guilt-attribution, and deidentification of German and Japanese civilians. Guilt was the most prominent theme. As the *Daily Mail* stated, in September 1940:

> The ruined homes and broken lives of Britain will be avenged. When Hitler has spent his fury in his useless effort to bring this country to her knees, the hour for attack will come. Then Britain must launch against Germany the most devastating offensive that has yet been seen.[252]

After Harris launched his first 1,000 bomber raid on Cologne, the *Express* declared 'The Vengeance Begins!', whilst the *Gaumont* newsreel service titled its coverage of the raid 'RAF Lets Hitler have it, right on the chin!'[253] Indeed, the media frequently criticized the government for not going far enough in punishing

---

[246] Sherry 1987, 129.
[247] Cited in ibid. 128.
[248] Cited in Hastings 2010, 216.
[249] Farr 2012, 147.
[250] Sherry 1987, 131.
[251] Cited in ibid. 139.
[252] Cited in Connelly 2002, 47.
[253] Cited in ibid. 50.

Germany, with the *Sunday Dispatch* opposing a government assurance that German civilians were not being killed wantonly with the assertion that: 'It is right that the German population should "smell death at close quarters". Now they are getting the stench of it.'[254] In the United States, likewise, *The New Yorker* celebrated the destruction of Berlin: 'it serves the bastards right … [and] it was a necessary action, efficiently and economically carried out …'.[255] But public discourse more broadly presented civilians as simply part of the enemy—as one contributor to the *New Republic* wrote: 'The natural enemy of every American man, woman and child is the Japanese man, woman and child.'[256] Frequently, Germans and especially Japanese were targets of blatant dehumanization—reduced to monkey-like, fang-toothed beasts, or stunted primitive hominids.[257] As Schaffer writes, 'during the war, hatred of the Japanese people was fed by accounts of enemy barbarism and by portrayal of the Japanese in American media as barely human creatures who were either comical, sinister, or grotesque'.[258] In June 1942, a float at a New York parade displayed 'a big American eagle leading a flight of bombers down on a herd of yellow rats which were trying to escape in all directions'.[259]

One important way in which public political discourse diverged from policymaker ideology, however, was the inclusion of a much heavier dose of outright misinformation. The British and American governments typically implied that civilians were not being directly targeted, even though they openly talked of the morale effect of bombing and retaliation against 'Germans' and 'Japs'. Adverts by Boeing 'stressed that its bombers destroyed factories without harming cities'.[260] General Arnold presented bombing as occurring 'with the care and accuracy of a marksman firing a rifle at a bullseye',[261] while journalist William Bradord Huie likewise contended that American bombers could strike 'with city-block accuracy without a bombardier's ever seeing anything on the ground [so as] not to destroy every building in the block; but to destroy the one building in the block where bearings for the Focke-Wulf planes are being made'.[262] This was completely false. Another *New Republic* issue deplored 'bombing defenceless people merely to instil terror in them' but assured its readers that 'so far as we are aware [terror bombing] is not the practice of the RAF and the AAF'.[263] Indeed, *The New Republic* was so confident in such assumptions that it responded to two major British papers that reported that American bombing had killed innocent civilians by suggesting they must have been 'misled'.[264]

---

[254] Cited in Hastings 2010, 216.
[255] Cited in Sherry 1987, 145.
[256] Cited in ibid. 141.
[257] Keen 1986; Sherry 1987. See also Russell 1996.
[258] Schaffer 1988, 155.
[259] Cited in Sherry 1987, 124.
[260] Ibid. 126.
[261] Cited in Davis Biddle 2002, 224. See also ibid. 212 & 245.
[262] Cited in Sherry 1987, 127.
[263] Cited in ibid. 140.
[264] Davis Biddle 2002, 258.

Dresden, whose bombing would ultimately cause enormous controversy, was illustrative of this interweaving of ideological justification and deception, but also highlights how the success of such legitimation efforts was not assured. The city was initially presented to both bombers and public as an important German military asset targeted as a necessary and beneficial act of warfare. 'Dresden', *The Times* reported, 'is a place of vital importance to the enemy ... the centre of a railway network and a great industrial town ... a meeting place of the main lines to eastern and southern Germany.'[265] Harris described it, falsely, as: 'a mass of munitions works, an intact government centre, and a key transportation point to the east'.[266] Dresden's controversy derived principally from the revelation that such presentations were false, and the widespread awareness that the Nazi threat in the last months of the war was nothing like what it had been before. At this point, the ideological justificatory narrative for area bombing broke down.

But this was rare. War alone made some sort of general shift in receptiveness towards greater violence against the Germans and Japanese probable whatever the state of political and media coverage. But it does not explain the close correspondence between dominant public attitudes and elite conceptions of area bombing, nor the specific support for such a massive assault on civilian populations. As in most security politics, it is critical to appreciate that ordinary British and especially American citizens could experience little of the bombing campaign directly, let alone assess the efficacy of the latest theories that guided it. They necessarily operated off information and meanings communicated to them via the government and media, both of which acted as ideological activists, strongly and consistently disseminating and authorizing justifications of violence against civilians. As Sherry summarizes: 'Few reporters were inclined to contest official claims. They saw themselves as enlisted in the war effort, their task that of establishing confidence in Allied virtue and victory and commanders.'[267] It is unsurprising that, in such a context, so many British and American citizens came to accept, or at least go along with, the apparently dominant consensus on the legitimacy of the bombing campaigns.

Ideological Restraints

Given that the Allies did engage in mass killing, it may seem perverse to talk of restraint in this case. Yet it is critical to realize that there were key limits to Allied bombing compared to more ultrahardline and/or genocidal campaigns of mass killing. These restraints reflected, on the one hand, limitationist elements of even

---

[265] Cited in ibid. 255.
[266] Cited in Overy 2005b, 293.
[267] Sherry 1987, 132–3.

the dominant Allied ideological narrative about area bombing. The Allies strictly instrumentalized violence against civilians to achieving a surrender from the German and Japanese government. They did not construct Germans and Japanese civilians as *intrinsic* threats, but as part of a wartime enemy with which future accommodation and coexistence was possible and desirable.

Once again, this conception of targeting civilians was not inevitable, but was consistent with prevailing liberal/conservative ideology in Allied states. It was, moreover, highly consequential. By the end of the war, the Allies had the opportunity to wipe out or punitively slaughter the Germans and Japanese on a much larger scale should they have wished to, and segments of their populations appear to have been willing to support such an act. Yet this was never remotely considered by policymakers, nor seriously lobbied for by publics. However horrific Allied bombing was, it was neither genocidal nor primarily a form of collective punishment. As the other cases in this book make clear, one cannot explain this merely by contending that the Allies had no strategic 'need' to engage in genocide or collective punishment against the Germans or Japanese, since there was no 'need' to annihilate Soviet peasants, Guatemalan Maya, or Rwandan Tutsi either. A fascist British or American government might well have been inclined to more extensively target German and Japanese civilians and could have perceived strategic and moral justifications for doing so.

Yet area bombing was not conceived in such terms. Indeed, it never even received—especially in Europe—the absolute support its leading bombing advocates demanded. When other objectives, such as the provision of air support to ground forces in the invasion of France, were more pressing, the bombing of cities was curtailed. With the war over, and its military rationale removed, it halted immediately. The Allies' hardline ideological conception of area bombing was critical in generating and enabling mass killing via the area bombing campaigns, but it also shaped the fundamental logics that determined those campaigns' scale, scope, and ultimate limits.

6.4 Conclusion

From Rudolph Rummel's quip that 'Power kills, absolute power kills absolutely' to Alex Bellamy's more recent contention that 'the idea that it is wrong to deliberately kill non-combatants in war is a pre-commitment of liberal politics',[268] mass killing has been primarily associated with autocratic, illiberal states. Yet cases like Allied area bombing in World War II inject a significant caveat into that association. Allied area bombing was clearly a case of mass killing, and was not simply a product of the strategic necessities of war. But nor was it primarily rooted in radical political

---

[268] Bellamy 2008, 43.

projects or nationalist hatreds. It was principally the result of specific hardline ideas about war and security—rooted in a particular historical pathway of ideological development in the United States and Britain—which were activated and intensified by armed conflict. Such ideas were sufficiently strong to underpin a military campaign of great cost, questionable strategic gains, and brutal human consequences.

The influence of such ideas did not cease at the end of World War II. Despite the fact that hard evidence on the effectiveness of bombing civilians was weak, Allied victory only reinforced many US air force leaders' confidence in such a strategy. In the immediate post-war years, US Air Force General Orvil Anderson found it reasonable to suggest initiating a nuclear war of aggression against the Soviet Union, asking: 'Which is the greater immorality—preventive war as a means to keep the U.S.S.R. from becoming a nuclear power; or, to allow a totalitarian dictatorial system to develop a means whereby the free world could be intimidated, blackmailed, and possibly destroyed?'[269] To a reporter he stated: 'Give me the order to do it and I can break up Russia's five A-bomb nests in a week! And when I went up to Christ, I think I could explain to Him why I wanted to do it—now—before it is too late. I think I could explain to Him that I had saved civilization.'[270] These comments were rejected by the US government, but they were not a complete outlier among senior military officers. In Korea, US-led forces killed between one million and three million civilians through bombing.[271] Yet LeMay still felt that victory was thrown away because bombing had escalated too slowly:

> We slipped a note kind of under the door into the Pentagon and said, 'Look, let us go up there ... and burn down five of the biggest towns in North Korea—and they're not very big—and that ought to do it.' Well the answer to that was four or five screams—'You'll kill a lot of non-combatants,' and 'It's too horrible.'[272]

The retired General Eaker responded in the same way to failure in Vietnam. 'How much better it would have been', he argued, 'if necessary, to destroy North Vietnam than to lose our first war. That would have saved us 50,000 American dead, 250,000 Allied dead, and, subsequently, the greatest genocide in this century ....'[273] For all the 'restraint' that frustrated Eaker, US bombing nevertheless killed around 600,000 Cambodian and 350,000 Laotian civilians.[274]

In all these cases, violence against civilians was not a natural consequence of objective political or military circumstances, crucial though those circumstances were. But it was a course that large proportions of liberal/conservative societies

---

[269] Cited in Schaffer 1988, 202.
[270] Cited in ibid.
[271] Bellamy 2012b, 166.
[272] Cited in Selden 2009, 93.
[273] Cited in Schaffer 1988, 212.
[274] Bellamy 2012b, 173.

believed justified, and which many more tacitly condoned. Such support for mass killing is a reminder that hardline ideological understandings of security politics, and the capacities for mass violence that result, are not only the province of manifestly 'evil' worldviews. They can come to look plausible and appealing to ordinary people, given the right ideological precedents, strong activism by hardliners, and conditions of crisis. Beyond aerial bombing, comparable ideas have influenced violence by liberal democracies in imperial adventures and efforts to suppress colonial rebellions, as in French Algeria, British Kenya and Malaya, and the westwards expansion of the United States.[275] Their influence is also visible in individual abuses that were encouraged by overarching policy, like the famous My Lai massacre in Vietnam and the Abu Ghraib abuses in Iraq—neither of which were as exceptional in those wars as is commonly assumed.[276]

To deny such campaigns is to sustain a sanitized and inaccurate self-narrative about liberal/conservative societies. Such societies do tend to be more restrained in many forms of political violence than the most brutal authoritarian regimes. But they also contain many people and institutions attracted to hardline ideas. When such hardliners have gained power, liberal/conservative societies have proved willing to perpetrate mass violence against civilians through the same basic justifications utilized by authoritarian states—construing such violence as an essential means for defeating their enemies and protecting their security.

---

[275] Blakeley 2009, 22; Gerlach 2010, ch.5; Bellamy 2012b, 81–91; Madley 2016.
[276] Glover 1999, 58–63; Doris and Murphy 2007; Bellamy 2012b, 186.

7
Mass Killing in Guatemala's Civil War

7.1 Overview

The Guatemalan civil war, running from the early 1960s until 1996, was not only one of the twentieth century's longest-running armed conflicts but also one of its most brutal for civilians. Guatemala's UN-sponsored Historical Clarification Commission (CEH), set up as part of the 1996 peace process, estimated that 200,000 civilians were killed or 'disappeared' over the course of the conflict, with mass killings prevalent in a peak period from 1980 to 1983 that encompassed over half of all civilian casualties.[1] While abuses were committed by all armed parties to the conflict, approximately 90% of recorded human rights violations were perpetrated by the Guatemalan government or groups under its control.[2] Fighting a left-wing guerrilla insurgency, Guatemala's military regime chose to target civilians en masse, especially via a campaign of terror waged against Guatemala's indigenous Maya[3] communities, who were perceived as providing vital support for the guerrillas. The CEH, several scholars of the civil war, and a second truth commission organized by the Human Rights Office of the Archdiocese of Guatemala have concluded that the military was guilty of genocide.[4]

---

[1] Garrard-Burnett 2009, 6–7. Precise figures are uncertain: the most detailed database, based on testimonies to the *Centro Internacional para Investigaciones en Darachos Humanos* (CIIDH), counts 36,906 killings and disappearances across the civil war, including 25,928 from 1980 to 1983. But as the CIIDH emphasizes, this 'presents only a fraction of the deaths attributable to the Guatemalan State during the years of armed conflict' (Ball, Kobrak, and Spirer 1999, 11).

[2] ODHAG 1999, 290; Rothenberg 2012, xxi–xxii.

[3] Identity labels in Guatemala are, as in all societies, complex. Using the historically common term 'Indian' to describe Guatemala's indigenous population is considered offensive and rejected as a self-identification. The indigenous population comprises 23 ethnolinguistic groups, of which 21 are considered 'Maya', and constitutes around half of Guatemala's population. During the civil war, however, neither ethnolinguistic subgroups nor 'Maya' were primary self-identifications for indigenous Guatemalans, who usually identified their communities with their Guatemalan *municipio* (municipality). Since the 1980s both indigenous activists and scholars have increasingly employed the term 'Maya', which I therefore employ (alongside the broader term 'indigenous'). For discussion see Arias 1990; Smith 1990, 3–5; Nelson 1999, esp. 5–7 & 24, fn.26; Garrard-Burnett 2009, 7, fn.16. See also footnote 6, below. Except when quoting other scholars, I follow recommended practice in using 'Maya' as singular and plural noun and adjective, with the exception of the 'Mayan' languages; see http://osea-cite.org/program/maya_or_mayans.php.

[4] ODHAG 1999; Rothenberg 2012; Salazar 2012; Brett 2016. See, however, Garrard-Burnett 2009, 13–16; Schwartz and Straus 2018, 231–3.

Both the truth commissions and specialist scholars emphasize the roots of Guatemala's civil war in the country's lengthy history of economic exploitation, political exclusion, and racism. One of the most unequal states in twentieth-century Latin America, Guatemala was governed for over a century after independence (from Spain in 1821) by dictatorial governments dominated by a small economic elite.[5] The economy was heavily reliant on what was practically forced labour by Maya and poor *ladinos*.[6] This period ended with the 1944 overthrow of the ruling dictatorship, producing a decade of democratic government under Presidents Juan José Arévalo (1945–51) and Jacobo Árbenz Guzmán (1951–54).[7] Arévalo and Árbenz both pursued substantial political, social, and economic reforms that incurred the hostility of the economic elite, and animated US fears of left-wing states flourishing in the Americas. The result was the 1954 overthrow of Árbenz's government in a CIA-backed coup, after which Guatemala was ruled until the final years of the civil war by a succession of highly repressive governments dominated by the military in alliance with far-right political parties.[8]

The Guatemalan civil war emerged from the fallout of this 1954 coup. On 13 November 1960, disaffected left-leaning army officers attempted to overthrow President Miguel Ydígoras Fuentes.[9] The attempt failed but morphed into a loosely organized left-wing guerrilla insurgency which linked rebel military officers, the Guatemalan Workers Party, and supporters from student, worker, and peasant organizations. Guatemala's military (with US assistance) largely pacified this phase of left-wing insurgency in 1966–68. But the subsequent military regimes of the 1960s and 1970s increasingly resorted to severe coercion and the creation of paramilitary death squads to quash political opposition.[10] While this solidified military control in the short run, '[t]he government's repression of and intractability toward even the most modest challenges to the status quo pushed moderate voices of reform from the political center to the radical left by the mid-1970s.'[11] The result was renewed violent opposition to the state.

The guerrillas gradually reorganized over the 1970s and shifted their strategy. Having previously relied on a rather small ladino support base in eastern Guatemala, the insurgents—principally the *Ejército Guerrillero de los Pobres* (Guerrilla Army of the Poor or EGP) and *Organización del Pueblo en Armas* (Organization of People in Arms or ORPA)—began to see a mass popular uprising

---

[5] Yashar 1997, 222.
[6] Like 'Maya', 'ladino' is a complex marker of identity, but is widely used in scholarship to refer to Guatemalans who do not identify as indigenous, so I broadly follow this practice. This excludes, however, a third category—*Garífuna*—denoting Guatemalans of African descent. All these categories are somewhat fluid cultural constructs. Five centuries after the Spanish conquest of Guatemala, most ladinos possess indigenous ancestry. See references in footnote 3, above, for discussion.
[7] Adams 1990, 141–2; Yashar 1997, 191–206; Schirmer 1998, 10.
[8] Yashar 1997, 206–8; Schirmer 1998, 13–17; Rothenberg 2012, xxvi–xxxiii; Grandin 2017, 3–5.
[9] ODHAG 1999, 190–2; Garrard-Burnett 2009, 26–7.
[10] ODHAG 1999, 194–201.
[11] Garrard-Burnett 2009, 24.

as the route to revolution. Crucially, this would involve the mobilization of long-exploited Maya communities in Guatemala's northern highlands.[12] While these communities were initially ambivalent towards the guerrillas, the insurgency gradually managed to consolidate a significant power base in the highlands. Blatantly fraudulent government elections in 1974 and a catastrophic earthquake in 1976 that devastated poorer and indigenous Guatemalans both intensified opposition to the regime.[13] The government's response was characteristically heavy-handed. Under the presidency of Romeo Lucas Garcia (1978–82), the military resorted to intensified state terror and mass killings against both indigenous communities and broader leftist social and political organizations.[14] This proved counterproductive: indigenous communities increasingly supported the guerrillas, either in search of protection or from outrage at government abuses.[15]

Lucas Garcia was deposed in an internal coup on 23 March 1982, having lost the support of the military high command and its right-wing allies. He was replaced by a triumvirate of military officers including General Efráin Ríos Montt, who became *de facto* President. Ríos Montt's 18-month presidency was the deadliest period of the civil war, but also involved a 'rationalization' of the violence from the regime's perspective: reducing killings in politically influential urban areas while extending military control over the highlands. Annihilating many indigenous communities, the army simultaneously concentrated those not killed in 'model villages' under tight military control, and forced their inhabitants to serve in weakly armed militia, disingenuously titled 'civil self-defence patrols' (*Patrullas de Auto-Defensa Civil* or PACs). These measures were strategically ruinous for the guerrillas, who lost most of their territorial presence in Guatemala by late 1983. While Ríos Montt—a born-again evangelical whose broader plans for Guatemala's moral and economic renewal found weak support in the military—was himself deposed in another internal coup on 8 August 1983, lower but still significant levels of state violence persisted throughout the 1980s.

The mass killing of civilians in Guatemala under Lucas Garcia and Ríos Montt offers an important 'hard test' for the arguments of this book, because it is often seen as a classic case for rationalist theories of counter-insurgent violence which make minimal reference to ideology. Ben Valentino, for example, presents mass killing in Guatemala's civil war as largely explained by the threat of the insurgency and the paucity of other available counter-insurgent strategies, rather than by the military regime's ideological worldview.[16] Stathis Kalyvas similarly emphasizes that, by contrast with 'findings that take ideology or ethnicity as the main causal variable of violence', the targeting of Maya communities:

---

[12] Ibid. 36–7.
[13] Ibid. 42–5; Gawronski and Olson 2013.
[14] ODHAG 1999, 211–15.
[15] Garrard-Burnett 2009, 45–50.
[16] Valentino 2004, ch.6.

could just as easily have been … because they were located in areas of guerrilla presence … the army's repression did not focus on areas where indigenous organizations (and presumably grievances) were strong and guerrillas had little presence, but rather in areas where the guerrillas were trying to organize the peasants despite weak indigenous organizations.[17]

Support for this portrayal can also be found in two leading quantitative analyses of state violence in Guatemala. While recognizing the potential relevance of racism, Yuichi Kubota offers statistical evidence that state violence was generally a response to rebel attacks.[18] Christopher M. Sullivan likewise offers a strategic explanation of the military's massacres, presenting statistical evidence that they were used for combating insurgency, suppressing ethnic groups, and extending the state's control over territory.[19] Though not explicitly declaring it irrelevant, Sullivan makes no mention of the military's ideology as a significant cause of the violence.

I fully accept that the army's mass killings in Guatemala were strategically motivated and employed as a method for defeating the insurgency.[20] While some genocide scholars portray the military's assertions of guerrilla threat as merely a rhetorical cover for an extermination campaign primarily motivated by racist hatred, this interpretation is not plausible for the reasons Valentino, Kalyvas, Kubota, and Sullivan emphasize. Though the mass killings did have a genocidal aspect, they did not target the Maya population uniformly, but were focused on zones of guerrilla activity and varied in intensity according to the military's perceptions of local civilian support and rebel threat.[21] The killings also targeted significant numbers of non-Maya ladinos who were considered 'subversives'.[22] At the rank-and-file level, moreover, a majority of perpetrators were not ladinos killing Maya out of racial hatred, but themselves indigenous Maya mobilized into the army through complex mixtures of coercion and military socialization.

Nevertheless, rationalist theories that neglect ideology fail to explain the occurrence and scale of mass killing in Guatemala. Strategic counter-insurgency aims were indeed central, but the decision to employ mass killing in pursuit of such aims only made sense within a certain set of ideological assumptions and narratives. When ideology is ignored, rationalist theories face multiple problems in explaining the occurrence and character of the killings.

For a start, such theories can only be employed with arbitrary selectivity for different phases of the mass killing. They look somewhat applicable to the period

---

[17] Kalyvas 2006, 136–7.
[18] Kubota 2017.
[19] Sullivan 2012.
[20] ODHAG 1999, 4–8 & 133; Oglesby and Nelson 2016, 139.
[21] Stoll 1993, 64–5 & 87; Garrard-Burnett 2009, 13–16; Rothenberg 2012, 49; Brett 2016, 128. This is not to deny that the mass killings were, in significant aspects, genocidal.
[22] Garrard-Burnett 2009, 17.

following Ríos Montt's 'rationalization' of the violence from roughly June 1982 onwards, since this period did prove strategically devastating for the guerrillas.[23] Yet scholars of Guatemala's civil war are almost unanimous in emphasizing that the strategy of mass killing under the preceding Lucas Garcia regime (and extending into the first month or so of Ríos Montt's tenure) was deeply counterproductive and strategically misguided, and served to fuel rather than defeat the insurgency.[24] As Garrard-Burnett summarizes:

> the insurgency had expanded quite dramatically in the late 1970s, as peasant opposition, pushed to the wall by Lucas's brutal and even irrational politics of violence, coalesced around the guerrillas and grassroots popular organizations ... it was the polarizing violence of the Lucas regime that actually did more than outright guerrilla recruitment to drive people outside the political system and into the armed opposition.[25]

There are two potential rationalist responses to this enigma of counterproductive mass killing. First, it might be suggested that mass killing, whilst an ineffective solution, was nevertheless the only real option the government had available to avoid a catastrophic military defeat. But this is not plausible for either the Lucas Garcia or the Ríos Montt regimes. The CEH report emphasizes how:

> The magnitude of the state's repressive response was totally disproportionate in relation to the military force of the insurgency ... at no time during the internal armed confrontation did the guerrilla groups have the military potential necessary to pose an imminent threat to the state ...[26]

Even when the threat posed by the guerrillas was relatively high—after the failure of the earlier Lucas Garcia killings and greater mass mobilization—alternative responses were available to the military that look *at least* as strategically rational as mass killing. Even ignoring the most laudable but unlikely alternatives, such as negotiations, the military could have pursued a far more targeted campaign to attack the guerrillas directly, engaged in more limited selective violence against

---

[23] As I emphasize below, however, this shift in strategy itself reflected a shift in power between two ideologically divergent factions of the military.

[24] Stoll 1993, 15, 96, 127, & 307; Schirmer 1998, 18; Kruit 2000, 15–16; Brett 2016, 115 & 131–2. Valentino also recognizes that the strategic effectiveness of the campaign was limited to the Ríos Montt era phase of the mass killings, and remained questionable in the long run; see Valentino 2004, 216–17. He leaves unaddressed, however, the problem of explanatory selectivity this creates for rational-strategic theories given the contrast between the Lucas Garcia and Ríos Montt era. Kubota similarly limits his quantitative analysis to the period after February 1982, on the grounds that only in early 1982 did the rebel forces unify to form the Guatemalan National Revolutionary Union; see Kubota 2017, 55.

[25] Garrard-Burnett 2009, 25–6. See also Davis 1988, 23; Stoll 1988, 94 & 103; Brett 2016, 109–11.

[26] Rothenberg 2012, 184. See also Nelson 1999, 11; Garrard-Burnett 2009, 18.

specific civilian collaborators, or relied more exclusively on the military's strategy of population control in so-called model villages.[27] Rationalist scholars might suggest that the military nevertheless resorted to indiscriminate or categorical violence because they lacked the intelligence or capacity to deploy more targeted and limited strategies. But this contention is also not supported by the evidence. The army often possessed detailed intelligence on guerrilla activities and areas of more specific support, often *avoided* available opportunities to engage or destroy guerrillas directly, and showed little interest in attempting to properly identify genuine civilian collaborators.[28] In short, civilians were not targeted en masse in Guatemala because no other options were available, but as a *preferred* strategy for devastating what was perceived as a sweepingly suspect population.[29]

A second potential rationalist explanation for counterproductive mass killings is that while mass killing was *in fact* counterproductive, the information needed to rationally conclude this was not available at the time. Once again, however, this explanation is not supported by the evidence. For a start, there was important elite dissensus over the rationality of mass killing—numerous officers within the Guatemalan military proved able to see the counterproductive nature of the policy, the inaccuracy of claims that indigenous communities were uniformly supporting the guerrillas, and the ultimately limited, non-existential threat posed by the guerrillas.[30] Moreover, actors of a less radically hardline ideological orientation—notably the US ambassador and CIA observers—likewise saw that mass killing was strategically asinine. A Lucas Garcia-era CIA report from February 1981, for example, observed how 'the percentage of the indigenous populations supporting the guerrilla movement apparently remains small and geographically limited',[31] and recognized that guerrilla recruitment was 'enhanced by the anti-government resentments provoked by the Army's chronic abuses of civilians'.[32]

In addition, while rationalist theories explain strategic patterns in how mass killing unfolded at the local level in Guatemala, there are several features of the violence they fail to account for. They cannot, for example, explain the intense targeting of Catholics in the highlands. This is a curious element of mass killing

---

[27] Massacres did facilitate the formation of such model villages, but there is no evidence that population control necessitated expansive mass killing. See Schirmer 1998, ch.4.

[28] Stoll 1993, 84; Schirmer 1998, 59; ODHAG 1999, 219; Brett 2016, 128; Schwartz and Straus 2018, 231. See also Rothenberg 2012, 184.

[29] As noted in Chapter 3, some rationalist theories point to the economic value of mass killing. This is also not a plausible explanation for Guatemala. The indigenous population had minimal resources to plunder, whilst being a critical source of cheap labour. It made little sense to kill such a population, and the mass killing intensified an economically ruinous armed conflict and left the agrarian economy devastated. Material self-interest did play a role in generating right-wing antipathy to left-wing challenges, but the link from such interests to mass killing was contingent on the form that ideology took. See Adams 1990, 149–50 & 153–7; Stoll 1993, 85–6; ODHAG 1999, 243; Garrard-Burnett 2009, 50.

[30] National Foreign Assessment Center 1981, iii–iv, 8, 12, & 14; Schirmer 1998, 37–8, 57, & 62; Garrard-Burnett 2009, 46 & 87; Grandin 2017, 6.

[31] National Foreign Assessment Center 1981, 8.

[32] Ibid. 19.

in Guatemala, because the regime was strongly supported by the Catholic church and many conservative Catholic sectors of society. Moreover, most Catholics in the highlands had minimal if any links to the guerrillas in the early 1980s. To explain the targeting of Catholics, I argue below, specific ideological assumptions of the military must be brought into focus: specifically, the association of Catholic activities in the poorest parts of Latin America with dangerous left-wing politics and ideologies of 'liberation theology'. Rationalist theories also struggle to explain the exceptionally cruel and abusive nature and scale of much of the violence.[33] As Greg Grandin notes: 'The killing was savage, markedly more brutal than similar repressive campaigns conducted elsewhere in Latin America during the same period.'[34] On many occasions the army burned unarmed civilians alive (most infamously in the notorious Spanish Embassy fire in January 1980) and directly targeted infants in needless, easily avoidable atrocities.[35] The roots of such gratuitous cruelty do not, of course, lie *solely* in ideology—endogenous processes of brutalization played a key role.[36] But hardline security doctrines were interwoven with such processes, promoting contempt for victims and the martial valorization of brutality.

Rather than supporting purely rationalist theories, the Guatemalan government's employment of mass killing is explained by the interaction of strategic circumstances with the distinctive ideological character of the military regime and its agencies: in particular, the brutal blend of *racist nationalism* and *ultraconservative anticommunism* which guided military elites' security politics. These ideological elements also explain why the military's strategy was never confined purely to counter-insurgency objectives, but more expansively aimed to devastate indigenous communities that were perceived as both intrinsically subversive and occupying threatening strategic locations.[37] In particular, the regime engaged in extensive efforts to convert those not exterminated into what Jennifer Schirmer has aptly termed the 'sanctioned Mayan', purged of traditional Maya culture and identity and tamed by encultured identification with the Guatemalan state.[38]

This synthetic neo-ideological emphasis of both counter-insurgency and ideology enjoys increasing support from specialist scholarship on Guatemala. The two truth commission reports on the mass killings firmly emphasize *both* strategic objectives *and* intense anticommunism and racism. Elizabeth Oglesby and Diane Nelson similarly emphasize how:

> Historically based racism contributed to conflating Mayas (or at least specific groups of Mayas viewed as especially 'rebellious') into the category of internal

---

[33] Brett 2016, 34. See also Garrard-Burnett 2009, 17–18.
[34] Grandin 2017, 1.
[35] Schirmer 1998, 55; ODHAG 1999, 29–36 & 73–80; Garrard-Burnett 2009, 93–7; Rothenberg 2012, ch.5; Grandin 2017, 10.
[36] Stoll 1993, 97. See also Kalyvas 2006, 55–8.
[37] Rothenberg 2012, 69.
[38] Schirmer 1998, 59; Brett 2016, 54 & 137.

enemy. The state's attack against the Ixils and other Maya communities was motivated by political and military criteria, rather than purely racial objectives ... [but] key here is to understand how racism and counterinsurgency were intertwined in specific places and moments.[39]

As with my other case studies, the remainder of this chapter first examines the evolution of the military's ideological worldview from the mid-twentieth century through to the mass killing campaigns, and then more directly assesses how ideology shaped the formulation of policies of mass killing, the mobilization of perpetrators, and the maintenance of support and legitimacy amongst public constituencies.

7.2 The Evolution of Military Ideology

The roots of the two key ideological elements underpinning mass killing in Guatemala—racist nationalism and ultraconservative anticommunism—can be traced back throughout the period since Guatemala's independence in 1821. In the late nineteenth century, Brett observes, 'the "indian" population ... was constructed through the predominant narrative as the unproductive vagabond, an outcast, not a citizen: the solution was to enforce conditions of cheap labour ... akin to modern slavery'.[40] Since Guatemala's society and economy depended in key respects on the immense inequality between the ladino and indigenous halves of the population, racist justifications of ethnic oppression coexisted with deep anxiety about the possibility of indigenous political mobilization—the fear, as one 1930s parish priest put it, that 'if organized and a bit educated the Indians might some night massacre all of the ladinos'.[41]

Such ideas continued to influence the country's political trajectory following the fundamental break with authoritarianism in 1944, and then the reimposition of dictatorship in 1954 after a decade of democratic reform. Examining political discourse in the mid-1940s at the start of the democratic opening, Richard Adams observes how: 'The dangerous and menacing behavior of Indians was well portrayed in the news articles of the time and was considered common knowledge of the dominant society.'[42] Adams' survey of such media publications provides a snapshot of political discourse prior to the military governments of the 1950s, 1960s, and 1970s, highlighting how the indigenous population was already consistently depicted as: (i) lazy and despicable; (ii) incapable of self-direction; (iii) nevertheless crucial for the national economy and national security; and (iv) in

---
[39] Oglesby and Nelson 2016, 139.
[40] Brett 2016, 58–9.
[41] Cited in Adams 1990, 148.
[42] Ibid. 147.

need of 'regeneration' led by ladinos so as to promote smooth integration into the national-democratic community as defined by ladino elites.

This combination of racist denigration of indigenous individuals with a perceived need to build a unified Guatemalan nation that would incorporate the obedient 'sanctioned Mayan' is visible across the media commentaries and editorials Adams analyses. The editors of leading newspapers in the 1940s lamented 'what an inferiority complex the Guatemalan suffers for his Indian blood, for the indigenous character of his nation! … The Guatemalan does not want to be Indian, and wishes his nation were not.'[43] Elite ideologies were not monolithic here. Indeed, one can discern two fundamentally opposed perspectives: a far-right standpoint which expressed unrestrained hatred and disgust of indigenous communities, and a second, more sympathetic, *'indigenista'* stance that preached the need to educate and civilize the indigenous. But these two perspectives overlapped in core racist notions and fears.

In particular, both shared a conception of indigenous people as dupes for political forces hostile to Guatemalan society (in this period, the main concerns were former members of the dictatorship deposed in 1944)—a notion that would remain crucial during the civil war. As one of the more hostile editorials argued:

[W]e are opposed to these [indigenous] compatriots, ignorant, filthy, lazy, sick, licentious, without consciousness. We have often felt ourselves rebel against their evilness. We have also found ourselves in agreement with those who would favour their gradual disappearance by whatever means that would progressively diminish their ranks. In addition: when we have witnessed row upon row of these robust but moronic beings bending low to kiss the bloody hand of their own unholy executioners [i.e. the former dictatorial Presidents and their allies], without the most minimal revulsion, we have wished that the earth would swallow them up, never to reappear.[44]

Though different in tone, the more sympathetic *'indigenista'* perspective fundamentally shared this conception of indigenous inferiority and threat—though sometimes extended to the rural peasantry as a whole:

The countryman, the illiterate, the labourer, Indian or ladino, continues in his ignorance and consequently continues to be a danger, to be manipulated by the perverse maneuvering of the enemy. These beings, because of their lack of consciousness, are a cloud in our sky of democratic liberties.[45]

---

[43] Cited in ibid.
[44] Cited in ibid. 148.
[45] Cited in ibid. 149.

In all such discourse, as Adams comments: 'No credibility was given to the possibility that the Indians may have had long and serious legitimate complaints within the system.'[46] This reflected not only the framing of Maya as ignorant and gullible, but also the fact that political interest in the conditions of indigenous communities was fundamentally instrumental—the indigenous were a concern not as citizens with rights but as either a potential threat against society or a supplicant element of the national economy.

This interest in making economic use of indigenous labour, through violence and coercion if necessary, is visible in debates over economic reform in the early democratic era. In defending Guatemala's 'vagrancy law', which effectively forced indigenous people who had lost land through debt accumulation into highly exploitative labour, one editor observed that:

> As long as there is no vagrancy law that accords with our needs, the [consequences] would deal agriculture a moral blow. It is argued that it is a harsh law, but our Indian requires harshness as long as he cannot meet his own needs ... Harshness, they might say; but harshness necessary to lift this child from the ignorance in which his father has lived.[47]

For the political elites of mid-twentieth-century Guatemala, then, the 'founding narrative' of the nation-state was built around virulent racism towards indigenous populations.[48] Indigenous persons were needed as economic instruments for the dominant elite, but fundamentally outside the primary political community represented by the state.[49]

As Brett notes, however: 'In and of itself, the racially motivated discourse of the indigenous population as filthy and lazy, as primitive, was not adequate to represent the mainstay of a sustainable, imminent threat to the primary political community: indolence, lascivity and drunkenness were an unlikely source of violent rebellion.'[50] Moreover, as noted, Guatemala's mass killings were not directed solely against the indigenous population—though they bore the bulk of the violence—but also left-wing parties, student movements, organized labour, and Catholic organizations engaged in development work. Racist ambitions to build a 'unified' Guatemalan nation purged of indigenous influences were insufficient: the rationale for Guatemala's mass killings also depended on anticommunist and ultraconservative ideological currents, which took hold at the end of the democratic period and radicalized progressively through the crises faced by the military regimes of the 1960s and 1970s. Such currents were also encouraged by

---

[46] Ibid.
[47] Cited in ibid. 160.
[48] Straus 2015a.
[49] Brett 2016, 57.
[50] Ibid. 59.

the regional hegemon as, with the onset of the Cold War, American notions of communist threat were propagated throughout Latin America.[51] Brutal political repression reflected the fact that the coalition of internal and external political forces that combined to topple the Árbenz government in the 1954 coup shared a paranoid impression of the threat of communist subversion in Guatemala, and consequently refused to tolerate *any* form of left-wing political activism.

For example, in opposing the labour reforms of the democratic period, the *Asociación General de Agricultores* (AGA) and *Comité de Comerciantes e Industriales de Guatemala* (CCIG), representing the business interests of agriculture and industry, portrayed any calls for improved workers' rights as carrying the imminent prospect of communist dictatorship. In a co-authored document, they complained that:

> The only reason [for a new labour code] is the union leaders' necessity to provoke economic disequilibrium in order to demand and to justify nationalization of land and the expropriation of the means of production ... Today we feel we are in a position to affirm that we are certain in having identified the existence of a ploy to paralyze the country's agrarian production; that will force private property at the very State, through strikes at harvest time, to accede to all types of pretension which occur to the union leaders; that they are trying to achieve total disequilibrium and complete insecurity for people and for good in order to easily impose a dictatorship of the masses led and sustained by professional leaders.[52]

Moreover, in a remarkable act of national redefinition, the AGA and CCIG explicitly argued that 'the Nation is not constituted by the persons who have been demanding the reforms'.[53] Hyperbolic ultraconservative assertions that any calls for economic reform reflected a global communist conspiracy and an existential threat to Guatemala intensified over the course of the early 1950s. As the AGA's *Buletin* apocalyptically emphasized in 1952: 'The first step is about to be taken: the expropriation of land. It will be followed by the nationalization of industry, and commerce. The Totalitarian State.'[54] This position was also supported by influential figures within the Guatemalan church. In April 1954, just months before the coup against the Árbenz government, Archbishop Rosell y Arellano authored a vitriolic letter decrying the imminent threat of communism in Guatemala, which was read out in churches across Guatemala. The Archbishop warned:

> anti-Christian communism—the worst atheist discourse of all time—is stalking our country under the cloaks of social justice. We warn you that those whom the

---

[51] Roniger 2010, 25–7.
[52] Cited in Yashar 1997, 194.
[53] Cited in ibid.
[54] Cited in ibid. 195.

communists help today, they will condemn to forced labor and terrible suffering tomorrow. Everyone who loves his country must fight against those who—loyal to no country, the scum of the earth—have repaid Guatemala's generous hospitality by fomenting class hatred, in preparation for the day of destruction and slaughter which they anticipate with such enthusiasm.[55]

Importantly, such apocalyptic narratives were increasingly expanded to incipient political activism in peasant and indigenous communities. As Yashar summarizes:

> The oligarchy claimed that labor organizing within indigenous communities would initiate race wars which in turn would exacerbate class conflict. The rise in rural organizing would threaten national production, lead to civil war, and allow agrarian committees to become [according to one 1953 newspaper] 'an unlimited, dictatorial, and absorbing power,' all of which was part of a communist plan. By 1952, the opposition's anti-communist discourse had become hostile and pervasive.[56]

By the time of the reinstitution of an authoritarian Guatemalan state in the mid-1950s, then, dominant political discourses already expressed many ideological features potentially conducive to mass killing, which in the short run fuelled the suspension of democracy and construction of increasingly militarized authoritarian governments in the 1960s and 1970s. Indigenous populations in particular, and to a lesser degree broader groups of poor peasants, workers, students, and left-wing activists, were deidentified from the primary political community and characterized as severe existential threats to society. The ultraconservative right consistently contended that such threats could only be addressed by an authoritarian state willing to employ large-scale repressive violence.

From 1954 onwards, the new military regimes (intersected by a single civilian but undemocratic presidency from 1966 to 1970) reproduced these ideological assumptions amongst the ruling elite, and entrenched and institutionalized them into the structures of the military and the state. This represented an important shift: the military had been relatively sympathetic to reform in 1944, but ultraconservative radicalization during the democratic era, coupled with the expulsion of more reform-minded elements after the failed left-wing coup attempt in 1960, left the military as the principal bastion of anticommunism and racism. The result was that '[a]nti-communism became an obsessive guiding principle for both the military and for Guatemala's economic elite ... any expression of opposition was condemned as communist-inspired and foreign-born.'[57] Under such an ideological infrastructure, as Christian Tomuschat notes, 'political

---

[55] Cited in ibid. 198.
[56] Ibid. 195.
[57] Ball, Kobrak, and Spirer 1999, 13. See also Plant 1978, ch.2.

claims that in Western Europe would have been characterized as belonging to a social-democratic perspective were stigmatized as endangering the security of the state'.[58] Such extreme intolerance took its strongest forms in the far-right paramilitary units that were given free reign or actively directed by military elements. One of the most infamous paramilitary organizations, the *Movimiento Anticomunista Nacional Organizado* (MANO), sent threatening letters to members of Guatemalan cooperatives stating that it 'knew by experience that all trade union and co-operative organizations fall under the control of communist leaders who are infiltrated into them'.[59]

Such inclinations were solidified through the Guatemalan state's incorporation of the 'National Security Doctrine': a set of ideas propagated throughout Latin America during the Cold War, which melded US anticommunist security thinking with military regimes' conceptions of themselves as saviours of civilized nationhood against alien Marxist corruption.[60] The US-run 'School of the Americas' (SOA)—attended by significant numbers of Latin American officers and soldiers, including future president Ríos Montt—was one important disseminator of such ideas. The school offered counter-insurgency courses which frequently advocated broad repression: 'training manuals did not differentiate between guerrilla insurgents and peaceful civilian protestors' and, perhaps unsurprisingly, '[g]raduates of the SOA have been implicated in massive human rights violations'.[61] The United States also trained over 1,000 Guatemalan police officers at its International Police Academy—eventually closed in the 1970s due to its association with dictatorial repression in Latin America.[62] But National Security Doctrine ideas dovetailed with existing features of the ideological environment of post-1954 Guatemala. As the CEH report notes: '[the] National Security Doctrine fell on fertile ground in Guatemala where anti-communist thinking had already taken root and, from the 1930s on, had merged with the defense of religion, tradition, and conservative values, all of which were allegedly threatened by the worldwide expansion of atheistic communism'.[63]

The combination of this ideological character of the political elite and state institutions—extreme even by the standards of Cold War Latin America—interacted with peaceful and violent opposition to the state to drive forward cumulative radicalization. Diagnosing all left-wing opposition as communist conspiracy, the state responded with severe but generally counterproductive state terror, which intensified opposition and further undermined state legitimacy, whilst also strengthening the institutional power of hardliners. The result was

---

[58] Tomuschat 2012, xvi.
[59] Cited in Plant 1978, 25.
[60] Pion-Berlin 1988; Roniger 2010, 31–3.
[61] Roniger 2010, 37.
[62] Plant 1978, 36.
[63] Rothenberg 2012, 182.

yet more expanded repression and an increasing reliance on paramilitaries and death squads over the 1960s and 1970s.[64] This was, to be sure, neither a linear nor homogenous process. The cycle was notably, though only temporarily, reversed following the successful defeat of the first phase of the guerrilla insurgency between 1966 and 1968, lessening the state's perception of crisis and permitting a period of relative stability and economic growth until roughly 1974.[65] Yet the basic ideological character of the elite remained largely unchanged, so that when the guerrilla threat re-emerged in the mid-1970s, the state's impulse was again to respond with severe repression.

This trajectory established the ideological foundations of the Lucas Garcia and Ríos Montt regimes, providing the dominant infrastructural influences on elite military officers and sustaining the critical hardline coalition between those officers, the broader military establishment, and core political and economic constituencies during the mass killings. As Manolo Vela Casteñeda summarizes:

> [T]he high command of the Guatemalan army, those serving in a planning capacity, were part of a generation developed in a climate of virulent anti-communism. They had begun their careers around 1954 … in a world formed by a coalition among the oligarchy, the Roman Catholic Church and a military high command strongly supported by the United States that was intent on undoing social reform, especially agrarian reform. Anti-communism cemented the alliance between the upper classes and the military, becoming the ideological mortar sustaining the entire political order.[66]

In combination with the new tactics of the guerrillas—specifically efforts to mobilize Guatemala's indigenous population—this ideological foundation created propitious conditions for mass killing.

Again, these cycles of cumulative radicalization were not simply 'natural' or 'rational' responses to the crises of twentieth-century Guatemala—indeed, they tended to worsen those crises. But once the pre-existing ideological infrastructure of the Guatemalan state and its constituencies is brought centrally into the picture in each crisis, radicalization becomes a predictable outcome. The pre-existing ideological history of twentieth-century Guatemala encouraged the formation of a catastrophic ideological synthesis in which, as Grandin puts it: 'Cold War anti-communism revitalized nationalist racism against Maya Indians and reinvigorated old forms and justifications of domination.'[67] This ideological character of the military regime was especially key in the eventual shift to mass killing under the Lucas Garcia regime. Faced with a growing guerrilla threat, the Lucas Garcia regime

---

[64] Brett 2016, 124. See also Yashar 1997, 223–5.
[65] Garrard-Burnett 2009, 31–5.
[66] Vela Castañeda 2016, 235–6.
[67] Grandin 2017, 4.

did not respond with a sustained effort to fight a limited antiguerrilla war. This would have been a plausible and strategically rational response to the crisis, but was precluded by the regime's ideological propensity to see the entire indigenous population as subversives and an entrenched belief that state terror would be effective in 'teaching the subversives a lesson'.

7.3 Ideology and Counter-Insurgent Mass Killing

The military regimes of Lucas Garcia and Ríos Montt were not ideologically homogenous.[68] There was an important ideological divergence between 'on the one hand, the more fiercely anticommunist counter-insurgent officers who believed "good works" to be far less effective than 100% brute force in countering the insurgency [and], on the other hand, civic-action counter-insurgent officers who believed in combining the two'.[69] While the former group's dominance lasted through the Lucas Garcia period, the latter 'developmentalist' group dominated under Ríos Montt, pushing for a rationalization of violence in which 'only' 30% of the civilian population in conflict zones would be targeted with mass killings (as opposed to the theoretical '100%' under Lucas Garcia), while 70% would be co-opted in 'model villages' and 'civil defence patrols'. A third ideological standpoint was Ríos Montt's own idiosyncratic blend of military ideology with evangelical Pentecostalism, which had significant appeal amongst the regime's constituencies but did not extend much within government circles beyond Ríos Montt and his close advisors. Beneath the policymakers, junior officers and rank-and-file soldiers were even more diverse, though often more conservative than the high command.[70] Far-right death squads, in clientelist but sometimes ambiguous relationships with the regime, formed an especially ultrahardline element.[71] Such ideological divergences were consequential for the nature of state violence under different governments. The ascendency of developmentalist officers under Ríos Montt and his successors, though initially increasing civilian fatalities, produced more restrained—though still violent—repression from the mid-1980s onwards, in the face of significant opposition and coup attempts from remaining *ultraderechista* (ultraright) elements.[72]

Guatemala's mass killings were not rooted, then, in some sort of monolithic ideology uniformly replicated across perpetrators, but in a dominant set of overlapping hardline doctrines that made massive violence against civilians appear justified and which sustained the necessary perpetrating coalition. The blend of

---

[68] Yashar 1997, 201–2 & 223; Schirmer 1998, 18.
[69] Schirmer 1998, 37. See also National Foreign Assessment Center 1981, iv & 9.
[70] Schirmer 1998, 207.
[71] National Foreign Assessment Center 1981, 14–15.
[72] Schirmer 1998, 157–85.

ultraconservative anticommunism and racist nationalism generated an understanding of the Maya population (and broader left-wing groups) as an extreme threat to society, sweepingly guilty of subversive collaboration with the guerrillas, and to a large degree external to the primary political community. In addition, a blend of militarism and nationalism led military elites to the conviction that mass killing was an ultimately effective, necessary, and valorous strategy in the fight against the guerrillas—disagreements on its exact form notwithstanding. Many also conceived of such a strategy as critical for a project of nation-building to forge, in Ríos Montt's term, *La Nueva Guatemala:* a new Guatemala in which social instability, developmental problems, and disunity (presumed to be rooted in the uncivilized and gullible character of the indigenous Maya) would be largely eliminated. Consequently, despite the ideological variations within and between the Lucas Garcia and Ríos Montt regimes, both willingly endorsed the use of mass killings.

The Military's Political Elite

It is widely accepted that Guatemala's political elites under Lucas Garcia and Ríos Montt perceived the indigenous population in its entirety as actively supporting the guerillas, and thus constituting an existential collective threat and criminally subversive sector of society. As Héctor Alejandro Gramajo Morales, an influential military officer who became Guatemala's Defence Minister in the late 1980s, observed: 'With the strategy of the EGP [insurgency] being played out in 1979 as whites fighting Indians, we [in the army] saw in every Indian an enemy.'[73] Towards the end of the Lucas Garcia period, the US Ambassador observed in a cable that: 'The well-documented belief by the Army that the entire Ixil Indian population is pro-EGP ... has created a situation in which the Army can be expected to give no quarter to combatants and non-combatants alike.'[74] This was, indeed, consistent with a broader Manicheanism amongst military leaders, as Dirk Kruijt summarizes: 'Whoever was considered not a hundred percent government loyalist was regarded as an enemy, an insurgent, a criminal, and thus a communist.'[75] Consequently, as the insurgency intensified, '[t]he Lucas government's reaction, led by the president's brother, Defence Minister Benedicto Lucas, was to respond with mass terror, convinced that only drastic and precipitous action could halt the guerrilla advance'.[76] While this proved, as observed earlier, hugely counterproductive in practice, the regime assumed that mass killing would ultimately defeat the insurgency and secure a prosperous future for Guatemala, with Lucas Garcia

---

[73] Ibid. 84.
[74] Cited in Garrard-Burnett 2009, 87.
[75] Kruit 2000, 16. See also Schirmer 1998, 209–12 & 251.
[76] Garrard-Burnett 2009, 49.

defending the violence on the ground that: 'When we have exterminated this social stain, those criminals, we will be able to advance more rapidly towards the collective common good.'[77]

On this, the leading figures of the Ríos Montt period were in full agreement. The threat of guerrilla insurgency was assessed to be extreme. As Harris Whitbeck, a personal advisor to Rios Montt in 1982–83, saw it: 'When we took power, the guerrillas controlled 90 percent of the antiplano. They really did. They absolutely could have taken over ... If the March 23 coup had not taken place, very possibly they would have won the war.'[78] Since this threat was perceived as deriving from indigenous support, the Ríos Montt regime—like the Lucas Garcia regime—continued to see mass killing as strategically necessary and justified self-defence and punishment. From an early stage, Ríos Montt had observed that: 'The problem of war is not just of who is shooting. For each one who is shooting, there are ten working behind him.'[79] Winning the war thus meant defeating these supporters—perceived as the overwhelming majority of the indigenous population. As Ríos Montt's press secretary later explained:

> The guerrillas won over many Indian collaborators ... Therefore the Indians were subversives, right? And how do you fight subversion? Clearly, you had to kill Indians because they were collaborating with subversion. And they [human rights advocates] would say, 'you're massacring innocent people.' But they weren't innocent. They had sold out to subversion.[80]

Ríos Montt consequently claimed that the mass killings reflected a policy 'not one of scorched earth but of scorched communists.'[81] In other words, he explained: 'we are killing people, we are slaughtering women and children. The problem is, everyone is a guerrilla there.'[82]

Such claims did not merely reflect a hugely exaggerated perception of how many indigenous Guatemalans were actually supporting the insurgency; they also reflected an identification of indigenous Guatemalans as *intrinsically* dangerous irrespective of any acts they had yet performed. As Brett observes: 'the very existence of indigenous communities was perceived as menacing, given their alleged propensity to oppose the values and practices of the dominant political community and thus their innate capacity to mobilize as subversives to overthrow it'.[83] Even children were readily targeted on the grounds that they would probably become

---

[77] Cited in Brett 2016, 127.
[78] Cited in Garrard-Burnett 2009, 25, fn.4.
[79] Cited in ibid. 90.
[80] Cited in Valentino 2004, 212.
[81] Cited in Garrard-Burnett 2009, 14.
[82] Cited in ODHAG 1999, 291, fn.4.
[83] Brett 2016, 42.

guerrillas in future.[84] Such conceptions imbued much of the violence in Guatemala with a genocidal character—targeting individuals on grounds of group membership and as part of a strategy of physical (or, through resocialization in model villages, cultural) elimination of Maya identity. But vast collective violence was also encouraged by the military's fundamental ideological intolerance to any form of opposition, and its extremely elastic conception of what a guerrilla or subversive was. As Garrard-Burnett observes: 'The vague definition of what being "with the guerrillas" meant eventually contributed to the deaths of many thousands of people, as the scorched-earth campaign called for the elimination of everyone who had seemed to have any connections with the guerrilla whatsoever.'[85]

As in many mass killings, moreover, these perpetrator perceptions often bore little if any relationship to how victims self-conceptualized. As survivors interviewed by Brett observed: 'They killed us because we were indigenous, but many of us only realized we were indigenous when the violence began.'[86] Garrard-Burnett similarly notes how: 'While the guerrillas had enjoyed explicit support in some villages, in many locations villages knew little about them or expressly did not want any association with them.'[87] For both the Lucas Garcia and Ríos Montt regimes, then, mass killing did not reflect especially esoteric ideological goals, but nor was it a rational response to the strategic realities of civil war that any incumbent government would have seen in similar terms. Instead, the strategic employment of mass killings reflected a particular interpretation of the guerrilla insurgency and its purported supporters rooted in the distinctive ideological worldview of Guatemalan military and political elites.

Other aspects of the mass killing were similarly rooted in the regime's ideological narrative of the civil war more than material or strategic realities on the ground. In particular, the military's counter-intuitive targeting of Catholics in the highlands, bearing in mind its support base in urban Catholic society, is comprehensible when the regime's extreme anticommunist and ultraconservative diagnosis of the conflict is unpacked. Catholic organizations had taken a leading role in forming economic cooperatives in the highlands since the 1960s, in the process promoting indigenous education and political awareness. This had, from a fairly early stage, presented potential frustrations for the future economic interests of the military and wealthy landowners in northern Guatemala. Over time, it began to provoke longstanding fears amongst Guatemala's elites of indigenous mobilization and animated suspicion of 'communist' economic activity. Stoll observes, moreover, that while military suspicion of Catholics was typically misplaced, 'the army's allegations against the Catholic Church did not lack plausibility

---

[84] Valentino 2004, 213; Brett 2016, 42.
[85] Garrard-Burnett 2009, 39, fn.82.
[86] Cited in Brett 2016, 46.
[87] Garrard-Burnett 2009, 97.

[since] guerrilla priests are part of Latin American folklore', due to prominent stories of Catholic priests joining revolutionary movements in the 1970s under the influence of new liberation theologies connecting Catholic religious belief and socialism.[88] Local Catholic priests—several of whom were US citizens engaged in development work—were also potentially troublesome for the military, given their local moral authority and ability to communicate with groups outside the highlands, and even Guatemala, through their ownership of aircraft and radio equipment.

Within the regime's intensely anticommunist security doctrine, then, Catholic activity in the highlands became diagnosed as part and parcel of the insurgency. As Falla concludes, the military 'considered religion to be merely a screen for the guerrillas',[89] with the result, as Frank Afflitto observes, that even 'attending Catholic Mass became a perceived communist threat to Guatemalan national security'.[90] Local inhabitants in the highlands testified that 'if you were a Catholic, the army said you were a communist',[91] and the army frequently recorded targeted killings of Catholics as the deaths of 'subversives' in combat.[92] These were not mere post-hoc rationalizations. On the contrary, army suspicion and hostility towards Catholic organizations in the northern highlands can be traced back to the 1970s.[93] This hostile ideological construction of Catholics was not sufficient on its own to generate massacres.[94] But *in tandem* with the growing intensity of the insurgency and the army's perception of extreme threat, it generated an otherwise puzzling current of repressive violence.

The specifically genocidal aspects of Guatemala's violence against indigenous communities also reflected the military's understanding of mass killing as necessary, functional, and morally laudable in service to the regime's core project of building a new, united, and secure Guatemalan nation. As suggested by the discourses of the mid-twentieth-century democratic period, indigenous communities—as well as left-wing ladino groups—had long held ambiguous status, at best, as members of that nation in the eyes of Guatemalan political elites. In the 1970s and 1980s, the military diagnosed the 'failure' of indigenous communities to integrate themselves—i.e. to become dutifully subservient Guatemalan citizens—as a central source of threat.[95] Though present under the Lucas Garcia regime, these themes were intensified considerably under Ríos Montt's much more explicit vision of national salvation. As Garrard-Burnett summarizes:

---

[88] Stoll 1993, 170
[89] Falla 1994, 29.
[90] Afflitto 2000, 122. See also Stoll 1988; Stoll 1993, ch.6.
[91] Stoll 1993, 177.
[92] E.g. Falla 1994, 28.
[93] Ibid. 24–8.
[94] Stoll 1993, 171–2.
[95] Grandin 2017, 9.

By the 1980s, this failure to subscribe to the national project carried, for the Guatemalan government, a strong suspicion of subversion ... Outsiders did not, in Ríos Montt's view, grasp the extent to which ethnic issues interfered with national unity ... At the heart of this Guatemalan uniqueness was the inchoate nation, based on problematic Indian roots, that still lacked definition and direction.[96]

Thus, in Ríos Montt's rather elliptic formulation: 'We are going to make Guatemala, we make Guatemala, we make it ... It will be grand, sovereign, and independent and when we have the strength, the consistency, and our own dignity, we will *be* Guatemalans.'[97] Anything outside the 'strong', 'consistent', and 'dignified' form of Guatemalan life as defined by Guatemalan elites was not actually Guatemalan at all, but a barrier to Guatemalan security and sovereignty. Brett likewise emphasizes how:

[T]he image of the indigenous population in Guatemala had, since the colonial encounter, been wrought through the figure of the abhorrent other within the discourse and imaginary of the primary political community. Of primitive and slovenly race, the 'indian' ... represented an obstacle to the country's modernisation, yet 'indian' marginalisation and invisibility, however much an impediment, had to be maintained indefinitely in order to preserve the economic interests and political power of the non-indigenous, primary political community.[98]

This specific understanding of indigenous otherness as incompatible with Guatemalan prosperity, security, and integrity again explains the genocidal elements of the campaign. Alongside the massacres, military and political strategists called for an 'intense, profound, and carefully prepared psychological campaign to impact the Ixil mentality and make them feel part of the Guatemalan nation.'[99] Most explicitly, the military's Civil Affairs division contended that such policies should involve 'intensifying the ladinization of the Ixil [Maya] population *until it disappears as a cultural subgroup foreign to the national way of being*.'[100]

This highlights a form of genocidal intent that problematizes current disputes over whether mass killing in Guatemala was 'really' counter-insurgency or genocide or 'cultural genocide.'[101] *The military simply perceived no mutual exclusivity here.* Since it did not act out of a purely biological, Nazi-style racial theory, it could employ a mixed genocidal strategy involving both physical killings and coercive cultural reorientation, which in turn was part and parcel of a strategy of

---

[96] Garrard-Burnett 2009, 71 & 78.
[97] Cited in ibid. 78.
[98] Brett 2016, 215.
[99] Rothenberg 2012, 68.
[100] Cited in Schirmer 1998, 104 (*my emphasis*).
[101] Somewhat contra: Schwartz and Straus 2018. See also, in general, Salazar 2012; Brett 2016.

counter-insurgency. As Schirmer has observed, the Ríos Montt regime's 'kill 30%, re-educate 70%' approach aimed to create a 'sanctioned Mayan', purged of genuine indigenous culture and identity, politically neutered, and unthreatening to national security as the military perceived it.[102]

This exclusivist nationalist vision was, moreover, coupled to the military's understanding of national prosperity as requiring an all-powerful security state, free from meaningful legal or political constraints. The virtues of ordinary citizens were understood in moralized and highly gendered terms as involving military-style masculine martialism and obedience to the state. As Ríos Montt expressed it: 'The poverty of our country is a poorness of men; Guatemala lacks men who have integrity, decency, honesty, truthfulness, honor, manliness, a manliness that builds its base from something very simple, which is obeying the law.'[103] At the same time, the military had long adhered to a deeply instrumentalist view of the law, which, in practice, meant that no legal constraints existed on what the state was allowed to do in the name of national security. As one of the military's senior lawyers later recalled:

> In the year of Watergate, when my American friends were considering whether Nixon had broken the law, I asked how it was that they believed that a country can be solid, structured, and powerful if it didn't have a system of [State] security? The security of the State is above all else, and the President must be in violation [of the law] for that security so that we can all be secure ... you can change the law to adjust it to State security ... [since] national security stands above all law. I don't believe that I have ever, in my thirty-five years of legal experience [here in Guatemala], come across a case in which one could say: 'Here the law supersedes national security.' It could be that the state of national security itself is violating rights, but one must see whether the person [whose rights are being violated] is doing something against the security of the State. One must always protect [the State].[104]

This demonstrates a highly distinctive ideological understanding of law and security. Its entirely instrumental vision of adherence to legal restrictions was frequently reflected in military planning documents. Those for the military's operations in the Ixil, for example, stated: 'We have all the structure of the Law which assists us and thus—even in a situation almost of war in which we live—we must attach ourselves to the law *up to where it permits us to fulfil the mission*.'[105] Nominal prohibitions on targeting civilians could thus be disregarded whenever they were

---

[102] Schirmer 1998, ch.6.
[103] Cited in Garrard-Burnett 2009, 79.
[104] Cited in Schirmer 1998, 141.
[105] Cited in ibid. 140 (*emphasis added*).

thought inconvenient. An American military attaché present at a key meeting between Ríos Montt and his senior military commanders in July 1982, just before the fresh intensification of mass killings, observed how the president:

> emphasized the fact that the plan was made very general to permit each commander as much freedom of action as possible in his assigned area ... He wanted each commander to take special care that innocent civilians would not be killed; however, if such unfortunate acts did take place he did not want to read about them in the newspapers.[106]

Similarly, Colonel Garcia, one of the military's senior field commanders, later explained:

> Under ideal circumstances, there should not be a moment when one acts outside the framework of the Law .... *But it was not us*, the military in this country, *not us who started this war*. This [war] came from an external aggression ... In accordance with its function, as its constitutional mandate, [the Army] fought [these guerrillas] on their terrain and under the conditions they offered. What happened was that our organisation proved to be more efficient, more efficacious.[107]

As Schirmer summarizes, the result was a situation where rights 'become "securitized": continuously subject to qualification or denial whenever they are deemed to be in conflict with security interests of the State ... Non-securitized and inherent rights not subject to the power of the state do not exist within the military's definition of the term in Guatemala.'[108] Since what was 'required' in the name of national security was left entirely to the discretion of military elites, this conception amounted to the flat denial of any constraints on the regime's waging of a total war against those civilian groups it saw as constitutive of the insurgent threat in civil war.

Rank-and-File Agents in Guatemala's Mass Killings

As in most cases, the drivers of rank-and-file participation in Guatemala's mass killings were diverse and cannot be reduced to ideology. There is considerable evidence of individuals participating for self-interested and essentially apolitical reasons, or guided by local rivalries and animosities disconnected from

---

[106] Cited in Garrard-Burnett 2009, 91.
[107] Cited in Schirmer 1998, 140 (*emphasis in original*).
[108] Ibid. 136.

the broader ideological justifications of the violence.[109] The Guatemalan military also employed 'hard' and 'soft' coercive pressures to mobilize direct killers. Many soldiers were forcibly recruited, and the military apparatus did sometimes threaten and resort to lethal violence against its own members, as well as using intensely brutalizing forms of military training.[110] Participants in the 'civil self-defence patrols' in indigenous communities under military occupation were also subjected to intense pressure from the army to collude in or perpetrate massacres.[111] As emphasized by situationist theories, military units and self-defence patrols became sites for intense intragroup pressures in which the more enthusiastic hardliners could impose participation in violence as the local social norm.

Yet ideology remained central to the mobilization and organization of rank-and-file perpetrators in Guatemala. The military's hardline security doctrine enabled and encouraged mass killing through both internalized justifications for the violence and institutionalized ideological structures that created social pressures and incentives for participation. While coercion was a factor, Manolo E. Vela Castañeda suggests, based on interviews with former rank-and-file perpetrators, that in most cases 'it seems clear that those who participated in the massacres did so voluntarily. No one forced them. Killing was a process of self-selection ... These groups of soldiers decided freely that the cruelty of the killing they carried out was something they had to do.'[112]

Ideology's central significance for rank-and-file perpetrators was not, however, as a form of 'special' motivation linked to esoteric political goals, but as a security-orientated justificatory narrative that drew together the military's values and interpretations of the conflict to promote extreme violence against civilians. While 'true believers' in regime ideology were in the minority, scholars of Guatemala's civil war emphasize that partial and selective internalization of justificatory ideology was widespread. Garrard-Burnett, for instance, concludes that:

> As members of civil patrols or indigenous conscripts—even ordinary Guatemalans who were not directly involved in the war—came to imbibe the doctrines of counterinsurgency ... [many] came to believe they were on some kind of higher mission: to save the nation, to root out the 'bad seed,' which morally justified the most extreme actions. In combination with the army's powerful esprit d'corps [sic], it was the regime's ability to give meaning to these actions that helped make them not only plausible but actually possible.[113]

---

[109] Brett 2016, 130–1; Vela Castañeda 2016, 241.
[110] Schirmer 1998, 155; ODHAG 1999, 121, 132, & 173; Vela Castañeda 2016.
[111] ODHAG 1999, 8; Garrard-Burnett 2009, 100–1.
[112] Vela Castañeda 2016, 238 & 241.
[113] Garrard-Burnett 2009, 111.

The Archdiocese of Guatemala truth commission likewise concluded that:

> During the training period, the army tried to instil in its soldiers an ideology that would serve as a frame of reference to justify their actions psychologically. The army fostered a sense of group unity and morale, and a preconditioned hostility toward anything that could be related to the guerrillas ... This portrayal of the conflict was designed to present the army as a victim. Poverty was blamed on guerrilla actions, and the mother country was glorified as a supreme entity that required unanimous cooperation to ward off the threat of communism.[114]

What evidence supports these conclusions? As in all cases, inferences about the precise role ideology played for different perpetrators can only be tentative. But two broad sources of evidence exist: first, a limited but revealing body of research involving direct interviews with perpetrators and their military and paramilitary peers; and second, the testimony of witnesses to and survivors of the military's massacres on the situational statements and behaviour of perpetrators.

In interviews, perpetrators repeatedly testify to the intense efforts undertaken by field commanders and training officers to inculcate soldiers on the ground with the military's conception of civilian communities as part of the guerrilla insurgency. As one former soldier attached to the army's intelligence division reported:

> When it was time to patrol, they told us, 'Okay, guys, we're going to an area where there are only guerrillas. Everyone is a guerrilla there. Children there have killed soldiers, and supposedly pregnant women have just come and thrown a bomb and killed; they have killed soldiers. And so you all must distrust everyone. No one is a friend where we are going. So, they are all guerrillas and all of them must be killed.[115]

Officers interviewed by Vela Castañeda likewise explained how: 'They made it clear that this whole area was under EGP [insurgent] control. From the Ixil Triangle north, you are fighting everything there. There was no differentiation. It was all enemy.'[116] Consequently, many commanding officers appear to have been pre-convinced of the guilt of local civilians irrespective of their actions or responses to interrogation. One soldier stated:

> [W]e heard a lecture from an officer and the thing I remember most strongly was when they showed us a map and said 'Look, this whole mountain range from the Cuchumatanes north, all of these people are guerrillas. They are collaborating, they are part of the guerrilla ... Meaning anyone you find there you have to kill them.[117]

[114] ODHAG 1999, 128. See also Vela Castañeda 2016, 230.
[115] Cited in ODHAG 1999, 31.
[116] Cited in Vela Castañeda 2016, 235.
[117] Cited in ibid.

Another explained:

> The civilian population was converted into the enemy, that which had to be fought ... I saw the young troops, twenty-three, twenty-two years old, firmly determined to fight and attack, they clearly saw the civilian population as an enemy. Because they were collaborating with the guerrilla they were against the army, against the country.[118]

Again, the threat posed by civilians was presented as so intolerable and deeply rooted in the population that even children had to be killed for fear that they might grow up to support the guerrilla. A former member of the civilian 'self-defence' patrols explains how perpetrators:

> told my sister ... that they had to finish off all the men and all the male children in order to eliminate the guerrillas. 'And why?' she asked, 'and why are you killing the children?' 'Because those wretches are going to come some day and screw us over.' That was their intention when they killed the little ones too.[119]

A detailed example of how officers mobilized soldiers' willingness to perpetrate comes from the especially brutal and lethal massacre in mid-March 1982 at the village of Cuarto Pueblo, in which an estimated 362 civilians were killed. One witness describes[120] how the commanding lieutenant exhorted the soldiers to continue with the massacre by claiming:

> They [the residents of Cuarto Pueblo] are friends of the guerrillas, but they aren't saying so. We have to finish them all off, to put an end to the guerrillas. The women are preparing their food. If we finish them all off, things will soon calm down. The men are helping them. But when we ask about that, they say they don't know anything. Lies! ... We're going to do in the lot of them. That way they won't be able to help out the guerrillas. If you finish off the people, there's no one left to help them ... We have a list of guerrilla villages. We already know which entire villages are going to be wiped out. We'll finish this town first and then we'll go to others ...[121]

Frequently such violence was in some sense retaliatory in nature—it was provoked by guerrilla attacks—but it took the form of *collective punishment* reflecting the army's assumption that whenever the guerrilla engaged in assaults, the local

---
[118] Cited in ibid. 233.
[119] Cited in ODHAG 1999, 31.
[120] The words are not a transcription, and allowance must be made for a testimonial gap of 1–2 years; see Falla 1994, 8, fn.1 & ch.7.
[121] Ibid. 89. See also Brett 2016, 168–9.

civilian population was equally guilty. The Cuarto Pueblo massacre itself may have been conducted in part because Cuarto Pueblo was the site of a particularly bloody guerrilla attack on the local army detachment a year earlier.[122] One widow whose (civilian) husband and son were executed by the military was told by an officer: 'What did you expect? ... They killed my boys. My boys had reason to be mad. Many of them died today, and you people killed them.'[123]

These accounts throw into relief the interwoven nature of ideology and strategic thinking at the leading edge of mass killing. Perpetrators did not appeal to abstract or utopian political ideals to explain their participation in the violence. But nor are they responding to self-evident realities or irresolvable ambiguities of the situation (as if it were rational to assume that children must be killed to win a civil war). Instead, Guatemalan perpetrators operated under a particular ideological construction of the security context, which reflected deeper racist and anticommunist elements of the military's distinctive political worldview. Whatever the civilians did or said was irrelevant: whether they admitted to collusion under torture and interrogation or denied it, the officers had generally already decided that they were guilty since they were on the military's 'lists'.

That such claims were sincerely internalized by at least some soldiers, and did not merely function as post-hoc rationalizations, is supported by much of the behaviour and discourse reported by witnesses from the massacres themselves. Some selection bias in memory is probable. But such testimony rarely portrays the most prominent front-line perpetrators as reluctant killers, grimly participating under coercive pressure. Significant numbers of rank-and-file perpetrators energetically drove on the violence, and 'many took the initiative to root out potential subversion on their own, often zealously'.[124] Organizing officers frequently appear to have fully endorsed the military's ideological rationale for mass killing. One local municipal leader observed how for the army's officers: 'The analysis was that they were all guerrillas. That was the analysis. The orders here were, "a known guerrilla is a dead guerrilla" and being indigenous meant being a guerrilla. They were the same for the Army; there was no difference.'[125]

When civilians took up a government amnesty in June 1982, they too faced officers who told them: 'You're all thieves. You're the ones that kill the Army. You're like cats that hide in the mountains. You're animals ....'[126] The brutality of the abuses by ordinary soldiers and their reported discourse during the massacres are also suggestive of the influence of military ideology. As the Archidiocese report concluded, citing witness testimony:

[122] Brett 2016, 167.
[123] Stoll 1993, 82–3.
[124] Garrard-Burnett 2009, 100.
[125] Cited in Rothenberg 2012, 147.
[126] Cited in ibid. 37.

This behavior, which was officially sanctioned by the army, was based on regarding people with contempt: 'They are shit; they don't deserve to live because they support subversion.' Extreme contempt is evident in the way the killing and razing took place. The army often used people's silence as evidence that they were guerrillas: 'We have to finish them all off, because those people don't say anything when we ask them questions.' ... According to many testimonies, the initiative behind the massacres came from higher levels. But witnesses also described the insensitivity of the soldiers who carried out the massacres: 'It made them laugh.'[127]

Numerous other witnesses testify to such evidence of enthusiastic brutality towards the population.[128] For at least a segment of the perpetrating forces, it seems that—in the words of one ladino member of the highlands cooperatives—'the military, they are proud of what they have done'.[129]

As the above examples indicate, not all testimony from front-line soldiers suggests a specifically *racialized* animosity as a direct motivator of the killing—perhaps unsurprisingly when it is remembered that so many of the perpetrators in Guatemala's mass killings were themselves indigenous. But soldiers and witnesses do suggest that a minority of perpetrators may have been motivated by racial hatred, and that broader racism was widespread, especially amongst superior officers. One of Vela Castañeda's interviewees reported how a senior officer declared that: 'It would be shameful if [an indigenous] name like Popsoc appeared on [the walls of the officer training school]. Indian names stain the race.'[130] Another explained how in the training of indigenous soldiers:

There was a sub-lieutenant who was just all the time, 'Indian this, Indian that, just Indian, Indian ...' He'd say to a soldier, 'Here, look, do this or that thing', and if the soldier didn't hear and ask him to say it again, the officer would start screaming, 'You disgusting dirty smelly Indian! Just do what I told you!' All the officers were like that, only 'Indian, dirty, repulsive, appalling, animal'.[131]

Undergirding the military's explanation of the indigenous population's purportedly universal support for the guerrillas was a set of racist claims about the stupidity and gullibility of the 'Indian'. The soldiers themselves often appeared to internalize such conceptions—over two decades later, one interviewee contended that: 'Most of the civilians who died were *naturales*, pure indigenous. All those people died because the guerrilla fooled them. The guerrilla deceived those Indians.'[132] Responsibility for the violence was thereby displaced onto the victims

---

[127] ODHAG 1999, 172.
[128] Ibid. 73–5 & 131; Rothenberg 2012, 51–2.
[129] Cited in Manz 1988, 85.
[130] Cited in Vela Castañeda 2016, 234.
[131] Cited in ibid.
[132] Cited in ibid. 233.

and/or the enemy, not the perpetrators. Vela Castañeda observes how: 'The constant repetition of the idea that "Indians allowed themselves to be deceived by the guerrilla" began to take on apocalyptic dimensions and to weave together the various strands that justified the state response.'[133]. Based on interviews with survivors, Brett reports that:

> [I]nterviewees stated how, during operations, massacres and torture, soldiers accused the indigenous of being inherently gullible and subversive, naturally untrustworthy. In the words of Casaus Arzu, the entire indigenous population was stigmatised as 'communists, infidels, idolators and sinners, as irrational and opposition' to facilitate the ethnic violence. According to a survivor of the massacres ... 'The soldiers screamed at us that we were only indians, that we were nothing, that we were only animals, that we didn't deserve the respect of a human being, that the violence was all our fault. By being indigenous, we were guilty.'[134]

Constructions of victims as threatening and guilty were thus interwoven with racist and dehumanized deidentification of the indigenous population.

Of course, there was considerable variation in how perpetrators internalized such justificatory claims and frames. Some became relatively committed to the campaign of violence. Vela Castañeda observes that: 'Within each unit there was always a more radical group. From within the platoons, among the troops, they would seek out the fiercest soldiers, those who were really violent, and they pushed them to become harder, even more radical.'[135] As in other cases, such enthusiastically hardline perpetrators were typically responsible for a disproportionate share of the violence, and appeared especially prominent in far-right death squads or the military's elite and vicious *Kaibiles* special forces. The identification and more extensive utilization of such individuals were encouraged by the military:

> [S]oldiers who stood out for their bloodlust and ferocity were encouraged to re-enlist, and undertake special training in the jungles of El Petén at a camp called the *Infierno* or hell. These troops embodied a whole different level of identification with the state project and an intensive esprit de corps, increased by their sense of being the elite of the elite fighting forces.[136]

Within these units, extreme anticommunist and/or racist ideological commitments were an object of pride and enthusiasm amongst members, whilst also serving to legitimate their actions in the eyes of the pro-military portion of Guatemalan society.[137] The practices and operating procedures of these units were

---
[133] Ibid.
[134] Brett 2016, 64. See also Rothenberg 2012, 66.
[135] Cited in Vela Castañeda 2016, 237.
[136] Ibid. 240.
[137] ODHAG 1999, 111.

also held up as a model by the army. As an officer explained: 'The techniques that the *Kaibiles* learned in the *Infierno* were developed throughout the whole army. They carried the seeds and transformed all army training in the same direction.'[138]

In other cases, internalization appeared real, but more selective. Perpetrators without deep underlying commitment to the military's seething anticommunism and broad racism nevertheless appear to have adopted key military frames of the conflict, the victims, and the violence that—at a minimum—legitimated the killings and may have also supplemented more situational motives. As one former intelligence officer testified:

> They put confusing ideas into your head there. For example, they tell you that, in Guatemala, 'We can't allow ourselves to be conquered. Nothing to do with communism. Communism comes to take away lands and everything. It comes to exploit; it comes to do this and it comes to do that.' They brainwash you; they brainwash you good ... And the soldier gets indignant and says: 'Yes, the ones who caused this are the guerrillas, and that's why Guatemala is poor.' ... So with a word they all become enemies of the people, of the whole country. And when you are in training, you say 'That's true.'[139]

Often what mattered was the inculcation in soldiers of a basic identification with their military units, which provided a foundation for ideologically adopting the military's broader narratives and claims. Vela Castañeda reports:

> For young indigenous men, total immersion in this world radically challenged the basic foundations of their identities ... Over and over again in the interviews, I saw how the men's indigenous identity had been displaced by a more powerful identification, one that would help them to survive the kind of daily battering they endured in the military—an ideological identity. Now, deeper than their indigenous roots, these young men identified as *soldiers*. It did not matter if back there in the mountains the enemy looked like them, or that those who commanded them did not. The enemy that had to be sacrificed was no longer fully human; their deaths were not crimes.[140]

Without becoming true believers, then, ordinary soldiers could come to sincerely accept the military's narrative of the civil war in ways that enabled relatively smooth participation in mass killing.

But the military's justificatory narrative also promoted participation in mass killing through structural pressures and incentives: generating norms of behaviour that reflected the military's hardline values and expectations, while funnelling

---
[138] Cited in: Vela Castañeda 2016, 240
[139] Cited in ODHAG 1999, 128–9.
[140] Vela Castañeda 2016, 234.

self-interested motives to generate instrumental employment of the military's ideological claims. Garrard-Burnett observes how, as military institutions penetrated the countryside, 'longstanding feuds, rivalries, land disputes, religious differences, and personal differences assumed a political veneer',[141] with local actors manipulating the ideological discourses of the military campaign to serve their interests, escalating the local scope of violence. Military socialization of soldiers likewise blended coercion and intragroup pressures with norms set by the military's ideology. The military espoused extreme demands for unquestioning and automatic obedience from subordinates and a maximal suppression of any critical consideration of army policy. As one officer explained: 'Let's say they told you to kill this person. You couldn't say, "I won't do it," because they had drilled into us that an order was to be obeyed without question.'[142] Another officer reported:

> Unlike other institutions where one might hold on to a certain independence or autonomy, or make decisions for yourself, the army just absorbs you. They put you in a small unit ... you become, well, like a number, a person whose identity has to mould itself to the organization's personality. You might not agree with something but you just have to bear it and you have to conform, and act like everyone else is acting. It's a system in which the soldiers almost stop, well, thinking, I mean you can't really stop thinking, but they were so regimented, every minute, every hour, every day, told what they had to do. You just don't have much alternative. You might desert or resist, but most just bear it, they stay inside the system.[143]

In such a system, self-interest and self-esteem overlapped with conformity to encourage compliance with the regime's ideology. Given the military's brutal moral ideals for soldiers' behaviour, a 'perverse system was created in which disregard for human life was a prerequisite for promotion'.[144]

To a degree, of course, such processes of authoritarian socialization and training are common to most militaries. What is distinctive about military units prone to perpetrate atrocities, however, is that such processes are taken beyond military professionalism, to create a hardline institutional culture that valorizes extreme martialism, disregard for human life, blind obedience, and the vigorous abandonment of potential restraints on egregious violence. Even relatively apathetic soldiers, coerced into serving, were subject to 'training in a series of values and customs that was completely devoid of any notion of human rights or international humanitarian law'.[145]

---

[141] Garrard-Burnett 2009, 102.
[142] Cited in ODHAG 1999, 129.
[143] Cited in Vela Castañeda 2016, 237.
[144] ODHAG 1999, 129.
[145] Ibid. 126.

This all illustrates the contrived nature of attempts to isolate whether perpetrators were 'primarily' guided by self-interest, social pressure, habituated practices, or ideology. In mass killing, such factors become interwoven and mutually reinforcing. What served a perpetrator's self-interest was to engage in the behaviour implied by the regime's ideology and the justificatory narrative for mass killing. Social conformity pressures were orientated towards military authorities and peer groups that espoused that narrative. More explicitly political motivations for violence were built on the foundations set by that narrative. Ideology mattered 'on the ground' as a framework, present in both sincere worldviews and institutional configurations and norms, that tied such diverse motives to the perpetration of state violence.[146]

Public Justification of the Killings

As in other cases of mass killing, the military's ideological justifications served (alongside outright denials about the extent of military atrocities) to legitimate the violence and maintain political support amongst key public constituencies. In this role too, the military's ideological justifications appealed to superficially prosaic claims of self-defence and national security rather than special ideological goals. The military presented all victims of army operations as guilty of collaboration with the guerrilla and thus part of an existential threat against the Guatemalan state—violence was, correspondingly, presented as a necessary and valorously patriotic defence of Guatemala against external enemies.

Such ideological constructions of the victims and the violence were consistently propagated through the media, through the speeches and statements of leading figures in the regime, and through the military institutions and political parties aligned to the ruling coalition. The central features of the publicly presented security doctrine emerged prior to the Lucas Garcia regime. Indeed, as soon as the insurgency began mobilizing indigenous supporters in the northern highlands, the previous government of President Kjell Laugerud Garcia had responded with human rights atrocities—notably the Panzós Massacre of 29 May 1978, in which Guatemalan soldiers shot several dozen peasant protestors. The Laugerud government's press statements immediately portrayed the peasant protestors as guerrillas. As the US embassy observed:

> According to the President, the incident was instigated by the Guatemalan Army of the Poor (EGP) ... Laugerud said that some of the '1,000' Indians assembled in

[146] Vela Castañeda 2016, 241.

the town square, armed with machetes and hoes, [and] attacked the military outpost of 22 soldiers from the rear ... Minister of Defence Otto Spiegeler [likewise] blamed guerrillas and 'religious' agitators ....[147]

As already observed, Lucas Garcia radicalized such claims by attributing guilty association with the guerrilla beyond specific protesting groups to the entire indigenous population of the highlands and all left-wing ladino political activism. The government also employed a broad futurizing narrative which presented violence as a necessary precondition for Guatemala's (military-led) economic development.[148] In certain important ways, however, the Lucas Garcia regime's legitimation efforts were not altogether successful. Both military insiders and large sections of the regime's core constituencies lost confidence in claims that the military violence would ultimately stabilize the security situation, and the heavy use of violence in politically influential urban areas was widely perceived as excessive and terrifying.

Scholars of the civil war generally agree that the Ríos Montt era marked an important shift in both the character and effectiveness of the regime's legitimation efforts. Stoll argues that:

> Until Ríos Montt arrived on the scene, the government had so disgraced itself that revolutionaries dominated the country's moral landscape. Ríos Montt challenged them 'morally as well as militarily,' providing a rationale for those who wished to support the army but had been stupefied by its behavior.[149]

Ríos Montt did this by layering both a richer moralized value-system, rooted in traditional religious values and narrative forms, and a more expansive vision of the future direction of the state and the function of wartime violence, on top of the Lucas Garcia regime's narrative of guerrilla threat and victim guilt. Though esoteric to some audiences, Ríos Montt's weekly 'Sunday Sermons', in which he offered lengthy speeches on the state of the civil war and Guatemala's moves to political reconstruction, were widely watched. Ríos Montt utilized such missives to entrench popular prejudices against the guerrilla and 'Indian' and to champion a confident moralized portrayal of the state and its actions. As Garrard-Burnett explains:

> Within this symbolic universe, Rios Montt's New Guatemala was built upon three fundamental principles. These, freely blending a kind of Pentecostalized liberalism with Cold War strategic interests, included: salvation, judgement and righteousness, each of which represented a specific strategy of political behaviour

---

[147] US Embassy, Guatemala, 'More on Campesino/Military Clash in Panzos,' 1 June 1978, DNSA.
[148] In, for example, the government's appeals to its 'thousand-day plan'; see ODHAG 1999, 217.
[149] Stoll 1988, 109.

and motivation … The second, judgement, was the darkest and perhaps the most elemental principle of the Rios Montt regime, in that it provided the necessary 'moral justification' for the Plan Victoria 82 [military campaign], including the massacres and the systematic effort to destroy Mayan culture, the 'fit punishment' for a wayward people seduced by the godless allures of communism and rebellion.[150]

Of course, this was heavily interwoven with dehumanization and racism—with the military frequently employing prejudicial stereotypes and implicit assumptions that the welfare of indigenous persons was not a priority, given their secondary status in the Guatemalan national community. As Brett summarizes:

Integral to the violence was the dehumanisation of the indigenous community … the creation of an ethnic hierarchy based upon invented criteria of biological, cultural and moral differences; and the attribution of responsibility to the indigenous for the crisis that the country was facing. In short, across the media and within the military doctrine, the narrative communicated *how indians are, and always have been, sub-human, untrustworthy, gullible, stupid and envious.*[151]

How important and successful were these processes of public legitimation? As in all cases, this is hard to assess with confidence, but two principal bodies of evidence suggest that legitimation efforts by the military significantly shaped public opinion and helped neutralize potential opposition to the mass killing campaign.

The first is the fact that, outside those longstanding left-wing organizations that had either effectively joined the violent resistance or been terrorized into submission, the main sectors of Guatemalan society remained largely quiescent in the face of mass killing by the state, even as severe repression in urban areas decreased under Ríos Montt. In part, this appears to have reflected belief in the military's dire threat constructions of the insurgency and genuine fear of a Marxist takeover. As the new US Ambassador to Guatemala, Frank V. Ortiz, noted on his arrival in September 1979:

Dominant Guatemalans see themselves as increasingly isolated and as the international target of a Marxist-inspired, hostile, coordinated campaign … It helps create a siege mentality. It strengthens the hand of those in Guatemala who favour repressive responses even to constructive criticism and to legitimate pressures for change.[152]

---

[150] Garrard-Burnett 2009, 83.
[151] Brett 2016, 64 (*emphasis in original*).
[152] Cited in Cangemi 2018, 629.

This significant hostility to and fear of the 'communist threat' was complemented by a general disinterest in the fate of people in the highlands amongst the more politically influential public constituencies. Alongside an acceptance of military claims that victims of violence were all guerrillas, this reflected a broad 'spectator racism', as Vela Castañeda labels it, which was 'found among those citizens who refused—despite the gravity of the events—to notice the terrible violations of human rights occurring between 1981 and 1982 in Guatemala. For them, the indigenous population was inferior, dispensable.'[153] Thus, while the Lucas Garcia regime's *urban violence* was widely perceived as excessive, contributing significantly to the collapse of support for the regime, dominant conceptions of the highland victims of mass killings as expendable in the fight against communism helped to diffuse potential opposition. The absolute peak of the violence, in April 1982, was sufficient to provoke some limited criticism, but was still 'not enough to jolt urban, middle-class Guatemala, comfortable in the city's newfound security ... even as violence in the countryside continued to unprecedented levels'.[154]

In addition, the regime's ideological justifications functioned as a form of negative legitimation—sowing confusion and frustrating potential solidaristic resistance to violence. Even where communities were not strongly supportive of the military and its broader ideological objectives, many presumed that its claims about those targeted contained at least a grain of truth. The CEH observed how:

> systematic accusations were created to ideologically co-opt the population while creating a negative and criminal image of social organizations and their representatives. This stigmatization was based on the idea that if something happened to someone it was because 'he was up to something.' In that way, the repression was justified, in many cases, relatives even blamed victims within their own families ... Phrases such as 'he must have done something' or 'why did he get involved with *babosadas* [guerrillas]' were repeated throughout the country ....[155]

In the highlands, similarly, the military managed to utilize its justificatory propaganda to generate animosity against the guerrillas—who were blamed for having brought down the army's repression upon local communities and failing to deliver their promises for revolutionary gains.[156] Such efforts helped fragment potential opposition to the military, even if it did not generate active public enthusiasm.

Again, this role of ideological justification in sustaining public quiescence in the face of the mass killings was especially effective under the Ríos Montt leadership, for three reasons. For one, Ríos Montt's effective cessation of urban violence

---

[153] Vela Castañeda 2016, 232. See also Brett 2016, 57.
[154] Garrard-Burnett 2009, 89.
[155] Rothenberg 2012, 147.
[156] Ibid. 152.

lent substantial plausibility—for those outside of the highlands—to the government's claims that it was restoring order to the security situation. Second, Ríos Montt's symbolic employment of a one-month 'amnesty' for the guerrillas in June 1982 (though applied with uneven consistency in practice) helped strengthen the government's claims that all supporters of the guerrillas after July 1982 were resolute traitors to the state and enemies of Guatemala, who had refused the opportunity to return to the community and thus deserved vengeful violence. Third, as noted above, the Ríos Montt government—and particularly the president himself—proved far more determined and capable in articulating and promoting a broad ideological vision of the government's activities by comparison to the Lucas Garcia regime. Garrard-Burnett contends that many Guatemalan citizens 'seem to have genuinely embraced the General's vision of a new Guatemala, formed, as it were, from a potent admixture of religion, racism, security, nationalism, and capitalism'.[157] 'By his astute appropriation and inversion of symbols, language, and meaning', she concludes, 'Rios Montt managed to engineer not only consent but even enthusiasm for the state's ideological reconquest of its people.'[158] While support for the regime was by no means uniform, then, pre-existing ideological sympathy amongst influential sectors, combined with the legitimation efforts of the government, succeeded in holding together the hardline coalition needed for the military to retain a tight grip on power.

The second body of evidence on the success of the regime's public legitimation efforts can be found in their enduring effects on Guatemalan politics and society years after the mass killings. The pursuit of justice and reconciliation after the end of the civil war in 1996 has been significantly undermined by the evident extent to which many sectors of Guatemalan society continue to adhere to the military's narrative of the conflict.[159] Survivors' organizations are attacked and discredited through the same slurs and accusations employed by the military regimes of the 1980s. The effect of regime propaganda is tragically illustrated by the difficulty many survivors have had in returning to their communities and reintegrating, due to the suspicion and accusations levelled on them by their neighbours. One, for example, reported how:

> my current neighbors are always pointing me out as a bad person ... We returned—and to hear the same problems again, to be singled out, to be threatened again, to be told we are murderers, that we are a bunch of guerrillas, that we are demons, that we have killed many people.[160]

---

[157] Garrard-Burnett 2009, 12–13.
[158] Ibid. 84.
[159] Rothenberg 2012, 187.
[160] Cited in ODHAG 1999, 50. The civil defence patrols proved to be a particularly effective means for the military to propagate its ideological account of the violence amongst targeted communities—with even members who broadly resented participation tacitly internalizing elements of the military's narrative; see Garrard-Burnett 2009, 104; Bateson 2017, 643.

The most striking evidence of all is found in the enduring popularity of Ríos Montt himself, even decades after he left office. It is this puzzle with which Garrard-Burnett opens her dedicated study of the General, discussing how:

> Ríos Montt's support came not only from predictable sectors—the conservative urban middle and upper classes—but also from many rural indigenous people, including, astoundingly, many who lived in areas most affected by the scorched-earth campaigns of 1982-1983 ... A 1989 survey—taken just before the second presidential election to follow military rule—posed the question, 'For whom would you vote if the elections were held today?' Ríos Montt won handily, even though he was not even a candidate for president ... It is a testament either to the enduring power of fear or the power of alternative discourses of reality that during the late 1980s and the early 1990s many indigenous informants who spoke with pollsters, human rights workers, and visiting anthropologists showered Ríos Montt with encomiums, describing him as a visionary leader, a champion of law and order, and a messenger of hope in the midst of despair.[161]

It is hard to contend, in light of such evidence, that the regime's ideological propaganda and justificatory narratives had little real impact. The capacity for concerted collective opposition amongst Guatemala's mass public was certainly limited in the face of the brutal state terror of the late 1970s and early 1980s. But, as in all mass killings, the extent to which the regime could legitimate the violence—positively or negatively—mattered, in influencing both the solidity of the regime's support base and the capacity (and willingness) for individuals and communities to co-ordinate acts of resistance and rescue on the ground. Indeed, the downfall of the Lucas Garcia regime—which, unlike the coup against Ríos Montt, reflected a broad collapse in support amongst influential social sectors as well as the military—is illustrative of the dangers of weak ideological legitimation. Although many factors contributed to that downfall, it was centrally rooted in the widespread sense that much of the violence, especially in urban areas, was 'out of control', and could not be plausibly construed as strategically and morally necessary.

The American Connection

Since the United States played a key role in the Guatemalan civil war as a supporter and influencer of the military governments, its own ideological relationship to the mass killing warrants some discussion. American military support had been integral to the restoration of Guatemala's authoritarian political structure in 1954, and to the solidification of its repressive national security structure thereafter.

---

[161] Garrard-Burnett 2009, 9-11. See also Kruit 2000, 21 & 24.

Ideologically, the US government promoted an intense brand of anticommunism that both influenced the worldview of the Guatemalan far-right and enhanced the elite's instrumental use of such an ideology. These two roles came together in the influence of US counter-insurgency doctrine—itself heavily coloured by anticommunism—on the Guatemalan military, bearing in mind the way 'U.S. advisors trained Guatemalan military and police officers in counter-insurgency tactics during the 1960s, including surveillance and interrogation methods that often involved torture and murder'.[162]

There were important divergences between prevailing US and Guatemalan ideologies. American security doctrines were mixed, and included limitationist notions of human rights that generated both normative and practical concerns about atrocities—especially under Jimmy Carter's presidency—that were largely disregarded by the Guatemalan military. US observers did press the Guatemalan military to exercise more restraint and discrimination, but this achieved little, since Guatemala's leaders were convinced that such killings were necessary and would prove effective. The United States was not, as such, the primary driver of Guatemalan mass killing. Indeed, discussing the United States' role in the region as a whole, Luis Roniger observes how:

> [I]n the view of several Latin American high officers, the US was passive and not aggressive enough in the fight against communism during the Cold War ... even taking into account the training and support of the US, one should look for a no less fundamental source of conviction of the [local] military commands in their shared, albeit nuanced, belief in doctrines of national security.[163]

In locating primary agency in Guatemala's military regime, and emphasizing its distinctive ideological worldview, I therefore side with the broader trend in recent historiography that, in Tanya Harmer's summary, 'emphasizes the importance of decentering Cold War narratives and recovering Latin American agency'.[164] Nevertheless, the *partnership* between the Guatemalan and US governments was built on an essential ideological alignment within a Cold War context in which human rights values were deemed secondary to the perceived geopolitical urgency of preventing the spread of communism in Latin America. In this sense, the United States was an element of the coalition on which Guatemalan mass killing depended, and ideology played a key role in explaining sustained US support.

Indeed, declassified documents from the Guatemalan Embassy and US State Department shed new light on the extent to which American officials' own anticommunist worldview blinded them to the scale of the military's mass killings. US Embassy cables throughout the late 1970s and early 1980s consistently and

---

[162] Cangemi 2018, 618–19.
[163] Roniger 2010, 31.
[164] Harmer 2019, 118.

erroneously attributed most atrocities to the guerrillas. As mass killings reached their peak in 1982, extensive reporting of army atrocities by Amnesty International and other human rights agencies based on witness reports and internal Guatemalan coverage were passed on to the US Congress, the White House, and State Department. In response, the US Embassy in Guatemala sought to systematically 'disprove' the non-governmental organization (NGO) allegations, in almost all instances by appealing to information it had received from the Guatemalan army (or Guatemalan press reporting that was, in turn, dependent on information from the army). The Embassy's cable to the State Department contended, in terms scarcely less paranoid than the Guatemalan government's, that:

> We conclude that a concerted disinformation campaign is being waged in the U.S. against the Guatemalan government by groups supporting the communist insurgency in Guatemala ... This is a campaign in which guerrilla mayhem and violations of human rights are ignored; a campaign in which responsibility for atrocities is assigned to the GOG [government of Guatemala] without verifiable evidence ... a campaign in which atrocities are cited that never occurred ... As Solzhenitsyn pointed out in his Nobel lecture: 'Anyone who has once proclaimed violence as his method must inexorably choose the lie as his principle.' ... It seems beyond question that the three [human rights] reports are drawing on many of the same sources ... most of which are well-known Communist front groups ... If the GOG were indeed engaged in massive extrajudicial executions—a 'mad, genocidal campaign'—in the highlands, one must wonder why Indians are joining Civil Defense Patrols in great numbers ...[165]

This document illustrates how far the US Embassy staff's ideological biases blocked any meaningful understanding of the military's mass killings on the ground (as well as the coercive character of the civil defence patrols). This was not a simple information problem. Evidence on the scale of government abuses was readily available, but embassy officials persistently assumed that human rights organizations were sympathetic to communism and that indigenous witnesses were guerrillas or guerrilla sympathizers and could not be trusted. Simultaneously, embassy staff displayed an astonishingly naïve credence in the military's reporting on its own alleged behaviour.

So severe was this problem that eventually US State Department officials within the Bureau of Human Rights and Humanitarian Affairs raised concerns with the credibility of embassy reporting. In an internal memorandum, Charles Fairbanks, Deputy Assistant Secretary for Human Rights and Humanitarian Affairs, sardonically observed that: 'All of the early Embassy responses to charges of Army massacres were based on asking the Army whether or not

[165] US Embassy, Guatemala, 'Analysis of Human Rights Reports on Guatemala by Amnesty International, WOLA/NISGUA and the Guatemalan Human Rights Commission,' 22 October 1982, DNSA.

they were true. There is an obvious problem here ….'[166] Indeed, Fairbanks observed that, despite widespread awareness that the army was committing human rights abuses (indeed, this was a period where the army was killing thousands of indigenous civilians every month), the US Embassy had 'not reported in any cable a single instance that it believes was done by the Army'.[167] 'I would conclude', Fairbanks wrote, 'that our Embassy does not really know who is responsible for the killings in rural Guatemala.' Still, ideological narratives die hard—despite these concerns, even Fairbanks continued to doubt the accuracy of human rights NGOs' reporting, and affirmed his belief that 'Ríos Montt is a reformer, who does not want his subordinates to be brutal' and that 'the guerrillas are doing more atrocities than the other side'.

The problem of US complicity in Guatemala's mass killings thus goes deeper than a cynical 'realpolitik' indifference to the fate of indigenous victims when weighed against US geopolitical interests. American policymakers were themselves profoundly (mis)guided by anticommunist ideology, which shaped their perceptions of the conflict. It cannot be assumed that if US policymakers had possessed a more accurate understanding of the scale of mass killings, they would have been able to halt them. While much has been made of US assistance to the Guatemalan government, the United States often struggled to leverage such assistance for significant influence—indeed, when the Carter administration pressured the Laugerud government over human rights abuses in 1977, Guatemala responded by unilaterally refusing further US military assistance, purchasing arms from other sources including Israel, Argentina, and private US contractors.[168] Nevertheless, the diplomatic, political, and material support of the regional hegemon and global superpower was hardly a marginal issue. Guatemalan elites generally, and correctly, gambled that the fundamental ties between the United States and Guatemala would prevail. What is clear from the US archives is that those ties were not purely geopolitical—American anticommunism, which only strengthened as control of the US executive passed to the Reagan administration, also served to align US perceptions of the conflict with the Guatemalan military's narrative, and thereby facilitated firm US backing even at the peak of genocidal atrocities.

7.4 Conclusion

While Guatemala's government was associated with severe human rights abuses for much of the twentieth century, a range of factors had to come together to produce mass killings between 1978 and 1983. Many of these are well understood by scholars of genocide and armed conflict. Guatemala's history of racism

---

[166] US Department of State, 'Credibility of Embassy Guatemala Human Rights Reporting', 23 November 1982, DNSA.
[167] Ibid. (emphasis in original).
[168] Cangemi 2018, 622–8.

and severe inequality, its authoritarian political institutions, and its conditions of severe socio-political crisis in the face of a guerrilla insurgency represent a familiar breeding ground for large-scale violence against civilians. In many respects, what is surprising about mass killing in Guatemala is not that it occurred, but that it has garnered so little scholarly attention in comparative conflict and genocide scholarship.

This is unfortunate, because Guatemala's appalling atrocities against civilians generate both evidence and puzzles for existing theories of genocide and mass killing. Guatemala highlights the fuzzy outer margins of a genocidal logic of violence, where killing targeted at broad ethnic groups can coexist with cultural assault and counter-insurgent tactics in an overlapping panoply of violent state initiatives. Guatemala also highlights the complexity of strategy and self-interest, and its ambiguous relationship to material 'realities', in civil war. Guatemalan mass killing is impossible to understand without appreciating its strategic logic for the military, and it is clear that local-level strategic factors that rarely appear in prevailing theories of genocide powerfully shaped the distribution of violence. At the same time, there was a profound illogicality to mass killing in Guatemala. Much of it was counterproductive, as many influential observers internal and external to the country pointed out, and many aspects of the killing were unrelated to any clearly identifiable incentives. The whole campaign was, at the very least, wildly disproportionate to the actual threats the military faced. Strategic circumstances *were* vital, but they were consistent with a range of other outcomes besides mass killing.

Attention to the military's ideology is consequently necessary to explain mass killing in Guatemala and solve multiple puzzles about its form and extent. Ideology was certainly *insufficient* to provoke mass killings in absence of crisis conditions and structural inequalities, and its radicalization was heavily intertwined with such factors. But without the military's possession of a powerful justificatory narrative rooted in an exceptionally hardline blend of anticommunism and racist nationalism, both the decision to employ mass killings and the form they took—targeting largely docile indigenous communities and making no meaningful effort at selective discrimination of guerrilla targets—made little sense. Had that ideology not been utilized to mobilize and organize rank-and-file perpetrators, moreover, it is unlikely that massacres would have occurred with the scale, ferocity, and willing local initiative that they did. Similarly, had the military not been able to draw on ideological justifications so effectively in public political discourse, its capacity to rely on strong backing from its key coalition, and enjoy such relatively quiescent passivity from influential internal and external actors, would have been questionable.

The emergence of a justificatory narrative for mass killing was guaranteed by neither Guatemalan history nor the conditions of crisis and conflict Guatemala faced. It emerged out of specific cultural and ideological antecedents, evolving and radicalizing through the agency of influential elites amongst Guatemala's far-right, especially within the military. Hardline ideology in Guatemala was in several respects classically belligerent—authoritarian, dismissive of individual rights, racist, and exclusionary. Yet the justificatory narrative for mass killing rested most centrally on the way such factors undergirded claims about self-defence, victim guilt, and strategic necessity in a time of crisis. It had relatively little to do with 'special ideological goals' involving utopian projects of political or racial purification, some of Ríos Montt's more far-reaching visions notwithstanding. Racism thus occupies a central but nuanced role: the perpetrators of Guatemala's mass killings did not start out with a genocidal blueprint which they then sought to implement, but arrived at a mixed policy of group-targeted mass killings and cultural elimination of Maya identity due to their racist ideological interpretation of the counter-insurgency task they set themselves.

Needless to say, a short comparative case study of this kind omits much. As in this book in general, I have not dwelt on the role of ideology in shaping how victims and survivors responded to the violence, though there is some evidence that it did play such a role in Guatemala. Falla, for example, highlights cases where individuals aligned to far-right parties or the Charismatic Catholic movement refused to flee because they believed their political loyalties would prevent them being targeted.[169] The paucity of available data also prevents detailed examination of practices of rescue,[170] and any precise unravelling of the nuances of internalized and structural ideological influence in the killings. More than any other case in this book, Guatemala's mass killings offer many avenues for future research and demand greater scholarly attention. My modest aims have been to contribute to a growing body of comparative research that is drawing Guatemala into genocide and mass killing scholarship, and to highlight the complex interconnection of ideology and strategy that drove the Guatemalan state's violence against its own people, both indigenous and ladino.

---

[169] Falla 1994, 68–9 & 81–4.
[170] But see Brett 2016, 204.

8
The Rwandan Genocide

8.1 Overview

Even in a phenomenally violent century, the Rwandan Genocide stands out as one of the most infamous, well-studied, and intense instances of mass killing. Rwanda, a country in Africa's Great Lakes region slightly smaller than Belgium, its former colonial overlord, has a population comprised of three principal ethnic groups: the Hutu (84–90% of the population at the time of the genocide), Tutsi (9–15%), and Twa (1%).[1] In just a hundred days from early April to mid-July 1994, forces organized by an extremist 'Hutu Power' government killed between 500,000 and 800,000 Tutsi victims, as well as tens of thousands of Hutu and Twa political rivals and opponents of the violence.[2]

The genocide and mass killing emerged out of a complex political crisis that escalated over the early 1990s. Following a period of relative stability under President Juvénal Habyarimana in the 1970s and early 1980s, Rwanda encountered significant economic difficulties in the late 1980s, primarily due to a collapse in the international prices of its key exports of coffee, tea, and tin.[3] With the end of the Cold War, Habyarimana also initiated limited democratization reforms under international pressure, which threatened elements of the prevailing political order. Most dangerously, Rwanda was in an uneven but recurrent state of civil war from 1990 to 1994 as a predominantly Tutsi insurgent force, the Rwandan Patriotic Front (RPF), invaded the country from neighbouring Uganda.[4]

Faced with this crisis, President Habyarimana agreed to internationally mediated power-sharing negotiations from 1992 onwards, despite bitter disapproval from hardline elements within his regime. As the negotiations were concluded, Habyarimana was assassinated—his plane shot down on 6 April 1994, as it

---

[1] Straus 2006, 19. Consistent with Kinyarwanda and United Nations practice, I do not add an -s suffix to pluralize these identity labels, except when quoting from other scholars.
[2] International Panel of Eminent Personalities 2000, para 14.2; Straus 2006, 51–2; McDoom 2021, ch.7. There are few estimates of Twa victims. For recent research on the estimates of the victims of the Rwandan Genocide see McDoom 2020a; Meierhenrich 2020; Verpoorten 2020. On broader patterns of violence in 1990s Rwanda see Straus 2019.
[3] Percival and Homer-Dixon 1996, 277–9; Uvin 1998, ch.4; Mamdani 2001, 147–8; Kimonyo 2016, 63–70.
[4] Prunier 1995, 114–20.

Ideology and Mass Killing: The Radicalized Security Politics of Genocides and Deadly Atrocities. Jonathan Leader Maynard, Oxford University Press. © Jonathan Leader Maynard (2022). DOI: 10.1093/oso/9780198776796.003.0008

approached Rwanda's capital, Kigali. The identity of the assassins remains unclear.[5] Hardliners rapidly took control of the government, mobilizing military forces, militias, and citizens to kill Tutsi and political opponents. The resulting genocide was ended only by the RPF insurgency taking over most of the country as 1994 progressed.[6]

As with other mass killings, multiple causes underpinned the Rwandan Genocide.[7] The civil war and political crisis were critical, threatening the self-interest of the dominant political elite. The Habyarimana government was highly authoritarian and corrupt, with power monopolized by an unofficial '*Akazu*'[8] network surrounding the family of the President's wife, Agathe Kanziga Habyarimana. This mafia-like cabal controlled political institutions and business interests across Rwanda, and its members devoted enormous energy to the pursuit of private self-enrichment and power. They stoked ethnic animosity and nationalism, but many of their efforts appear cynical rather than reflective of ideological conviction. For rank-and-file perpetrators of the genocide, local social pressures and the Rwandan government's dense coercive apparatus were central drivers of violence.[9] Genocide was also facilitated by the international community, which generally avoided exerting meaningful pressure on Rwanda to respect human rights,[10] blocked the local United Nations mission from preventing massacres,[11] and helped develop the regime's capacity for violence.[12]

Ideology's role in the genocide therefore remains heavily debated. Leading studies published in the decade following the genocide, such as those by Alison Des Forges, Gerard Prunier, and Mahmood Mamdani, often placed ideology at the centre of their analysis, emphasizing the intense racism of a supremacist 'Hutu Power' regime and the influence of hate propaganda in driving the violence.[13] Yet influential subsequent analyses of the genocide, such as those by Scott Straus and Lee Ann Fujii, have cast doubt on ideology's centrality, at least for rank-and-file

---

[5] Many suspect that Hutu extremists assassinated the President (see, e.g., Article 19 1996, 10 & 65; Percival and Homer-Dixon 1996, 275 & 284; Gourevitch 1999, 113; Wallis 2019, 374). But several scholars now identify the RPF as likely perpetrators, and French and Spanish judges have issued arrest warrants against members of the RPF. For discussion see Straus 2006, 44–5; Guichaoua 2015, 169–73.

[6] Combined with a late, and highly controversial, French-led intervention; see Prunier 1995, 281–311; Wallis 2006/2014, ch.7.

[7] On broader structural features of Rwanda conducive to mass violence see McDoom 2021, ch.2.

[8] Typically translated as 'little house', the term *Akazu* originally denoted the inner courtly circle surrounding the *Mwami* (king) of Rwanda. For the most detailed study of the *Akazu* network see Wallis 2019.

[9] Straus 2006; Fujii 2009.

[10] Even though on two notable exceptions, in 1990–91 and 1993, such pressure may have been consequential; see Uvin 1998, 96.

[11] The UN force commander, Roméo Dallaire, lobbied for authorization to use force to protect civilians, but was overruled by the UN hierarchy; see Dallaire 2003.

[12] Uvin 1998.

[13] Prunier 1995; Forges 1999; Mamdani 2001. See also Des Forges 1995; Thompson 2007; White 2009.

perpetrators.[14] Other scholars express broader scepticism, suggesting that ideology was largely 'a mask or pseudo-justification for the more fundamental goal of regime survival',[15] that ideological propaganda played little role,[16] and that the genocide was simply a ruthlessly self-interested elite response to political opposition.[17]

I argue that neither traditional-ideological accounts of mass killing nor sceptical perspectives which largely ignore ideology convincingly explain the genocide. Against traditional-ideological accounts, there is little evidence that special ideological goals or long-standing mass hatreds were primary motives for most participants. In Straus's detailed survey of Rwandan perpetrators, 86.5% reported that their relations with Tutsi were positive prior to the genocide, two thirds had Tutsi family members, and only 6.5% suggested that Hutu in general hated Tutsi.[18] Similar pictures emerge from other scholars' interviews with former perpetrators,[19] and several elite figures who orchestrated the genocide showed little evidence of deep ethnic animosity.[20] Rwanda was also much more ideologically diverse than traditional-ideological accounts of mass killing typically imply. There was considerable opposition to hardline ethnonationalism, and in many respects 'the moderate centre ground was strong and popular' prior to 1994.[21]

Yet accounts that sideline ideology are also unconvincing. There is no question that Rwanda's political elite were ruthlessly self-interested, but this does not explain their choice to kill civilians on such a scale. The genocide made no sense as a *military* strategy. There was no evidence of mass support for the RPF 'enemy' from Tutsi within Rwanda, nor did the RPF's insurgent strategy depend on mobilizing such support.[22] On the contrary, as André Guichaoua reports:

---

[14] E.g. Straus 2006; Fujii 2009. Straus still emphasizes that without 'pre-existing [ethnic] categories and without a history of ethnic or racial political ideologies, the call to attack Tutsis would not have resonated, and genocide would not likely have happened' (Straus 2006, 226). Straus places more emphasis on ideology in Straus 2015a, ch.9.

[15] Hintjens 1999, 242.

[16] Danning 2018.

[17] de Figueiredo and Weingast 1999; Lemarchand 2002.

[18] Straus 2006, 128–30. See also the interviews in Lyons and Straus 2006, 39–96. Jean-Paul Kimonyo alleges that Straus's data is too unreliable to yield general conclusions, since a statistical analysis by David Backer suggests that the data underestimate the scale of participation; see Kimonyo 2016, 5–6. While the testimony of former perpetrators must be treated with caution (as Straus emphasizes; see Straus 2006, 123), Kimonyo's critique is unconvincing. Backer's analysis suggests, at most, that Straus's perpetrators unsurprisingly understate *their level of involvement* in killings; see Backer 2008. Even if true, this does not render their testimony about the broader dynamics of violence useless, nor imply that it should be ignored. Kimonyo appears partly concerned here with rejecting the argument, which he imputes to Straus, that the Rwandan Genocide resulted from provocation by the RPF's invasion. But Straus makes no such argument, merely emphasizing the context of war as critical. For an actual assertion that the RPF provoked the genocide see Kuperman 2004.

[19] I draw heavily on and am indebted to such interview-based research—in particular Mironko 2004; Hatzfeld 2005; Lyons and Straus 2006; Straus 2006; Mironko 2007; Adler et al. 2008; Fujii 2009; Scull, Mbonyingabo, and Kotb 2016; Anderson 2017a; Jessee 2017; McDoom 2021.

[20] Wallis 2019, 16 & 29.

[21] Article 19 1996, 107. See also McDoom 2021, 145–62.

[22] Mamdani 2001, 188.

Tutsi notables, as well as a substantial segment of the Tutsi population in Rwanda, reacted to the news of the [RPF invasion] with great misgivings ... There was hardly a voice that deviated, and most Tutsi intellectuals, many of whom felt little affinity with the potential 'returnees,' joined in the various initiatives in support of 'national unity.'[23]

Much of the genocide, moreover, targeted parts of Rwanda distant from the front line of the civil war, while diverting militia and combat battalions away from the war effort.[24]

A more popular rationalist explanation of the genocide is that it nevertheless represented a brutally logical *political* strategy for consolidating the incumbent elite's control over the government and mobilizing popular support. Some scholars suggest that the military campaign against the RPF was doomed anyway—genocide therefore represented a 'gambling for resurrection' strategy designed to make the country ungovernable for the RPF after it took power.[25] There is evidence of such strategic logics in the genocide. But even for a cynically self-interested and morally bankrupt elite, genocide was strategically indeterminate and probably self-destructive. In the most detailed study of the genocide's organizers, Guichaoua emphasizes how:

> We would have expected a more 'realistic' approach, especially in light of the two unavoidable questions: With inevitable military defeat looming on the horizon, was it not possible to negotiate and to salvage or conserve matters, and if so, how? And if nothing could be negotiated or salvaged in the short term, how could options or resources be preserved for the future? There was none of that ... The final defeat was all the more decisive when unanimous moral outrage and condemnation capped the military debacle. The futility of the strategic option thus appeared overwhelming ....[26]

Some of the earliest massacres, moreover, targeted Tutsi groups that were politically inactive and had no history of opposition to the state,[27] and elites *themselves* were divided on the rationality of genocide, with many favouring a less hardline course.[28]

Theories focused on the economic incentives for mass killing are likewise inadequate. Aside from its overpopulation,[29] Rwanda's economic problems were not exceptional: many other states face similar challenges, even coupled with contexts

---

[23] Guichaoua 2015, 28.
[24] See also Gourevitch 1999, 156–7; Guichaoua 2015, 228, 260, & 281–2; Wallis 2019, 428.
[25] de Figueiredo and Weingast 1999. See also Wallis 2019, 315–16.
[26] Guichaoua 2015, 333–4.
[27] Verwimp 2011, 415.
[28] Guichaoua 2015, 54, 71–6, 163, 167–8, 188, 207–10, 223, 231, & 263.
[29] Prunier 1995, 353–4.

of armed conflict, without descending into genocidal violence.[30] There was looting in the genocide, but the genocide's rank-and-file perpetrators rarely mention material self-interest as a motive for participation,[31] and the intensity of violence did not correlate with the parts of Rwanda with the greatest economic deprivation and grievance.[32] As Timothy Longman observes, moreover, studies that emphasize the roots of the genocide in economic difficulties fail to explain 'how economic frustration was channelled away from those with the greatest wealth and power, mostly Hutu, to the Tutsi, who were generally as poor as their Hutu neighbors.'[33]

To effectively explain why key political elites pursued genocide and were able to implement it so effectively, a powerful ethnonationalist ideological infrastructure within Rwanda, produced by the country's specific history of ideological development, has to be brought into focus. As in other mass killings, that ideological infrastructure was sustained by (i) real but uneven internalization of ethnonationalist ideas amongst both elites and the broader population, and (ii) ethnonationalist norms, institutions, and discourses through which patterns of conformity were generated and self-interested political strategies pursued. Elites were not guided by longstanding goals to ethnically transform society. But they did develop a hardline ethnonationalist interpretation of the intensifying security crisis that made mass killing look like a strategically and morally justifiable response. That narrative was key in informing the elite's initiation of genocide, and in mobilizing and coordinating supporters amongst local political elites, the militias, and the broader population.

8.2 The Evolution of Hardline Hutu Ethnonationalism

Hutu and Tutsi Identity in Rwanda

The identity categories of 'Hutu' and 'Tutsi', around which most violence in the genocide revolved, have not been historically stable.[34] In the precolonial period of the Rwandan monarchy, Hutu and Tutsi were defined by a somewhat shifting range of caste, class, kinship, and cultural criteria. 'Hutu' may have originally been a demeaning term denoting 'rural boorishness', before becoming attributed to the arable farmers who filled support roles in the monarchy's army, by distinction with 'Tutsi' warriors drawn from pastoral herders.[35] Tutsi and Hutu increasingly became associated with economic activity: the Tutsi depending on animal husbandry

---

[30] Mamdani 2001, 149. See also Percival and Homer-Dixon 1996; Straus 2006.
[31] Only 5% of those surveyed by Straus mention such a motive; see Straus 2006, 136. Fujii's study is an exception; see: Fujii 2009, 96–8.
[32] Percival and Homer-Dixon 1996, 282.
[33] Longman 2004, 37.
[34] Mamdani 2001, 34. The pre-eminent study remains Newbury 1988.
[35] Kimonyo 2016, 15.

and cattle ownership, the Hutu on arable agriculture. Some distinct migratory waves may have contributed to the formation of such social groups, although this is disputed.[36]

Ultimately, Hutu and Tutsi became hierarchal status categories, with the Tutsi associated with the elite aristocracy that dominated the expanding Nyiginya kingdom that would become the pre-colonial Rwandan monarchy. Yet several factors complicate this picture. Most Tutsi were still not aristocrats, and certain regions of Rwanda, incorporated relatively late into the central monarchy, held onto distinct identities and may only have been subsumed into the Hutu/Tutsi schema over time.[37] Unlike conventional 'ethnic' identities, the categories were also not entirely fixed for individuals, with Hutu occasionally able to achieve the status of Tutsi—a process known as *kwihutura*—and Tutsi capable of losing property and becoming Hutu—a process of *gucapira*.[38] Tutsi and Hutu identities were also maintained through patrilineage, so the fact that children inherited the identity of their father should not obscure the reality of widespread intermarriage across groups.[39] Nor did Hutu and Tutsi identities appear to be significant bases for conflict prior to the twentieth century. Prunier observes that: 'there is no trace in [Rwanda's] precolonial history of systematic violence between Tutsi and Hutu as such'.[40]

But ethnic relations within Rwanda changed dramatically under first German (from the 1890s to 1916) and then Belgian (1916–62) colonial control. These states gained dominion over Rwanda at a time when pseudoscientific racial theories were of huge intellectual influence in Europe, and the colonialists rapidly reinterpreted the ethnic makeup of Rwanda (and neighbouring Burundi) according to such theories. German and Belgian occupiers declared that Hutu and Tutsi were rigidly distinct biological *races*. Moreover, encountering the relatively well-developed administrative practices of the Rwandan monarchy, European colonialists reasoned that such 'civilized' institutions could not have been developed by 'primitive Africans'. Under the so-called Hamitic hypothesis, the Europeans theorized that the stereotypically taller and finer-featured Tutsi aristocracy were biologically closer to European whites, must have originated outside of Africa in the Caucasus, and then brought civilization to Africa's inferior natives.[41] This racist pseudoscience shaped colonial ruling ideology in Rwanda and neighbouring Burundi, with Europeans supporting the Tutsi (through the monarchy) as local agents of imperial rule.[42] Schools, increasingly controlled by the church, segregated education in favour of the Tutsi and propagated the racialized European ideology.[43]

---

[36] Uvin 1998, 14.
[37] Mamdani 2001, 69–70.
[38] Ibid. 70.
[39] Uvin 1998, 29; Mamdani 2001, 4 & 54; Straus 2006, 69 & 72.
[40] Prunier 1995, 39. See also Mamdani 2001, 105.
[41] See, in general, Prunier 1995, 5–11; Taylor 1999, ch.2; Mamdani 2001, ch.3; Kimonyo 2016, 16–19.
[42] Mamdani 2001, 87.
[43] Ibid. 89–90.

In 1933, the Belgians further introduced a national census and compulsory identity cards which would legally define every Rwandan's singular ethnicity in immutable, biological terms.[44]

This ideological infrastructure of colonial Rwanda would significantly shape the country's post-colonial politics. By the 1950s, Belgium acknowledged the need for eventual independence for Rwanda-Burundi. By contrast with Burundi (which would continue to be controlled by a Tutsi-dominated political order), Rwanda's independence saw the Tutsi monarchy overthrown in favour of a Hutu ethnonationalist[45] majoritarian regime. Two interrelated processes were crucial here. First, Belgian orientations shifted, with a new generation of missionaries and colonial administrators displaying markedly greater public sympathy for the Hutu as an oppressed majority.[46] Second, a new Hutu counter-elite, partly generated by gradually expanding educational opportunities, emerged and agitated against Tutsi privileges.[47]

This Hutu counter-elite was diverse, but its most powerful factions articulated revolutionary demands in racially exclusive terms. They were encouraged by the new generation of Catholic clergy, who helped fund Hutu revolutionary activities and handed over key media publications to Hutu politicians—in particular, the relatively widely published *Kinyamateka*, of which the future President of Rwanda, Grégoire Kayibanda, was editor.[48] The most influential statement of the revolutionary movement, the March 1957 'Bahutu Manifesto' signed by Kayibanda and other prominent Hutu elites, emphasized that: 'The problem is above all a problem of political monopoly which is held by one race, the Tutsi.'[49] Most ominously, the manifesto embraced the mythical colonialist account of the Tutsi as Hamitic conquerors of Rwanda by portraying the Tutsi as foreigners who had oppressed the true Rwandans—the Hutu. It thus called for liberation of 'Hutu from both the "Hamites" [Tutsi] and the "Bazungu" [whites] colonization',[50] and demanded the retention of ethnic labels on identity cards in order to 'keep a close check on this racial monopoly'.[51]

Kayibanda proceeded to create a formal political party in 1959, the *Parti du Mouvement de l'Emancipation Hutu* (PARMEHUTU), to push for Hutu national liberation. The party was racially exclusive, with Kayibanda stating:

---

[44] Ibid. 98–102.
[45] Mamdani emphasizes how the post-colonial Rwandan regime conceptualized the Hutu–Tutsi divide as *racial* and not *merely ethnic*; see ibid. 28–35. But these notions overlapped, so rather than continuously shifting between notions of race and ethnicity, I broadly refer to ideologies that posit an essential division between Hutu and Tutsi, and contend that politics must be dominated by Hutu, as *Hutu ethnonationalism*.
[46] Lemarchand 1970, 106–11; Mamdani 2001, 113; Kimonyo 2016, 23–4.
[47] Mamdani 2001, 106–14.
[48] Lemarchand 1970, 108–9; Prunier 1995, 45; Mamdani 2001, 114.
[49] Cited in Mamdani 2001, 116.
[50] Cited in ibid.
[51] Cited in Straus 2015a, 279.

'Our movement is targeted at the Hutu, who have been despised, humiliated and regarded with contempt by the Tutsi ... We need to enlighten the people that we are here to restore the country to its real owners, as this is the country of the Bahutu.'[52] Other major parties formed around the same time: the *Union Nationale Rwandaise* (UNAR), largely aligned to the Tutsi monarchy, the *Association pour la Promotion Sociale de la Masse* (APROSOMA), a Hutu-dominated but less ethnically divisive party, and the *Rassemblement Démocratique Rwandais* (RADER), a moderate cross-ethnic and largely intellectual party.[53]

Violence between members of UNAR and PARMEHUTU broke out in November 1959. Reflecting the shift in Belgian attitudes, the local Belgian military commander, Colonel Guy Logeist, sided with the 'Hutu' parties, interpreted the revolt as a 'revolution', and refused to support the Tutsi monarchy, instead recognizing or installing Hutu local chiefs in place of Tutsi incumbents. PARMEHUTU went on to dominate local elections in June 1960 (helped by a misguided boycott by UNAR), and organized a congress on 28 January 1961, which proclaimed the dissolution of the monarchy and the creation of a new republic under interim President Dominique Mbonyumutwa, with Kayibanda as Prime Minister.[54] Kayibanda quickly took over as President, and on 1 July 1962 Rwanda officially gained independence from Belgium, while tens of thousands of Tutsi fled the country.

The result was an ethnonationalist authoritarian Rwandan regime—the 'First Republic'—which adopted, in Omar McDoom's words: 'an ideology that framed the [1959] revolution as the moment of Hutu emancipation and equated democracy with Hutu majority rule.'[55] This regime repackaged the racial ideology of the colonial era with an idiosyncratic ideological blend of egalitarianism, conservatism, and xenophobia,[56] portraying Tutsi as a foreign race that had oppressed the Hutu majority for centuries.[57] Some Tutsi who had fled Rwanda during the revolution immediately organized violent raids on the new republic between 1961 and 1963, calling themselves '*inyenzi*' ('cockroaches').[58] Kayibanda's regime repelled them and initiated massacres of Tutsi, killing approximately 10,000 between December 1963 and January 1964.[59] The genocidal potential of the regime's orientation was explicit, with Kayibanda warning in 1964 that if Tutsi refugees continued to seek political power, this risked 'the complete and precipitous end of the Tutsi race'.[60]

---

[52] Cited in Kimonyo 2016, 29.
[53] Ibid. 29–30; Jessee 2017, 7.
[54] Kimonyo 2016, 32–3.
[55] McDoom 2021, 29.
[56] Lemarchand 1970, 102; Prunier 1995, 57–9.
[57] Prunier 1995, 80.
[58] This superficially curious self-appellation is variously claimed to reflect the raiders' tendency to attack at night or their hard-to-eradicate nature; see Kuperman 2004, 63; Fujii 2009, 83, fn.3.
[59] Straus 2006, 186–7.
[60] Cited in Straus 2015a, 285.

This anti-Tutsi ideological stance appears to have enjoyed significant public support. The Rwandan human rights activist André Sibomana recalls how many Rwandans in the period believed 'that they represented the original people who had inalienable and eternal rights on Rwandan soil. Suddenly they viewed the Tutsi not only as a race, but as a foreign race, a race of conquerors who had imposed domination over the Hutu and had to be driven out of the country.'[61] Faced with growing public dissatisfaction and the mass killing of Hutu in neighbouring Burundi in 1972, the regime intensified its ethnic appeals in an effort to mobilize support, but also seemingly out of sincere belief, with state intelligence privately blaming political opposition on the Tutsi minority.[62] Either way, 'Radio Rwanda was used to broadcast appeals inciting the Hutu to rise up and avenge themselves. Some extreme proponents of Hutu Power began publicly to call for a "final solution" to the Tutsi question.'[63] The regime's renewed discrimination efforts again received a positive response from significant sectors of the population.[64] When Kayibanda sought to contain the resulting growth in political violence, moreover, hardliners within the regime denounced him for going soft on Tutsi.[65]

This all testifies to the establishment of a powerful ethnonationalist ideological infrastructure within Rwandan society between the years preceding independence and the eventual fall of Kayibanda's regime in 1973. It was not an especially obvious direction for Rwanda to take, nor a necessary consequence of objective material, political, or structural conditions.[66] Despite Belgium's ideological elevation of Tutsi in colonial Rwanda, the overwhelming bulk of the Tutsi population lived in the same conditions as the Hutu.[67] The average incomes of Tutsi and Hutu families were barely different.[68] Many other states in sub-Saharan Africa were governed under the rubric of similar racial ideologies in the colonial period, yet decolonized under independence movements that were at least formally cross-ethnic and in some cases pan-African.[69] In Rwanda itself, the political environment in the years surrounding independence was a diverse one, in which 'two political tendencies—one accommodationist, the other exclusionist—vied for supremacy between 1959 and 1964'.[70]

Hardline ethnonationalism was thus a contingent and consequential feature of post-independence Rwanda. It emerged from the complex interaction of Rwanda's

---

[61] Sibomana 1999, 94.
[62] Wallis 2019, 47.
[63] Mamdani 2001, 137.
[64] See, for example, Wallis 2019, 49–50.
[65] Ibid. 50.
[66] Straus 2015a, 278. Such conditions did facilitate ethnonationalism, however; see McDoom 2021, ch.2.
[67] Wallis 2019, 22.
[68] Prunier 1995, 50.
[69] Straus 2015a, chs.5–7. See also Getachew 2019.
[70] Mamdani 2001, 126. See also Lemarchand 1970, 99–102; Mamdani 2001, 117; Kimonyo 2016, 28.

existing racial politics under colonialism, the genuine obstacles and insecurities posed by Tutsi opponents of independence, ideological activism by Kayibanda and other Hutu intellectuals, and specific interventions by Belgium. Other paths were possible: Kayibanda could have supported a less rigid racial hierarchy, and other organizations, notably the less divisive APROSOMA party, could have provided the core of a more inclusive post-independence regime.[71] Moreover, other paths *remained* possible: important ideological changes under the subsequent 'Second Republic' illustrate that extreme ethnic violence was not 'locked in' by the ideological direction taken at Rwanda's independence.[72] What is crucial, therefore, is how ideology developed under Kayibanda's successor, particularly during the escalating political crisis Rwanda faced in the early 1990s.[73]

The Second Republic Under Habyarimana

Facing internal criticism and widespread disorder, Kayibanda was overthrown in a coup by the head of the army, Major-General Juvénal Habyarimana, on 5 July 1973. Habyarimana ushered in a 'Second Republic', with himself as President, that continued to invoke the legitimacy of the 1959 revolution, while making important ideological revisions. These were clearest in the new regime's more ambiguous ideological characterization of Tutsi. In many respects, 'the ethnic politics of the Second Republic appear contradictory'.[74] By contrast with the Kayibanda regime, Habyarimana's government condemned ethnic divisionism, framed the Tutsi as genuine Rwandans, and promoted national unity in the pursuit of economic development. Yet Habyarimana still endorsed the established narrative that the 1959 revolution had overthrown centuries of brutal Tutsi oppression, and he maintained ethnic categories on identity papers and ethnic quotas in schools, universities, and state institutions. This complex mix of inclusion and prejudice was articulated throughout Habyarimana's major speeches. On a 1975 tour of the country to promote the new ruling party, the *Mouvement Révolutionnaire Nationale pour le Développement* (MRND), Habyarimana proclaimed:

> We know that for more than four hundred years, the Tutsi harshly oppressed the Hutu and did everything to deny the Twa confidence in their humanity. Our revolution of 1959 was the first to have overthrown Tutsi power ... However, there was at some time [under Kayibanda] a program to ignore the minority and even expel

[71] On APROSOMA's ideology and key figures see Kimonyo 2016, 113–23.
[72] Uvin 2001, 79.
[73] Somewhat contra: Mamdani 2001, 36.
[74] Kimonyo 2016, 59.

them from the country. Our movement must support and implement our commitment to abolish ethnic discrimination. The movement will give all Rwandans the opportunity to build their country in peace, unity and integrity ....'[75]

In 1987, Habyarimana similarly announced that: 'Our elders revolted against a feudal, minority power ... the majority of our youth know today that our struggle was led against an outdated and hegemonic feudal system, but not against an ethnic group as such ... We are all Rwandans.'[76]

This undermines traditional-ideological portrayals of genocide in Rwanda as an output of long-standing ideological ambitions to eliminate the Tutsi from society. Some suggest, for example, that 'in the years between 1959 and 1994, the idea of genocide, although never official recognized, became a part of life',[77] or that 'in Germany as in Rwanda ... the elimination of the Jew, the Gypsy, or the Tutsi had been openly part of the regime's political agenda from the moment it took power.'[78] In truth, there is little if any evidence that the Habyarimana regime contemplated genocidal violence in the 1970s and 1980s. The regime remained discriminatory, but, as Straus comments, 'Rwandan and outside observers attest to the ways in which anti-Tutsi sentiment decreased, at least until 1990'.[79]

Yet an ethnonationalist ideological infrastructure continued to sustain attitudes, practices, and institutions that, as the regime faced intensified political crisis in the 1990s, would lend themselves to hardline radicalization. Most obviously, the regime sustained a society-wide awareness of ethnic categories as unchangeable and consequential identities. Tutsi were largely excluded from local government and barely represented amongst army officers and parliamentarians.[80] Widespread discrimination along ethnic lines remained. Andrew Wallis reports how: 'Despite Habyarimana's "unity" rhetoric, lists that had appeared on shop and office doors banning Tutsis from entering under Kayibanda continued under the new regime.'[81] In schools, students continued to have to identify themselves in class as Hutu or Tutsi,[82] and recalled being taught that Tutsi were 'wicked' and 'bad people'.[83] A university student reported: 'When I arrived at the Ruhengeri university campus, I faced a lot of abuse from classmates who called myself and other Tutsi students "cockroaches" or "snakes" ... The authorities were afraid we were working for Tutsis in exile and somehow would ferment unrest.'[84] Another observed

[75] Cited in ibid. 54.
[76] Cited in Straus 2015a, 290.
[77] Melvern 2004, 8.
[78] Hatzfeld 2005, 52.
[79] Straus 2015a, 287. See also Wallis 2019, 109.
[80] Prunier 1995, 75.
[81] Wallis 2019, 74. See also Kimonyo 2016, 55–9.
[82] Wallis 2019, 75.
[83] Scull, Mbonyingabo, and Kotb 2016, 338.
[84] Wallis 2019, 123.

how: 'The Dean who had been appointed by Habyarimana kept asking why "such people", meaning Tutsis and those from the south, were in the faculty at all.'[85]

Similarly, while those surrounding Habyarimana were not uniformly wedded to anti-Tutsi ideology, the most influential elite networks supported hardline ethnonationalism. Wallis observes that: 'Hardliners around Habyarimana, such as [Alexis] Kanyarengwe, condemned any attempt at a policy of ethnic equilibrium.'[86] One such faction associated with Major Théoneste Lizinde, Habyarimana's intelligence chief, was 'considered to be violently anti-Tutsi and unhappy with Habyarimana's efforts to reconcile the Hutu and the Tutsi'.[87] Kanyarengwe and Lizinde were eventually pushed out of office, accused of plotting a coup (the former fleeing, the latter arrested and executed).[88] But other hardliners remained, including Colonel Théoneste Bagosora, who would play a central role in organizing the genocide, and Ferdinand Nahimana, who helped create the hate-radio station RTLM, both of whom displayed long-standing animosity towards Tutsi.[89] Others simply exploited ethnicity instrumentally—one Tutsi businessman noted how the President's brother-in-law, Protais Zigiranyirazo, was 'constantly pushing a negative stereotype about Tutsis, less I think from any real ethnic prejudice but more because someone like myself took a position he had his eyes on for one of his own men'.[90]

The regime's ideology contained other features that would become consequential for later violence. 'Authoritarian but somewhat debonair', in Prunier's words, the Habyarimana government ideologically emphasized mass citizen participation in collective social activities organized by the state, and a conception of state officials' authority as absolute. Several scholars consequently characterize the regime as totalitarian.[91] Certainly it had such aspirations, with the MRND stating explicitly:

> The movement intends to be popular and expects unreserved support. In other words, the action of the people, of the whole society, is modelled on one sole pattern, producing unity of purpose, harmony, and cohesion from the cells at the base of the movement up to the tip of the pyramid—in other words, the entire nation. No individual or group of individuals can escape the total social control at work here.[92]

---

[85] Cited in ibid. 121.
[86] Ibid. 64.
[87] Cited in Mamdani 2001, 141.
[88] Kanyarengwe's hardline stance may have been opportunistic—after fleeing Rwanda he joined the RPF, eventually becoming its (largely titular) president.
[89] Article 19 1996, 24 & 41; Wallis 2019, 29.
[90] Wallis 2019, 109.
[91] Kimonyo 2016, 59–62; Wallis 2019, 124.
[92] Kimonyo 2016, 60.

Such total control was more limited in practice, but the same can be said for less contentiously totalitarian regimes such as Stalin's Soviet Union, as seen in Chapter 5.

Crucially, though, it was not totalitarian *goals* of radical social transformation that mattered here but, as in the Soviet Union, the regime's ideological conception of political order and state power. The MRND party-state vision encouraged a dense authoritarian system involving high citizen mobilization and a deep penetration of (officially unquestionable) state authority in local communities.[93] This system inculcated intense interaction between state officials and citizens through the *umuganda* programme, a form of compulsory labour in which once a week the population were required to contribute to state-directed communal projects.[94] The regime's ideological emphasis on developmentalist citizen mobilization created, as a by-product, a powerful infrastructure for mobilizing violence, and *umuganda* became infamous since it was euphemistically used to describe participation in the genocidal killings in 1994. I will suggest, below, that this link between ideology, state authority, and citizen mobilization better explains Rwandans' responsiveness to local authorities' orders to kill in the genocide than the somewhat essentialist assertions of a Rwandan 'culture of obedience' that can be found in much commentary on the genocide.

Hardline Radicalization in the 1990s

For most of its lifetime, then, the Second Republic under Habyarimana resembled neither an ethnic powder keg of seething animosity, nor a benign environment of toleration. The dominant ideological infrastructure was mixed: highly controlling of society and encouraging significant ethnic discrimination, but lacking in explicit justificatory narratives for mass violence. Moreover, the regime's ideological dominance was not total: there were many other ideological currents in Rwandan society, which expanded as Rwanda partially democratized in the early 1990s.[95] The emerging 1980s to 1990s crisis was, moreover, partly rooted in growing discontent with the regime. Despite regime propaganda, surveys in July 1990 suggested that 96% of the population affirmed there was famine in the country, 91% complained of poor security, and 90% said corruption was out of control.[96] By 1992, the political opposition was able to mobilize between 30,000 and 100,000 protestors in demonstrations against the government in the capital (then a city of only 250,000).[97]

---

[93] Mamdani 2001, 143–4; McDoom 2021, 258–66.
[94] Mamdani 2001, 146–7; Kimonyo 2016, 61–2.
[95] McDoom 2021, 145–57.
[96] Wallis 2019, 203–4.
[97] Ibid. 263.

Four interrelated developments would escalate the political crisis and encourage significant hardline radicalization between 1990 and 1994: first, a shift to multiparty politics that threatened the incumbent elites; second, the onset of civil war against the Rwandan Patriotic Front insurgency, which invaded in October 1990;[98] third, the Arusha peace and power-sharing negotiations between the regime, the opposition parties, and the RPF between 1991 and 1993, which infuriated hardliners; and finally, the assassination of the moderate Hutu President of Burundi, Melchior Ndadaye, by the Tutsi-controlled Burundian Army in 1993, which further inflamed anti-Tutsi fears and rhetoric within Rwanda.

This escalating crisis was neither a sufficient basis for mass killing that operated independent of existing ideology, nor merely a trigger for pre-existing genocidal ambitions. In combination with existing hardline ethnonationalism, it generated a radicalizing narrative of the worsening situation which eventually made mass killing look like a strategically and morally justifiable response.[99] Two interrelated radicalizing dynamics were critical.

First, the onset of the civil war, the negotiations of a power-sharing arrangement with the RPF, and the Ndadaye assassination in Burundi activated and radicalized *sincere hardline sentiments* within Habyarimana's MRND[100] party, most of the opposition parties, and sections of the population.[101] The regime immediately associated the RPF with the entire Tutsi population within Rwanda, and construed them as monarchists determined to overturn the 1959 revolution.[102] The Rwandan Army produced an internal report in December 1991, written by a commission headed by the long-standing hardliner Bagosora, which identified the country's 'principle enemy' as 'Tutsi inside or outside the country, who are extremist and nostalgic for power and who have never recognized and still do not recognize the realities of the 1959 Social Revolution and who want to take power by any means necessary, including arms'.[103] Within a few months of the invasion, the President's wife reportedly argued that 'Hutu youth needed military training to confront the "Tutsi enemy" and that every Tutsi peasant should be included in such a definition'.[104] The party chairman of the MRND wrote privately to Habyarimana, recommending the formation of armed civilian militia loyal to the President, explaining: 'In my opinion, the young must be trained urgently (secretly of course). It is clear that the initial [Tutsi] plan to conquer Rwanda, Burundi and eastern Zaïre is under way. The only way to stop this is the participation by all of the people.'[105]

---

[98] On the origins of the RPF see Prunier 1995, 67–74; Mamdani 2001, 172–84.
[99] In this sense, a pre-existing genocidal ideology did not simply lie dormant in the 1970s and 1980s, waiting to be activated in the 1990s (cf. Uvin 1998, 36–7).
[100] In the multiparty period, the MRND added a further word 'Democracy' to its title, to become the MRND(D), but for simplicity I continue to use the four-letter acronym. See Mamdani 2001, 154.
[101] On popular attitudes see McDoom 2021, 93–115.
[102] International Panel of Eminent Personalities 2000, para 6.3.
[103] Cited in Straus 2006, 25.
[104] Wallis 2019, 311.
[105] Cited in Guichaoua 2015, 140.

The regime also showed its willingness to respond to the invasion with violence against civilians. While not immediately engaging in mass killings, it rounded up around 10,000 'RPF sympathizers', of whom 90% were Tutsi—despite the initial RPF attack being easily rebuffed.[106] The regime also instigated a massacre of around 350 civilians in Kabilira, having 'incited the population under the fabricated story that Tutsi had come to exterminate Hutu'.[107] Further massacres followed, notably of the Bagogwe Tutsi in Ruhengeri prefecture.[108] Pamphlets from the Minister of the Interior instructed the population to: 'Go do a special "*umuganda*". Destroy all the bushes; and all the *Inkotanyi* [a label for the RPF insurgency] who are hiding there. And don't forget that those who are destroying weeds must also get rid of the roots [a euphemistic instruction to also kill children].'[109] Families who fled to the local church were apparently refused entry, with the priest telling them the 'church of God was not able to house cockroaches'.[110]

Sincere hardline radicalization extended to, and split, the political opposition. There were three principal opposition parties: the *Mouvement Démocratic Républicaine* (MDR), a revival of Kayibanda's ruling party from the First Republic; the *Parti Social Démocrate* (PSD), a cross-ethnic left-wing party; and the *Parti Liberal* (PL), a predominantly Tutsi and intellectual party. Initially, these formed a tacit common front against the Habyarimana government. Yet most members of the MDR embraced the ethnonationalist ideology of the First Republic, and therefore opposed the RPF even more vehemently than Habyarimana. Consequently, when peace negotiations raised the prospect of power-sharing with the RPF, hardline elements of the MDR rejected the peace process and sided with hardliners within Habyarimana's government. As Kimonyo shows, this appears to have reflected predominant support for a hardline position within the MDR, since when an internal stand-off between its more moderate and more hardline factions was placed before the party's congress, 70% of party representatives backed the hardline faction.[111] A fourth party, the *Coalition pour la Défense de la République* (CDR), was an ultrahardline racist party formed, with the connivance of Habyarimana, out of extremist MRND members.[112] Small but violent, the CDR declared the Arusha negotiations 'an act of high treason'.[113] Thus, 'by 1993, almost all opposition parties had split between radical, so-called Hutu power wings that were close to the CDR and its discourse, and moderate wings'.[114]

---

[106] Wallis 2019, 224.
[107] Verwimp 2011, 409.
[108] International Panel of Eminent Personalities 2000, para 6.23; Wallis 2019, 247–8.
[109] Cited in Article 19 1996, 15.
[110] Cited in Wallis 2019, 247.
[111] This factional dispute was not *purely* about hardline–moderate disagreements, however—for details see Kimonyo 2016, 92–3.
[112] Wallis 2019, 256.
[113] Cited in ibid. 337.
[114] Uvin 1998, 65.

A hardline 'Hutu Power' coalition between the regime and elements of the opposition parties was consequently able to crystallize following the assassination of the Hutu President Ndadaye in neighbouring Burundi by the Tutsi-controlled Burundian Army in October 1993.[115] The assassination was seized on by Rwandan hardliners as evidence of a transnational Tutsi conspiracy to oppress and exterminate the Hutu (despite the lack of meaningful links between the Burundian Army and the RPF). Just days after the assassination, the MDR held a rally in which its Vice-President, Froduald Karamira, declared that the RPF had 'taken democracy away from the Burundians' and that 'every Hutu living in Rwanda must rise up in turn, so that we do what needs to be done'.[116] He went on:

> the unspeakable act that has just been committed in Burundi will also be committed in Rwanda, if we do not take care ... the enemy ... is among us, right from now ... the enemy is already in our midst ... Let us avoid attacking each other, while we are being attacked. Let us prevent the traitor from infiltrating our ranks and stealing our power.[117]

Calling for a hardline 'Hutu Power' alliance between the MRND, MDR, and CDR, Karamira's cries of 'Hutu Power', 'MRND Power', 'CDR Power', and 'MDR Power', were echoed by the crowd shouting 'Power! Power! Power!'.[118] Actual and rumoured massacres of Hutu in Burundi further intensified this process, so that, as Guichaoua summarizes: 'a defensive ethnic chauvinism crystallized around the Hutu Power movement and paved the way for a broad alliance among the parties opposed to a military takeover of power by the RPF'.[119]

The second radicalizing dynamic revolved around two *instrumental political strategies* the government and opposition parties pursued to enhance their influence in the context of civil war and political competition: first, ideological activism via propaganda and grass-roots mobilization, and second, political militarization, principally through the creation and use of militias. Catastrophizing ideological rhetoric was utilized as a means for building public support and neutralizing opposition. Immediately after the RPF invasion, the regime staged fake RPF attacks on Kigali and other areas to magnify public fears and mobilize support for the regime.[120] Those parties (principally the PL and PSD) who represented Tutsi constituencies or advocated cross-ethnic politics were branded as RPF 'accomplices' by the hardliners.[121] Such hardline rhetoric easily became an object of competitive

---

[115] Guichaoua 2015, 97.
[116] Cited in Straus 2015a, 300.
[117] Cited in Wallis 2019, 346–8.
[118] Straus 2015a, 301; Wallis 2019, 348.
[119] Guichaoua 2015, 98–9. See also Hintjens 1999, 277.
[120] Article 19 1996, 16; Hintjens 1999, 266.
[121] Wallis 2019, 291.

'outbidding' between the MRND, MDR, and CDR.[122] Propaganda proliferated—Uvin reports how in the early 1990s, 'more than twenty papers regularly published racist editorials and cartoons; the official Radio Rwanda often produced similar material'.[123] In May 1990, the CDR founding member Hassan Ngeze set up *Kangura*, an extremist newspaper with a wide circulation in urban areas, under the slogan: 'the voice which seeks to awaken and defend the "majority people"'.[124] Long-standing hardliners helped create the infamous radio station *Radio Télévision Libre des Milles Collines* (RTLM), which targeted youth with more irreverent content than the national broadcaster *Radio Rwanda*, and eventually became a central organ of genocidal propaganda.[125]

Hardliners also used rallies and other forms of face-to-face activism to mobilize those who were either sympathetic to or willing to go along with extremist ethnonationalist claims. Most infamously, a leading hardliner within the MRND, Léon Mugesera—who had been involved in anti-Tutsi ideology and the organization of massacres for years—made a vitriolic, rambling, and explicitly genocidal speech to a major MRND rally in November 1992. Branding opposition parties 'accomplices of the *Inyenzis* [cockroaches] and calling for moderate politicians who had 'demoralized' the armed forces to face the death penalty, he declared:

> Our movement is also a movement for peace. However, we have to know that, for our peace, there is no way to have it but to defend ourselves ... In the country, you know people they call '*Inyenzis*', no longer call them '*Inkotanyi*', as they are actually '*Inyenzis*'. These people called *Inyenzis* are on their way to attack us ... Are we really waiting until they come to exterminate us? ... I am telling you, and I am not lying ... they only want to exterminate us ... We should not allow ourselves to be invaded ... Do not be afraid, know that anyone whose neck you do not cut is the one who will cut your neck. Let me tell you, these people [Tutsi] should begin leaving while there is still time to go and live with their people, or even go to the '*Inyenzis*,' instead of living among us and keeping their guns, so that when we are asleep they can shoot us.[126]

Attacks on Tutsi in the area followed within hours of Mugesera's speech.[127]

But alongside such ideological activism, the regime and the opposition parties also progressively militarized the domestic political space in ways that further fuelled ideological radicalization. Death squads such as the *Amasasu* ('bullets') were created within the army, consisting of: 'extremist officers who felt that the fight against the RPF was not being carried out with the necessary energy ... [and who

---

[122] Hintjens 1999, 261. See also Uvin 1998, 53; Mamdani 2001, 209; Guichaoua 2015, 55.
[123] Uvin 1998, 64.
[124] Cited in Article 19 1996, 35.
[125] Ibid. 41–5.
[126] Mugesera 1992.
[127] Wallis 2019, 288.

began] to hand out weapons to the militias'.[128] The MDR had created a youth wing *Inkuba* in 1991, the MRND followed suit by creating the *Interahamwe* in early 1992, with the CDR creating the *Impuzamugambi*.[129] Between late 1992 and late 1993, these youth wings were converted into militias through military training and arms distributions and, enjoying powerful political backing, they acted with virtual impunity.[130] Simultaneously, the government sought to create 'self-defence forces' of civilians across the country, which would eventually be organized to carry out the genocide.[131]

Though party militias were illegal, Habyarimana publicly authorized them and encouraged their political violence, telling a 1992 rally (a week prior to Mugesera's more infamous speech): 'I personally believe that political rallies have not yet really started [applause]. When they start, I will call upon the *Interahamwe* and we will then actually descend ... What is wrong with that?'[132] Lest the implication was unclear, the speech was followed by an *Interahamwe* 'dramatic tableau' depicting savage attacks on the militia of other political parties. A former *Interahamwe* leader reported how, shortly after the speech:

> I heard instructions had been given to us [*Interahamwe*] to kill some people after the Ruhengeri rally. It seemed that the president of the *Interahamwe* in a nearby commune had been attacked by the population and beheaded. The *Interahamwe* leadership then told us to take revenge, which we did ... I remember how [Habyarimana's] speech was all about the need to stop 'the enemy'. We took this to mean both the Tutsi generally and anyone who opposed the party. We liked Habyarimana a lot and he was very popular as a man who had promised us jobs and good times ahead.[133]

Militarization was radicalizing through several mechanisms. Most obviously, violence was utilized to intimidate and eliminate moderates. Defence Minister James Gasana left Rwanda in fear of his life in July 1993, allowing hardliners to consolidate control over the military.[134] After Habyarimana's assassination, the hardliners immediately set about using the *Interahamwe* and Presidential Guard to kill moderates, most notably Prime Minister Agathe Uwilingiyimana.[135] But militarization also served as a vehicle of hardline propaganda. Militias were not initially recruited from especially ideological individuals, but they became

---

[128] Prunier 1995, 168–9.
[129] Article 19 1996, 30–1; Mamdani 2001, 204; Wallis 2019, 269–71. Later in the genocide, the term 'interahamwe' became applied more generally in some areas to denote bands of local killers.
[130] Guichaoua 2015, 133; Wallis 2019, 223.
[131] Mamdani 2001, 206.
[132] Wallis 2019, 284.
[133] Cited in ibid. 285.
[134] Guichaoua 2015, 88.
[135] Prunier 1995, 230–2; Mamdani 2001, 216.

instruments of indoctrination.[136] Elaborate ideological commitments were secondary, here, to strong identification with the parties in question and adoption of their ideological narratives. As one member explained: 'With our smart new uniforms, being part of such a militia made us very proud and people noticed us, we had status and we repaid it by being totally loyal to the [MRND] party ....'[137] The militia thereby became visible supporters of extremist ethnonationalism, chanting genocidal slogans and songs. Even the less institutionalized civilian 'self-defence forces' served to communicate regime assertions of infiltration and 'promote a fear that all Tutsi were probable RPF sympathisers'.[138]

The escalating political crisis of the early 1990s was thus crucial in explaining why hardliners, only ever one faction of the Rwandan ideological landscape, came to dominate over moderates. Crisis intensified hardline sentiments, while also presenting opportunities and incentives for instrumental hardline propaganda and violence. But crisis alone did not force elites to choose a radicalizing path—the impact of crisis depended on the existing ideological infrastructure. As an Organization of African Union study of the genocide observed: 'even many Tutsi were initially unsympathetic to [the RPF] invasion. Unexpectedly, the government had a perfect opportunity to unite the country against the alien raiders. They rejected it.'[139] Indeed, radicalization was ideologically contested.[140] As Guichaoua emphasizes, just prior to Habyarimana's assassination: 'the government was headed by a prime minister from the opposition and contained a good number of ministers with known democratic convictions; in the armed forces, the general staff and high command had not been won over by genocidal thinking.'[141] Mass killing and genocide emerged in Rwanda, not as a uniform response by elites to clear rational incentives or irresistible social pressure, but as a hardline factional project organized by 'the nucleus of loyalists around the presidential clan' and its supporters within the state, the militias, and Rwandan society.[142]

8.3 Ideology in the Genocide

The Hutu Power Elite

Around 8.20 pm on 6 April 1994, the aircraft containing President Habyarimana, as well as several senior Rwandan officials and the President of Burundi, was shot

---

[136] See also Wallis 2019, 355.
[137] Cited in ibid. 274.
[138] Ibid. 244.
[139] International Panel of Eminent Personalities 2000, para. 7.7.
[140] Straus 2006, 63; McDoom 2021, ch.4.
[141] Guichaoua 2015, 232.
[142] Ibid. 327.

down on its approach to Kigali. Over the following 48 hours, key members of the *Akazu* power network surrounding the President's family organized killings of both Hutu political opponents and Tutsi in the capital. To prevent her leading the government, the relatively moderate Prime Minister Agathe Uwilingiyimana was assassinated. Habyarimana's wife, Agathe Kanziga, her brother, Protais Zigiranyirazo, and the chief of staff at the Defence Ministry, Théoneste Bagosora, took control in the hours after Habyarimana's death.[143] Bagosora organized a new 'Interim Government', built from the hardline 'Hutu Power' factions of the various political parties. Several other figures played key roles in organizing the killings, including Defence Minister Augustin Bizimana and Joseph Nzirorera, national secretary-general of the MRND, who coordinated the *Interahamwe* militia alongside its leader, Robert Kajuga.[144] Eventually, the new Prime Minister, Jean Kambanda, would play an increasing role. Initial resistance from moderates in the military high command was outmanoeuvred and such figures were largely replaced by mid-April.[145]

As noted in the last section, the regime had already employed limited massacres of Tutsi since 1990, and it had prepared lists of key opponents and suspect figures, perhaps in anticipation of more wide-ranging killings. But the genocide emerged unevenly over the following days, as Bagosora and leading figures in the interim government dispatched orders, military units, and *Interahamwe* or *Impuzamugambi* militias to areas outside the capital to organize killings of Tutsi and political opponents.[146] Anti-Tutsi violence did not emerge 'spontaneously' around the country, but in response to government orders and mobilization efforts propagated by apex elites (although local intermediary elites loyal to the regime played a crucial role and often benefited from significant local support).[147] Again, these elite instigators of the Rwandan Genocide, centred on the kleptocratic *Akazu* clique, were unquestionably motivated by material and political self-interest in a context of political crisis.[148] But, as already explained, this combination of elite avarice and crisis/war did not implicate mass killing in any straightforwardly 'rational' sense. Instead, the genocide emerged from self-interested elite manoeuvring within a particular ideological infrastructure. A political strategy of eliminating the Tutsi and the regime's political opponents en masse reflected a particular ideological interpretation of the crisis, and relied on existing ideological structures that rendered such a strategy viable.

[143] Ibid. 156 & 166.
[144] Prunier 1995, 240–1.
[145] Ibid. 229.
[146] Contra common suggestions that the genocide was 'meticulously' planned, e.g. Article 19 1996, 10, 102, & 105; Kellow and Steeves 1998, 116–17; Gourevitch 1999, 104–5; Wallis 2019, 576. See also International Panel of Eminent Personalities 2000, paras. 7.3 & 14.3.
[147] Occasionally local elites were pressured by locals into violence, e.g. Straus 2006, 72–6.
[148] Guichaoua 2015, 202. Personal revenge dynamics in the hours following Habyarimana's assassination may have played a role; see ibid. 149, 152, & 155–6; Wallis 2019, 375–8.

This is not a wholly novel interpretation: several scholars of Rwanda endorse some form of this explanation. As Prunier, in one of the earliest accounts of the genocide, summarizes:

> The actual organisers of the genocide were a small tight group, belonging to the regime's political, military and economic élite who had decided through a mixture of ideological and material motivation radically to resist political change which they perceived as threatening. Many of them had collaborated with the 'Zero Network' killer squad in earlier smaller massacres, and shared a common ideology of radical Hutu domination over Rwanda.[149]

What remains somewhat more contentious is (i) what ideological content was most crucial in shaping elite choices, and (ii) how such ideological content mattered—whether as a cynical tool for manipulating the masses, a more thoroughgoing institutional basis for state power, or a substantive worldview sincerely adhered to by the elite.

As I have argued is true in mass killings more generally, genocide did not really emerge in Rwanda because of special ideological goals long held by the regime. While Philip Verwimp has linked violence against Tutsi in Rwanda to the regime's agrarian social engineering projects, there seems little evidence that Tutsi pastoralists were killed because they 'did not fit within the agrarian order of the Second Republic'.[150] Nor is it accurate to explain the genocide with reference to the way, as Gourevitch puts it, 'a vigorous totalitarian order requires that people be invested in the leaders' scheme ... [so that the] spectre of an absolute menace that requires absolute eradication binds leader and people in a hermetic utopian embrace'.[151] It is hard to find utopian arguments in the justificatory narrative for mass killing in 1994, and if totalitarian order required the spectre of an absolute menace requiring eradication, why did genocide emerge only in 1994, over two decades after Habyarimana's political system was established? In reality, as Straus argues: 'the leaders who instigated genocide did so primarily to win a civil war, not to radically restructure society'.[152]

Hardline ethnonationalist ideology mattered, *not* because of its long-term goals, but because of how it radicalized Rwandan security politics in response to crisis. As in other mass killings, two ideological conceptions of security were critical here.[153] First, and most obviously, implementation of mass killing only made sense given

---

[149] Prunier 1995, 241–2.
[150] Verwimp 2011, 398. Verwimp's argument is sophisticated and focuses on massacres in 1990–92. I don't deny that transformative agricultural projects mattered to some degree, but evidence for their centrality is highly circumstantial: the fact that many of the early massacres were in areas of high population density or agricultural resettlement.
[151] Gourevitch 1999, 95.
[152] Straus 2006, 11.
[153] See also Prunier 1995, 169–70.

an expansive definition of the enemy so as to include broad civilian groups, and primarily the Tutsi ethnic group.[154] There was nothing 'natural' about this conception on the face of it. But it could look plausible in a context where genuinely alarming features of the security environment (invasion by a 'Tutsi'-dominated insurgent force, assassinations of Ndadaye and Habyarimana, and killings of Hutu in Burundi) interacted with an existing elite ideology that identified *Tutsi as a category* with former historical oppressors, and represented Rwanda as, in the private remarks of the President's brother-in-law, 'the republic of the Hutus'.[155]

This conception of the enemy relied on familiar justificatory mechanisms: all Tutsi were construed as an enormous existential *threat*, as *guilty* of terrible machinations against the Hutu majority, and as a fundamentally *separate* group to which Hutu owed lesser if any obligations. Bagosora, who took the lead role in initiating the initial wave of massacres after Habyarimana's assassination, would subsequently explain his view that:

> This is the Rwandan drama ... The Tutsi minority wants to seize power at all costs and the Hutu majority disagrees ... In effect, the Tutsis, at once conceited, arrogant, conniving, and perfidious remain convinced that the good Tutsi is a Tutsi in power and that the good Hutu is the so-called moderate Hutu who serves the interests of the Tutsis unconditionally.[156]

Not only does Bagosora thus associate danger with *the Tutsi as a whole* rather than simply the RPF invasion, but the danger is also understood in genocidal terms. In this tract, written in 1995, Bagosora continues to assert that the Hutu are a 'people demonized by its Tutsi detractors and their allies' who '*risk disappearing* if nothing is quickly done to help them.'[157] Similarly, Robert Kajuga, President of the *Interahamwe* militia and himself of mixed ethnic parentage, told an English journalist during the genocide: 'It's a war against the Tutsis because they want to take power, and we Hutus are more numerous. Most Tutsis support the RPF, so they fight and they kill. We have to defend our country. The government authorizes us.'[158] Likewise Stanislas Mbonampeka, once a liberal politician, would later defend his involvement in the genocide with the claim:

> This was not a conventional war. The enemies were everywhere. The Tutsis were not killed as Tutsis only as sympathizers of the RPF ... There was no difference between the ethnic and the political ... Ninety-nine percent of Tutsis were pro-RPF.[159]

---

[154] Straus 2006, 7; Guichaoua 2015, 215.
[155] Cited in Wallis 2019, 216.
[156] Cited in Straus 2015a, 318.
[157] Cited in ibid. 316 (*my emphasis*).
[158] Cited in Mamdani 2001, 212.
[159] Cited in Gourevitch 1999, 98.

Whether sincerely, unreflectively, or opportunistically, elites explicitly conceptualized Tutsi, as well as Hutu/Twa political opponents or resisters, as an enemy in a military sense.

Second, the genocide depended on a representation of mass killing as *necessary and valorous warfare*, demanding uncompromising participation by the whole of society to dutifully follow the leadership, without which the war, and Hutu society in general, would be lost. As Guichaoua summarizes, the genocide's organizers 'did not view their crimes as 'violations' but rather as acts of war ... [that] could be deemed a functional necessity wholly derived from vital, strategic decision making'.[160] As is always true of collective security politics, moreover, strategic choice was moralized and occurred within particular conceptions of virtue—specifically, ones in which anything short of vast uncompromising violence was a sign of weakness and a betrayal of the Hutu Republic and the social revolution of 1959. 'In a war, you can't be neutral', Mbonampeka would later state. 'If you're not for your country, are you not for its attackers?'[161] At internal security meetings, the Interim Government's Prime Minister, Jean Kambanda, told officials: 'Each leader should reveal his colours because the bell for a decision has tolled. Whoever is not with us is against us. Nobody should say, "I am going to seek refuge somewhere" ... They should make up their minds. Are they for the RPF or are they defending the people's cause?'[162]

As in other cases, these twin ideological conceptions mattered not only in the central elite's initiation of mass killing and genocide, but also in how the violence was implemented by *intermediary elites* across the country. While the army and militia units provided the Interim Government with coercive instruments, the most significant power of the Rwandan state was the density of its 'administrative machinery' in local communities and the general dependability of local officials—prefects and burgomasters—in implementing central directives.[163] As Prunier observes: 'If the local administration[s] had not carried out orders from the capital so blindly, many lives would have been saved.'[164] But the responses of local intermediary elites were not uniform. Many of the most intense and/or early massacres were driven forward by local leaders either known for long-standing ethnic extremism or who displayed marked enthusiasm and determination in hunting down Tutsi to be killed.[165] Straus summarizes how: 'In Gafunzo, political party leaders, reservists and armed youth took the initiative; in Kayove, a well-educated man from a powerful family stepped forward; in Kanzenze, the burgomaster eventually took control; in other places, a *conseiller* led the call to attack

[160] Guichaoua 2015, xxviii.
[161] Cited in Gourevitch 1999, 97.
[162] Cited in Guichaoua 2015, 272.
[163] Mamdani 2001, 144.
[164] Prunier 1995, 244. See also International Panel of Eminent Personalities 2000, para. 14.47; Wallis 2019, 410–11.
[165] See examples in Prunier 1995, 136–9; Gourevitch 1999, 310; Wallis 2019, 243–4, 291, 305, & 415.

Tutsis; in still other places, aggressive young men did.'[166] Militia leaders also played a key role, and frequently avowed hardline ideology. Wallis reports how: 'when *Interahamwe* leader Cyasa Habimana, having just slaughtered 1000 Tutsi men, women and children, was asked why he had done it, he pointed to his lapel badge picturing Habyarimana and simply replied, "they killed him"'.[167]

Conversely, local elites unsympathetic to the hardline justificatory narrative for violence could act to delay the onset of violence and/or save lives. Early in the civil war, the government responded to the RPF capture of the local prison in Ruhengeri by ordering the town's commandant to destroy the prison together with everyone inside it. The commandant later explained, 'I immediately replied that killing prisoners during war was a criminal act', and refused to carry out the order.[168] Famously, the most serious resistance to the genocide came in Butare prefecture, where the local Tutsi prefect worked energetically, helped by the strength of local moderates and positive interethnic relations, to prevent violence spreading to his area of authority.[169] He largely succeeded for two weeks, until the central government managed to replace him with a hardline extremist who immediately initiated local killings.[170] Likewise in Musambira commune (Gitarama prefecture), the local MDR burgomaster Justin Nyandwi was able to mobilize members of the population to prevent violence: 'Together with our prefect, we took the decision to thwart the killings, to repulse them, and each commune organized to do this.'[171] Similarly in Giti, in Byumba prefecture, the local burgomaster—feeling that 'One cannot fight for one's country by killing people'—decided to try to prevent violence.[172] This ultimately helped prevent genocide in Giti, since the burgomaster was able to maintain his resistance until the RPF took control of the commune less than a week after Habyarimana's assassination. Ideology was not the only factor shaping such behaviour by local elites, but their lack of sympathy for hardline ideas, and rejection of the justificatory narrative, was evident and crucial. As Straus concludes: 'Had the moderates controlled the balance of power in various local areas most Rwandans would probably have accepted a moderate position.'[173]

What causally linked ideology to the behaviour of apex and intermediary elites? Again, any attempt to reduce ideology to either deep conviction or purely an instrumental tool provides little purchase here. Hardline ethnonationalist conceptions of Tutsi, political opponents, and state violence were infrastructural:

---

[166] Straus 2006, 88.
[167] Wallis 2019, 423.
[168] Cited in ibid. 246–7.
[169] Prunier 1995, 244; Kimonyo 2016, ch.4.
[170] Prunier 1995, 244. See also the example of Mayor Callixte Ndagijimana in Mugina, in Scull, Mbonyingabo, and Kotb 2016, 337.
[171] Cited in Straus 2006, 80.
[172] Ibid. 85–7.
[173] Ibid. 13. See also Klusemann 2012; McDoom 2021, 32–4.

shaping elite political behaviour through mutually reinforcing blends of sincere internalization and the incentives and pressures of structural norms and institutions.

On the internalized side, I share Uvin's view that explanations narrowly focused on the Rwandan elite's pursuit of power and wealth have 'a tendency to underestimate, if not entirely neglect, how much the elite themselves believe in the ideological messages they broadcast'.[174] Again, Bagosora had long-standing ethnic prejudices, while the new Prime Minister, Jean Kambanda, had chosen to participate in anti-Tutsi persecutions in the early 1970s.[175] As Wallis summarizes: '*Akazu*—both the family group and its "outer" adherents—included some ethnic-based fanatics and ideologues. For these individuals, the genocide was a chance to wreak revenge on the Tutsi it accused of historical abuses suffered when the minority group had held power under Belgian colonial rule.'[176] Beyond these ideologues, many hardliners within the elite were somewhat opportunistic, but still seem to have sincerely interpreted the crisis according to ethnonationalist ideas. Most saw Tutsi as objects of contempt and suspicion, even if not outright hatred, and extreme uncompromising violence was consistently presumed to be the most effective way of dealing with the elite's enemies. 'In their minds …', as Guichaoua puts it, 'personal ambition and the "overriding national imperative" always overlapped in one way or another.'[177]

But hardline ideas were important to the elite's behaviour in ways that go beyond sincere belief. Whatever individual members of the elite privately thought, hardline ethnonationalism had become, by April 1994, a public ideological structure on which they depended, and which provided both instrumental incentives for manipulation and conformist pressures for compliance. Racist stereotypes were part of the social fabric, a mutually understood 'fact of life' even when they were not sources of intense hatred.[178] They sincerely resonated for some segments of society, but were also focal points around which expectations of behaviour coalesced. By 1994, social norms had sufficiently radicalized to the point that when someone invoked extremist claims about Tutsi, others would tend to assume that such claims might generate popular and official support rather than condemnation or sanction. This public power of ethnonationalism was constitutive of the strategic context within which elite strategies of public legitimation and mobilization took place.

I suspect that hardline ideology also exerted structural pressure on the *private* deliberations of elites, although this is difficult to prove. In the days following Habyarimana's assassination, the most extreme hardliners had created 'a climate

---

[174] Uvin 2001, 81.
[175] Melvern 2004, 10.
[176] Wallis 2019, 17.
[177] Guichaoua 2015, xxvii. See also Prunier 1995, 227.
[178] Uvin 1997, 94.

marked by a spirit of revenge and by a necessary realism in confronting a common adversary' amongst the senior networks of the government.[179] Some elites may have thought the hardline ideological rhetoric exaggerated, but nevertheless felt intense pressure created by the apparent dominance of the hardliners to comply with it. One of Habyarimana's confidents recalled: 'Calling an officer, whatever his rank, a traitor or an *inyenzi* was enough at that time to destabilize him. You had to do everything to make sure you were not classified with the traitors.'[180] A mix of genuine sympathy for some hardline claims and the assumption that others within the elite supported these ideas made the hardliners' justificatory narrative difficult to contest, and easy to go along with. As Guichaoua comments:

> A good number of ministers, both the veterans and the newly installed, then felt that the demands of the nation, as expressed by extremist officers from the north [Habyarimana's traditional support base], were simply not up for discussion ... Everyone had to play the card of military preparedness and popular mobilization. The fate of the nation was at stake.[181]

Again, these hardline ideas were fundamentally *factional*—other parts of the elite did not adhere to them, and alternative ideological conceptions, implying alternative responses, were available.[182] But hardline ideology won out because 'the hardliners had the balance of power among Hutus in the country at the time. They dominated the central state, key army units, militias, and radio broadcasts.'[183] That dominant power position was itself, moreover, partly constituted by ideology—by real sympathies for hardline ideas, and the perception that they defined a prevailing normative consensus in state and society.

Rank-and-File Agents in the Genocide

While both Stalinist repression and the Guatemalan civil war involved degrees of public involvement, 'the massive participation of the population in the Rwandan Genocide', as a post-genocide international conference in Kigali reported, 'is virtually without historical precedent'.[184] The most influential estimates suggest that between 10% and 20% of the Hutu male population committed an act of violence during the genocide—around 200,000 perpetrators (although a disproportionate amount of the violence was carried by a minority of the most enthusiastic

---
[179] Guichaoua 2015, 201.
[180] Cited in Wallis 2019, 268.
[181] Guichaoua 2015, 196.
[182] Straus 2015a, 313–14 & 320–1.
[183] Straus 2006, 66.
[184] Cited in Mamdani 2001, 199.

or professional killers).[185] This creates a particularly strong overlap between the ideological attitudes of those mobilized to kill, discussed in this section, and the broader public dissemination of ideology, discussed in the next section.

Though this mass mobilization introduces some distinctive patterns into the character of rank-and-file perpetrators in Rwanda, the picture remains similar to other mass killings. Those who implemented the genocide 'on the ground' were overwhelmingly 'ordinary', not psychopathic, and were guided by a range of private motives beyond ideology. Participation in the killing itself brutalized perpetrators and habituated/desensitized them to violence,[186] and a context of impunity was an important enabler.[187] Most obviously, coercion probably was a key factor, stressed by most perpetrators in subsequent interviews.[188] While such testimony is obviously self-exculpating and likely exaggerates the true extent of coercion, it appears credible in many instances. But two points about such coercive pressures require emphasis.

First, the exact nature of coercive pressure varied, and often left significant room for agency.[189] A few perpetrators offer plausible accounts of being forced to kill on pain of death. Yet others suggest that coercion was absent, or limited to softer forms of threat, intimidation, authoritative orders, and peer-pressure.[190] Some appeal to coercion whilst simultaneously suggesting that they willingly accepted the violence as justified. Prunier cites a former perpetrator who claimed: 'I regret what I did ... I am ashamed, but what would you have done if you had been in my place? Either you took part in the massacre or else you were massacred yourself. So I took weapons *and I defended the members of my tribe against the Tutsi.*'[191] As Prunier observes: 'Even as the man pleads compulsion, in the same breath he switches his discourse to adjust it to the dominant ideology.'[192]

In their softer forms, coercive dynamics were partly sustained by prevailing ideological conceptions of state authority and the valorization of dutiful participation in violence.[193] The common suggestion that Rwandans possessed a sweeping 'tradition of unquestioning obedience'[194] or blindly adhered to 'norms of rote obedience'[195] is overstated, since Rwandans ignored state directives on a range of

[185] Straus 2004, 95; McDoom 2021, 5; Nyseth Brehm, Edgerton, and Frizzell 2022[forthcoming].
[186] Hatzfeld 2005, 31, 42, 44–5, & 68; Lyons and Straus 2006, 64; Scull, Mbonyingabo, and Kotb 2016, 340.
[187] Lyons and Straus 2006, 93.
[188] 64.1% in Straus's survey, and 16 of the 17 interviewees in the Scull, Mbonyingabo, and Kotb study; see Straus 2006, 136; Scull, Mbonyingabo, and Kotb 2016, 339–40. See also Gourevitch 1999, 136; Mamdani 2001, 219; Hatzfeld 2005, 70; Lyons and Straus 2006, 41, 45, 48–50, 61, 64–5, 69–71, 74, 79–80, 86, & 94; Jessee 2017, 168–73.
[189] Prunier 1995, 247.
[190] E.g. Gourevitch 1999, 307; Hatzfeld 2005, 33, 65, & 67; Lyons and Straus 2006, 91.
[191] Cited in Prunier 1995, 247 (*Prunier's emphasis*).
[192] Ibid.
[193] McDoom 2021, 33–4.
[194] Prunier 1995, 245; Gourevitch 1999, 243.
[195] Kellow and Steeves 1998, 116.

issues, and a minority did seek to evade participation in genocide.[196] But again, societies are ideologically heterogenous. *Large numbers* of Rwandans appear to have ideologically conceptualized state authorities as generally unquestionable, and not given disobedience serious consideration. As one of the ministers in the Interim Government explained: 'The party's militants were trained in such an ethos [of strict unity of thought], and the multiparty system changed nothing as regards their mentality of absolute respect for authority.'[197] Many perpetrators stated, for example:

> I could not [refuse]. They were authorities. I respected them. If you come and order me, can I refuse? ... It was not my will. It was because of the authorities who asked me to do it.[198]
> We did it because it was the Law. The authorities told us we had to do it.[199]
> [The local *conseiller*] said we had to look for the enemy, that the Hutu's enemy was the Tutsis ... It was a law. Whenever a leader gives you a command, you do it.[200]
> Yes, you obey your leaders. Otherwise you must have your own country, if you don't follow your leaders.[201]

Many also attested to the way violence itself had been ideologically valorized, generating a moral environment in which one faced intense peer pressure to conform. 'When the killings begin', one perpetrator proclaimed, 'you find it easier to ply the machete than to be stabbed by ridicule and contempt.'[202] Guichaoua observes that *Interahamwe* leaders who failed to show up to participate in killings in the 24 hours after Habyarimana's assassination 'were viewed as "cowardly" or "deserters" the following day.'[203]

Second, coercive pressure existed in large part because of the *willing participation and initiative* of large numbers of rank-and-file agents who appeared to sincerely embrace the regime's justifications for mass killing. As in other cases, the regime could not coercively micromanage violence across Rwanda directly—they depended on vanguards, primarily local elites (as discussed above), army units, and the energetic *Interahamwe* and *Impuzamugambi* militia to both implement the violence and pressure other less-committed Rwandans to take part. In this sense, as Mamdani emphasizes: 'The genocide was not simply a state project. Had the

---

[196] Uvin 1998, 67 & 215; Straus 2006, 63; Fox and Nyseth Brehm 2018.
[197] Cited in Guichaoua 2015, 219.
[198] Cited in Lyons and Straus 2006, 88.
[199] Cited in McDoom 2021, 1.
[200] Cited in Lyons and Straus 2006, 78–9.
[201] Cited in Anderson 2017a, 43.
[202] Cited in Hatzfeld 2005, 213.
[203] Guichaoua 2015, 216. See also Gourevitch 1999, 26.

killing been the work of state functionaries and those bribed by them, it would have translated into no more than a string of massacres perpetrated by death squads.'[204]

Coercion therefore only explains a fragment of rank-and-file perpetration in the genocide. Against stronger situationist accounts, moreover, perpetrators in Rwanda do not appear to have typically been reluctant or unthinking conformists. Instead, most appeared to fit into two broad categories. First, a sizeable but nevertheless minority group enthusiastically embraced the genocide and were crucial in driving the violence forward across the country.[205] They were especially prominent amongst the militia: riding through the streets of Rwandan cities singing songs and slogans such as 'let us eliminate the enemies of Rwanda, they have provoked us'.[206] A Rwandan general recalls: 'One could hear [the *Interahamwe*] returning from political meetings, atop of trucks and buses, singing and screaming at the top of their lungs without the least bit of concern for anyone at all: "*Tuzabatsembatsemba*"—"We have come to exterminate."'[207] Based on amateur video footage, Wallis describes the atmosphere of killing roadblocks in Kigali as 'jovial', with *Interahamwe* members celebrating killings and 'excited, laughing and filled with stories of their latest kills'.[208] A Canadian physician in Rwanda likewise recalled: 'The *Interahamwe* were terrifying, blood-thirsty, drunk—they did a lot of dancing at roadblocks.'[209] Many observers also testify to ordinary groups of killers enthusiastically chanting 'Eliminate the Tutsis'.[210] Such perpetrators were rarely intellectual ideologues, but they appeared to fully identify with the regime and accept its justificatory narrative, and this not only legitimated but often seems to have actively motivated their participation in violence. As one of Anderson's interviewees stated simply: 'I supported the Hutu power movement because it called for the killing of Tutsis.'[211]

The justificatory narrative such enthusiastic killers employed broadly matched the central conceptions articulated by elites. The Tutsi were an existential threat to the Hutu, determined to oppress or even commit genocide against them. They were collectively guilty for Habyarimana's assassination. Violence against them was a patriotic duty to the state, and a necessary measure to save the Hutu. A leading perpetrator of multiple massacres in Gisenyi prefecture reported how, after Habyarimana was assassinated:

> In my mind, I understood right away that the Tutsis were responsible. I was angry, and I said to myself, 'It is true. The Tutsis are mean.' And I said everything people

[204] Mamdani 2001, 225.
[205] Ibid. 219.
[206] Cited in Wallis 2019, 403.
[207] Cited in Guichaoua 2015, 139.
[208] Wallis 2019, 401.
[209] Gourevitch 1999, 134.
[210] Ibid. 29.
[211] Cited in Anderson 2017a, 103.

say about them is true … We said … we are going to kill them before being killed by them … I did not remember that I was educated by Tutsis, or that I had once lived with priests. I did not even think about the fact that my older brothers had Tutsi wives or that my mother-in-law was Tutsi.[212]

An army reservist who joined the extreme racist CDR party and later confessed to killing ten people likewise explained:

I saw that the CDR party was for the Hutus and that the CDR party would defeat the PL, which was the Tutsi party … [In the 1990s] I saw ID cards and tattoos saying 'Vive Rwigema' [an RPF rebel commander] … I saw that the Tutsi were accomplices … All Tutsis. I thought everyone was the same way … I thought that the *Ubwoko* ['category', i.e. Tutsi] was the enemy. We were angry about the death of the president … And if the children were not killed, would the ethnic group be exterminated? … We had to kill accomplices so that they would not increase their forces.[213]

Many of these figures firmly retained this conception of the violence years after the genocide.[214] One perpetrator suggested that of those in prison: 'most of the killers are sorry they didn't finish the job. They accuse themselves of negligence rather than wickedness.'[215]

These seemingly strong believers in the regime's justificatory narrative for genocide were crucial for two reasons. First, they appear to have been responsible for a disproportionate share of the killing. As Straus observes from his survey of perpetrators: 'those respondents who claimed that war-related anger and fear drove them were consistently the most violent. The most violent perpetrators claimed that Hutus had to fight for the country and they mobilized the less violent to join them.'[216] Straus elsewhere estimates that the roughly 10% of perpetrators who were 'soldiers, paramilitaries and extremely zealous killers' may have accounted for 75% of all genocide deaths.[217] Second, as Straus suggests, such perpetrators played a vital role in projecting the policy of genocide across the country, generating the coercive pressures noted above and mobilizing the less enthusiastic. Against the image of mass 'spontaneous' participation in violence, Fujii notes how 'leaders of the violence could not take for granted that the population would automatically obey or follow orders to kill their Tutsi neighbors.'[218] They frequently had to bring

---

[212] Cited in Straus 2006, 74.
[213] Lyons and Straus 2006, 41–3.
[214] Mironko 2004; Jessee 2017, 1–3 & 155–67.
[215] Cited in Hatzfeld 2005, 154. By contrast, see Scull, Mbonyingabo, and Kotb 2016, 341.
[216] Straus 2006, 150.
[217] Straus 2004, 95. See also Nyseth Brehm, Edgerton, and Frizzell 2022.
[218] Fujii 2009, 82.

in relatively committed *Interahamwe* militia, and senior politicians, to pressure the population to get involved.

Most perpetrators, however, appear to have fallen into a second group, who did internalize regime justifications of the violence but in a far more selective fashion. Such perpetrators still seem to have sincerely thought that the killing was necessary, but they were often unenthusiastic, rather apolitical prior to the genocide, and accepted justifications for violence only in response to active mobilization efforts. For many, such justifications may have functioned primarily as a legitimation for violence committed under social pressure, and only secondarily as a motive. Nevertheless, they readily espoused justifications that remained centred around hardline security politics. As excepts from Anderson's interviews with former perpetrators assert:

> The MRND was telling people to open their eyes and do something because the Tutsi are planning.
> People used to say the RPF would vivisect us.
> They taught us that if the RPF took power they would destroy the country.
> We thought if the RPF took power they would kill all the Hutus.[219]

Since these more selective adherents to the hardline narrative tended to lack strong pre-existing political views, Habyarimana's assassination was often crucial in lending such portrayals plausibility. As a fisherman involved in killings in Cyangugu prefecture recalled:

> The leaders came, those who led us on the patrol ... They told us that Habyarimana's plane had been shot down, that we had to find a way to defend ourselves, and that the Tutsis had killed him. That is when we understood that the Tutsis were the enemy. They told us, 'Find a way to defend yourself. You are with the enemy.' We became angry because they told us that everywhere the *inkotanyi* went they killed Hutus.[220]

A farmer from the southern prefecture of Gikongoro likewise explained:

> The authorities made the population believe that the Tutsis would kill us ... Hearts changed, little by little, because of what these authorities said ... In this period everyone was angry because the president had been killed and everyone said that one had to protect oneself ... I understood that the enemy was the Tutsis because the president had died. But it was not prepared. Our minds changed because of the events in the country.[221]

---

[219] Cited in Anderson 2017a, 73–4. See also McDoom 2021, 95–115.
[220] Lyons and Straus 2006, 91.
[221] Ibid. 39–40.

Again, what mattered here were not long-standing ideological goals or hatreds, but hardline interpretations of security, as manifested in the prevailing justificatory narrative for mass killing. In this, ideological constructions of threat, guilt, valour, and necessity loom largest. Thus, perpetrators explained how:

> [In the war] the Tutsis attacked Rwanda from Uganda and they killed Hutus ... When the Hutus saw that their people were finished, they attacked their Tutsi neighbours ... Because the *inkotanyi* killed Hutus; when Hutus saw their own killed by the *inkotanyi*, they killed the Bagogwe [Tutsis] ... We, the peasants, believed that the person who had killed the president was an enemy, that is what we believed ... The *inkotanyi*, they were Tutsis; they were Tutsis, so we believed the solution was to kill the Tutsis ... We said we were defending ourselves against the enemy ....[222]

Another perpetrator emphasized more simply: 'I went as someone who defends his country that was attacked. I thought that if the enemy came, he would kill me. I went as someone who loves his country, but after having received a command.'[223]

The more selective way in which such perpetrators internalized this justificatory narrative often imbued ambivalence into their viewpoint. Some suggested that they accepted the general claim that the Tutsi were culpable for the RPF invasion or the killing of the president, while nevertheless retaining a sense that their Tutsi neighbours were innocent. One farmer, whose wife was Tutsi, affirmed that: 'I thought the accomplices were in the city. But with my neighbors, I didn't think about it.'[224] Another stated: 'They were telling us that the *Inyenzi* were the Tutsi who were in the forests. For us, we had the impression that the *Inyenzi* did not look like us. We thought that they were not like our Tutsi neighbors.'[225] Nevertheless, some such perpetrators remained unrepentant years after the genocide, continuing to espouse the justificatory narrative rather than simply claiming they had been forced to kill. One asserted that: 'It was not only Tutsi who died ... RPF did worse.'[226] Another went as far as to claim that: 'These killings were begun by the *inkotanyi* after they had killed the head of state ... The *inkotanyi* made these things happen. The *inkotanyi* went into the peace negotiations with the president. Instead of accepting, they killed him; that is why these things happened.'[227]

A third category of rank-and-file agents showed genuine reluctance to participate in violence, and scepticism about the justificatory narrative for genocide, participating only under intense pressure. But they appear to have been a relative minority, *despite* the incentives one would expect, after the genocide, to

---

[222] Ibid. 81–2.
[223] Ibid. 78–9.
[224] Ibid. 47.
[225] Cited in Fujii 2009, 83.
[226] Cited in Mironko 2004, 56.
[227] Lyons and Straus 2006, 89–91.

self-exculpate by stressing coercion and disavowing willing belief. Even in this group, there was a spectrum of attitudes and behaviour in the genocide.[228] Some perpetrators suggested that they participated despite thinking the killing wrong and felt 'bad' about the violence even at the time.[229] Many Rwandans engaged in mixed behaviour, moreover, participating both in killings and in acts of rescue or resistance.[230]

Others more consistently sought to evade participation in the violence, or to resist it or rescue potential victims. Because many factors shaped individuals' willingness and capacity to resist or rescue, there is no simple one-to-one relationship between ideological attitudes and behaviour here. Yet research on resistance and rescue in Rwanda again suggests that individuals' adherence to limitationist ideas, inconsistent with the regime's justificatory narrative, was key. Based on 31 interviews with rescuers, Paul Di Stefano concludes that two factors were crucial: 'First, the influence of positive parental role models; and second, the possession of an ethnically blind worldview grounded in a "common humanity".'[231] When asked to explain why they rescued Tutsi, interviewees testified, for example:

> Other students used to harass [a Tutsi girl in my school] and persecute her in front of others in the schoolyard. It pained me a lot and I was very devastated when I heard that she was being abused just because she was a Tutsi ... This whole experience changed my life and I decided never to associate myself with these divisive ideas.[232]
>
> A human being is a human being. All people are the same; may his blood be red; may he be Twa; may he be Hutu; may he be someone else; he is still a human being.[233]
>
> I never discriminate [against] people. I know that we are all one ... We all have one blood.[234]

Similarly, having interviewed 45 rescuers, Nicole Fox, Hollie Nyseth Brehm, and John Gasana Gasasira emphasize the importance of both rescuers' situational positions and their socialization into particular worldviews, ideologies, and values, especially certain religious orientations.[235] In part, such orientations shaped their sincere interpretations of the violence: 40% of the interviewees stated that the primary motivation for rescuing was a moral code, while over half mentioned religious faith as an important influence on their actions.[236] But in addition, distinct

---

[228] Fujii 2008; Fujii 2009.
[229] Scull, Mbonyingabo, and Kotb 2016, 340. See also Lyons and Straus 2006, 50–8.
[230] Gourevitch 1999, 130–1; Mamdani 2001, 221.
[231] Stefano 2016, 199.
[232] Cited in ibid.
[233] Cited in ibid. 201.
[234] Cited in ibid.
[235] Fox and Nyseth Brehm 2018, 1641; Fox, Nyseth Brehm, and Gasasira 2021.
[236] Fox and Nyseth Brehm 2018, 1637 & 1643, fn.14.

ideological worldviews tied such individuals into somewhat segregated social networks, which offered different structural opportunities for avoiding violence and engaging in rescue.[237]

This again suggests that ideological worldviews and the acceptance or rejection of the regime's hardline justificatory narrative were important sources of variation in behaviour during the genocide. More broadly, levels of popular ideological sympathy towards the regime in different parts of Rwanda seem to have been associated with support for the violence. In analysing regional variation in the onset of the violence, Straus stresses that:

> the strongest relationship is between onset and support for the ruling MRND party. In areas where Habyarimana's party had the strongest support, violence materialized quickly; in areas where opposition had the strongest support, there was the most resistance to violence … in some way political commitments and alliances drove early participation.[238]

The OAU panel on Rwanda likewise observed how 'there were different levels of preparedness [for genocide] around the country, depending on local attitudes to Tutsi'.[239]

Despite evidence of sincere acceptance of the regime's justificatory narrative for genocide, evidence of long-standing ethnic hatred to Tutsi *as such* surfaces relatively rarely in the testimony of former perpetrators, survivors, and witnesses. More abstract claims about the Tutsi being foreign Hamites or desiring to reimpose the monarchy also appear to have garnered limited belief, and rank-and-file perpetrators make little mention of utopian ideals or broader political projects.[240] Occasionally perpetrators aver to dehumanized conceptions of Tutsi. 'I thought they were like animals deserving to be slaughtered', reports one. 'Nothing else.'[241] Charles Mironko observes that local killings were often organized with the use of hunting terminology that may have implicitly dehumanized victims.[242] Overall, though, evidence on the extent of dehumanization in rank-and-file perpetrators' conceptions of the violence is mixed, and it's unclear how commonly Tutsi were actually referred to as '*inyenzi*' ('cockroaches') by rank-and-file perpetrators. Dehumanization was probably a facilitator of violence, but it generally appears secondary to more familiar securitized claims about Tutsi as threatening, guilty, and deidentified from the national in-group.

---

[237] Fox, Nyseth Brehm, and Gasasira 2021.
[238] Straus 2006, 61.
[239] International Panel of Eminent Personalities 2000, 14.3.
[240] Straus 2007b, 627.
[241] Cited in Scull, Mbonyingabo, and Kotb 2016, 341. See also Gourevitch 1999, 141; Hatzfeld 2005, 42, 93, & 136; Lyons and Straus 2006, 49; Scull, Mbonyingabo, and Kotb 2016, 340–1.
[242] Mironko 2004, 51–3.

Moreover, these varying degrees of ideological internalization should not obscure how the hardline justificatory narrative for genocide also functioned as an ideological structure—both an apparent social consensus and an official normative line authorized by the state. Across fieldwork research on the genocide, perpetrators, survivors, and witnesses frequently refer to core justifications of the violence with phrases like 'people were saying', 'people used to say', 'everyone was angry', 'the Hutus believed', and so forth. Whatever people's private views, hardliners' claims appear to have constituted the dominant public political discourse during the genocide. One former perpetrator explained:

> When [the RPF] came, *we were told* that they hunted and beat Hutus, and the refugees from Byumba confirmed that that was true ... That morning [after Habyarimana's assassination] *everyone you saw said*, 'We have been saying ... for a long time that the Tutsis will exterminate us and, *voilà*, they just killed Habyarimana, who was protected. You, the simple peasants, you are finished.' CDR and MRND members said this ... Heads became hot. *The youths began to suspect everyone who was Tutsi* ... [The MRND and CDR members] were in the roads, in people's homes, and they even took vehicles by force, saying 'Unite! We will take revenge!' ... *We were people convinced the Tutsis would kill us* ... I protected my ubwoko. People said *they* would exterminate us.[243]

As this transcript suggests, perception of a dominant social consensus could easily overlap with sincere acceptance. But even when individuals possessed limited belief in the ideological narratives of the regime, they were still presented with a set of notions and a language of perpetration—orientated around *ibyitso* ('accomplices'), *igitero* ('attack'), *umuganda* ('coming together in common purpose'), *intambara* ('war'), and other somewhat banalized concepts—that constituted collective expectations about such violence in ways that encouraged participation and made opposition appear risky.[244] For many, such language may have been picked up and used instrumentally—as a pretext for legitimating private agendas. But the power of such language to serve such instrumental functions depended on ideological structures: on the fact that other individuals either sincerely accepted these ideas or themselves sensed that there were instrumental benefits to going along with them and significant risks in failing to do so.

So rank-and-file perpetrators of the Rwandan Genocide cannot be generally characterized as either zealous killers or as blandly unideological. Most rank-and-file agents lacked deep and long-standing commitments to anti-Tutsi ideology or hatreds, but accepted, to varying degrees, the regime's hardline justificatory narrative that framed the political crisis in ethnic and catastrophic terms and cast

---

[243] Lyons and Straus 2006, 75–8 (*my emphasis*).
[244] See also Li 2004, 18; Mironko 2004.

genocidal violence as a necessary solution. At a minimum they understood and reproduced that narrative as a discursive framework for rendering their actions socially acceptable and meaningful. A critical but smaller group, prevalent within the party militias and the most enthusiastic local killers in rural areas, seemed to have much more thoroughly embraced hardline ideas. Even for this group, ideology operated alongside a panoply of other motives, and was often a vernacular set of tropes, assertions, and frames rather than intellectual 'high doctrine'. But it is hard to see how the genocide could have occurred, at least on anything like the same scale, in the absence of such a hardline ethnonationalist infrastructure.

Disseminating the Hardline Narrative

This leaves the question of why and how the regime's hardline justificatory narrative for genocide was successfully established. The context of crisis was clearly vital, and many scholars recognize the importance of prior ethnonationalist ideological legacies in lending the regime's justificatory narrative plausibility. There is more dispute, however, concerning the role of propaganda and ideological activism.

Many commentaries strongly emphasize propaganda's role—almost always identified with the notorious RTLM 'hate radio' station—as an instrument for promoting the genocide. An influential French study, for example, stated: 'Two tools, one very modern, the other less so, were particularly used during the genocide of the Tutsi in Rwanda: the radio and the machete, the first to give and receive orders, the second to carry them out!'[245] An early analysis by Christine Kellow and Leslie Steeves endorses Western journalists' contentions that 'the Rwandan people spend all their time with a receiver stuck to their ear'[246] and 'When the radio said it was time to kill the people opposed to the government, the masses slid off a dark edge into insanity'.[247] Kellow and Steeves also emphasize 'strong traditions of hierarchy and authoritarianism, which increase the likelihood of blind obedience to the orders of officials on the radio'.[248] Philip Gourevitch's influential account of the Rwandan Genocide claims that 'The Hutu Ten Commandments', a notorious piece of hate propaganda in the hardline newspaper *Kangura*, was 'widely circulated and immensely popular ... Community leaders across Rwanda regarded them as tantamount to law'.[249] It is unclear, however, what evidence lies behind this claim—and less than 3% of the genocide perpetrators actually surveyed by Straus reported that they had even heard of the Hutu Ten Commandments.[250]

---

[245] Chrétien, Dupaquier, Kambanda, and Ngarambe cited in Li 2004, 11.
[246] Kellow and Steeves 1998, 118.
[247] Ibid. 124.
[248] Ibid. 116.
[249] Gourevitch 1999, 88.
[250] Straus 2006, 130.

Like most recent scholars, I push back against these exaggerated accounts of propaganda's role. The balance of evidence does suggest that hardline ideological activism was an important factor in the genocide, but the traditional image of radio propaganda whipping up pliant Hutu masses into ethnic hatred is inaccurate on three counts. First, *face-to-face activism* in the form of local speeches by hardliners, briefings by local officials, and exhortations by military units and militia groups appears to have been equally or more important than radio as the medium for disseminating the hardline justificatory narrative. Second, the impact of ideological activism was, as always, *conditioned by the existing attitudes and orientations* of those it targeted. This did not mean that propaganda merely encouraged what people were going to do anyway, but its effects were varied and contested. Third, propaganda's impact is not best understood through notions of blind obedience to authority or the triggering of ethnic hatred. Instead, hardline propaganda was influential because of the dynamics of *epistemic dependence* and *motivated reasoning* that are typical of security politics in conditions of crisis.

The broad ethnonationalist ideological infrastructure of the Habyarimana regime had long been sustained through a mixture of state education, the national broadcaster Radio Rwanda, the loyal support of the Catholic church (which largely continued into the genocide itself),[251] and the dense party networks of the MRND across the country. The hardline justificatory narrative for genocide gained plausibility from this long-standing ideology and its account of past Tutsi oppression, which many perpetrators appear to have sincerely accepted.[252] But, as already noted, the justificatory narrative itself was principally disseminated over the 1990–94 period, in which the media landscape in Rwanda transformed dramatically. The most infamous hardline newspaper, *Kangura*, was set up in May 1990 by Hassan Ngeze, supported by members of the *Akazu* network, including the ethnonationalist ideologue Ferdinand Nahimana. By 1992, it operated alongside several other important hardline anti-Tutsi papers such as *Interahamwe*, the paper of the MRND militia, and *Zirikana*, the paper of the extremist CDR party, again launched with assistance from the *Akazu*.[253] Radio Rwanda was used to explicitly call for violence in the Bugesera massacre in March 1992,[254] but as the Arusha negotiations progressed, its role and content became increasingly contested by the opposition political parties. In this context, hardliners from within the MRND, CDR, and *Akazu* decided to set up a private station in July 1993: RTLM.[255]

Many of these media produced racist and alarmist propaganda from 1990 onwards, with a dramatic uptick in hardline output from late 1993 which intensified

---

[251] International Panel of Eminent Personalities 2000, para. 14.66–71.
[252] Jessee 2017, 155–63; McDoom 2021, 94 & 157–60.
[253] Wallis 2019, 260–1.
[254] Gourevitch 1999, 93–4; Wallis 2019, 305.
[255] Forges 2007, 44.

further after Habyarimana's assassination.[256] Straus and McDoom both observe that much of the most virulent exterminatory media rhetoric was not produced until late in the genocide, suggesting significant radicalization as a *result* of the violence.[257] This should not obscure, however, the way a much more prosaic hardline justificatory narrative for violence was espoused earlier and disseminated either prior to or during the opening phases of mass killing. *Kangura* asserted in 1990, for example, that: 'After the infamous massacres of Hutus in 1965, 1972–73 and 1988 by minority Tutsi regimes in Burundi, the Hutu people are today, more than ever before, threatened with extermination.'[258] In July 1991 *Kagura* declared: 'If the Hutu continue to bicker amongst themselves, in different political parties, the *Inkotanyi* and their accomplices will exterminate us.'[259] In November it asked: 'what weapons shall we use to conquer the cockroaches once and for all?'.[260] In September 1991, the paper *Le Médaille Nyiramcibiri* covered massacres of the Bagogwe Tutsi by blaming the victims, asserting: '[The Tutsi] are the ones who have sparked the violence in the Mututa region by provoking the population, beating and killing a soldier just as … the *Inyenzi* had stepped up their attacks … the fury of the Bagogwe is stronger than that of lions … Why do certain Tutsi like blood?'[261]

In a context with relatively low literacy, radio stations had more influence outside educated circles and urban areas. RTLM escalated its hardline content over late 1993 and 1994.[262] In December 1993, RTLM asserted: 'Tutsi are nomads and invaders who came to Rwanda in search of pasture, but because they are so cunning and malicious, the Tutsi managed to stay and rule. If you allow the Tutsi-Hamites to come back, they will not only rule you in Rwanda, but will also extend their power throughout the Great Lakes Region.'[263] In January 1994, RTLM warned: 'Masses, be vigilant … Your property is being taken away. What you fought for in' 59 is being taken away.'[264] A month prior to the genocide, it warned: 'At RTLM, we have decided to remain vigilant. I urge you, people of Biryogo, who are listening to us, to remain vigilant. Be advised that a weevil has crept into your midst. Be advised that you have been infiltrated, that you must be extra vigilant in order to defend and protect yourself.'[265]

From the early massacres of the civil war through to the end of the genocide, such print and radio propaganda was accompanied by more focused face-to-face dissemination of hardline claims justifying violence against Tutsi as a form of defensive warfare. Prunier observes how:

[256] Ibid. 45–50.
[257] Straus 2007b, 621–2; McDoom 2021, 8 & 117–22.
[258] Anderson 2017a, 79.
[259] Cited in Article 19 1996, 36.
[260] Cited in Wallis 2019, 261.
[261] Cited in Article 19 1996, 40.
[262] Ibid. 51; Forges 2007, 45.
[263] Cited in Mironko 2007, 127.
[264] Cited in Li 2004, 13.
[265] Cited in Anderson 2017a, 81.

A common feature of all the [1991–92] massacres is that they were preceded by political meetings during which a 'sensibilisation' process was carried out. These seemed to have been designed to put the local peasants 'in the mood', to drum into them that the people they were soon to kill were *ibyitso*, i.e. actual or potential collaborators of the RPF arch-enemy. These meetings were always presided over and attended by the local authorities with whom the local peasants were familiar; but they also usually featured the presence of an 'important person' who would come from Kigali to lend the event an aura of added respectability and official sanction.[266]

In addition, the justificatory narrative spread through what Debra Spitulnik labels 'lateral communication': the reproduction of claims heard on radio or from leaders amongst ordinary citizens.[267] As Li summarizes: 'RTLM explicitly informed conversations that took place away from the physical contexts of listening as well. Broadcasts were often reincarnated elsewhere as rumour, where the possibilities for exaggeration or reinterpretation could only expand.'[268] Many perpetrators reported hearing justifications of violence from neighbours who had listed to the radio, or from roving bands of other perpetrators during the early stages of the genocide. As one reported:

> At that time you would see many small groups of people ... you would hear them saying that Tutsis dug holes where they will throw Hutus. You go to another group and you hear them saying that Tutsi have bought big metal tanks in which they will fry Hutus when *inyenzi* win the battle.[269]

Another explained:

> I don't own a radio because I am a cultivator ... But in fact, [some people] used to say that RTLM was urging us to kill people ... It said: 'the enemy is Tutsi.' And when a peasant hears that, that person has no choice ... When he meets a person he doesn't know, he says to himself: 'this is the one who came to eradicate us, he came to fight us.'[270]

Evidence on the exact reach of radio propaganda is mixed. In Straus's survey of perpetrators, of the 52% of respondents who said they owned a radio, 60% said they did not listen to RTLM, the most intense producer of hardline rhetoric.[271] Anderson, by contrast, reports that: 'Among my 80 respondents in

---

[266] Prunier 1995, 138.
[267] Spitulnik cited in Mironko 2007, 128. See also Jessee 2017, 164–7.
[268] Li 2004, 19.
[269] Cited in Scull, Mbonyingabo, and Kotb 2016, 339.
[270] Cited in Mironko 2007, 131.
[271] Straus 2007b, 628–70. See also Mironko 2007, 129–31.

Rwanda, about two-thirds listened to RTLM ... while a third read *Kangura* ... A former *Interahamwe* argues: "All Hutus listen to RTLM; it was their own radio." Another perpetrator argued that RTLM was "unavoidable".[272] In McDoom's survey, meanwhile, around 60% of respondents reported listening to RTLM.[273]

Ultimately, radio propaganda and face-to-face activism played a mutually reinforcing role in Rwanda, rendering the major elements of the hardline justificatory narrative socially pervasive. As an *Interahamwe* member reported, the radio stations 'were always telling people that if the RPF, the rebel Rwandan Patriotic Front, comes, it will return Rwanda to feudalism, that it would bring oppression'.[274] Another recalled: 'I heard it so many times ... that the Tutsis were from Ethiopia ... during the genocide they killed them [Tutsis] and put the dead bodies in the river so that they go back to Ethiopia quickly.'[275]

Precisely assessing the effects of such hardline activism is difficult, but the balance of evidence suggests that it played a significant role in the violence. For a start, there are several specific instances of RTLM directing and coordinating specific attacks against certain areas and targets, and militias immediately responding to those instructions and carrying out such attacks.[276] Statistical analysis of RTLM's broader influence presents somewhat mixed results. David Yanagizawa-Drott's study, based on an analysis of local variation in violence in relationship to the likely strength of available radio broadcasts given local topography, suggests that RTLM broadcasts increased overall participation in the genocide by 10%,[277] although his method of utilizing radio broadcast strength as 'proxy' variable for estimating the actual exposure to radio has been critiqued.[278] Hollie Nyseth Brehm, by contrast, finds no statistically significant impact of RTLM broadcasts on local variation in the *level* of killing, but does suggest that greater RTLM coverage was associated with earlier *onsets* of violence—consistent with the thought that radio propaganda was key to the early mobilization of violence, which then spread and escalated through other channels.[279] Straus finds that radio may have affected the form and intensity of an individual perpetrator's involvement, since: 'the perpetrators who said radio incited them were more likely to commit more violence and to be leaders of the killing than those who said radio did not incite them'.[280]

Direct interview data more strongly suggest that exposure to hardline propaganda, whether by radio or in face-to-face mobilization, was a central factor

---

[272] Anderson 2017a, 76.
[273] McDoom 2021, 89.
[274] Cited in Mamdani 2001, 191.
[275] Anderson 2017a, 83.
[276] Article 19 1996, 55–6 & 74–9; Gourevitch 1999, 14; Li 2004, 20; Wallis 2019, 358.
[277] Yanagizawa-Drott 2014.
[278] Danning 2018.
[279] Nyseth Brehm 2017, 21.
[280] Straus 2007b, 628.

shaping participation in attacks. Perpetrators repeatedly emphasize that they accepted or complied with justifications for violence because either they heard them from local authority figures and radio broadcasts, or 'everyone said' them. Only a small subset, by contrast, suggest long-standing pre-existing belief in such ideas. Across different studies, perpetrators claim, for example:

> I did not believe the Tutsis were coming to kill us, but when the government radio continued to broadcast that they were coming to take our land, were coming to kill the Hutus—when this was repeated over and over—I began to feel some kind of fear.[281]
>
> They were using songs and propaganda to drive hate into our heads—to make the Hutus believe that the Tutsis were just there to take their things ... We did not think it was hate—just reality.[282]
>
> After Habyarimana's death, our national leaders made it understood that the enemy was none other than the Tutsis, because the RPF opposed the government in place. It was made clear that the RPF wanted the Tutsis to be in power. It was necessary that the people for whom the RPF was fighting be killed ... [This message was transmitted by] the radio, RTLM, which said that our enemies were not far ....[283]
>
> On the radio stations, Tutsis were accused accomplices. The refugees told us that they grind children and that they open pregnant mothers' wombs to see the baby's position. All those accusations changed the way we viewed them.[284]

Such claims were representative. Around 70% of the Rwandans surveyed by Mc-Doom reported that, during the civil war, people assumed that if the RPF were to be victorious 'the Hutu would be killed'.[285] Straus similarly observes that:

> about half the respondents countrywide reported being afraid of the rebel Rwandan Patriotic Front (RPF) before the genocide. They said that they had heard that RPF soldiers killed Hutu civilians, that the rebels disembowelled pregnant women, that the rebels were 'snakes,' and that they had tails. When I asked respondents how they learned this information, most answered radio broadcasts.[286]

Moreover, the consistency with which Rwandans, during or after the genocide, invoked the specific discourses of ethnonationalist hardliners, despite the seem-

---

[281] Cited in Kellow and Steeves 1998, 123.
[282] Cited in Anderson 2017a, 76.
[283] Lyons and Straus 2006, 96.
[284] Cited in Scull, Mbonyingabo, and Kotb 2016, 339.
[285] McDoom 2021, 80–3.
[286] Straus 2006, 124. See also Li 2004, 15.

ing mismatch between these claims and the positive local ethnic relations typically testified to by perpetrators, is hard to explain without reference to ideological activism. Mironko observes how the perpetrators he interviewed reproduced 'the euphemisms and code words used during the genocide campaign in speeches, radio broadcasts and other media'.[287] During attacks, moreover, perpetrators often repeated or acted according to quite specific propaganda claims—for example, beginning to target churches after hardliners asserted that the RPF was using churches as military bases.[288]

However, the fact that the mix of face-to-face and print/radio propaganda was a powerful means of spreading hardline justifications of the genocide should not obscure the unevenness of its impact, and the way it influenced rather than eliminated the agency of ordinary Rwandans. Many areas exposed to radio propaganda at an early stage did not immediately see outbreaks of violence,[289] and individuals who expressed familiarity with hardline ethnonationalist claims sometimes suggested that they nevertheless did not believe them.[290] Li emphasizes that his interviewees 'recalled actively debating, comparing, and doubting broadcasts from different radio stations, including RTLM'.[291] While it is clear that propaganda was often effective in encouraging perpetrators to accept hardline justifications of the violence, for many its role may have been more structural: shifting perceptions of social norms, creating the impression of a shared social consensus on the legitimacy of violence, and signalling to ordinary Rwandans what was expected of them by the authorities.[292]

While commentaries on propaganda in Rwanda often focus on the most grotesque expressions of ethnic hatred and dehumanization, moreover, the bulk of propaganda revolved around a more prosaic hardline justificatory narrative for violence: building a perception of pervasive Tutsi support for the RPF and dangerous infiltration, asserting that the RPF would visit mass atrocities and extermination on ordinary Hutu, valorizing participation in violence as simply defending the country, and denigrating opponents as traitors.[293] As the OAU report summarizes: 'Always the language underlined the image of a country under siege, calling for the Hutu to exercise "self-defence" by using their "tools" to do their "work" against "enemy accomplices."'[294] Examples of RTLM broadcasts in the immediate days after Habyarimana's assassination on 6 April include:

---

[287] Mironko 2004, 51.
[288] Article 19 1996, 78.
[289] Straus 2006, 63.
[290] Straus 2007b, 627.
[291] Li 2004, 12.
[292] See also Fujii 2004; Anderson 2017a, ch.4.
[293] Article 19 1996, 52–4; Straus 2007b, 623–5.
[294] International Panel of Eminent Personalities 2000, para. 14.53.

8 April: We know that the *Inkotanyi* are now dispersing, in fact, they are spreading out amongst the inhabitants.[295]

10 April: Citizens are asked to remain vigilant, to stand up like real men, to defend themselves. Roadblocks must be maintained. They really must be maintained during the day so that they can halt these *Inkotanyi*. Because there are some coming ... dressed as civilians and unarmed ... apparently seeking reinforcements.[296]

13 April: continue to keep your eyes open, remain vigilant ... and give them the punishment they deserve.[297]

15 April: If you do not want to have Rwandans exterminated, stand up, take action ... without worrying about international opinion.[298]

Often the propaganda also adopted a passive, deagentifying language that helped cast violence as inevitable or the responsibility of the Tutsi.[299] As one RTLM announcer put it: 'these people [the Tutsi], they are going to continue to commit suicide ... to engage in a suicidal battle against a large group ... won't they be exterminated?'.[300] On another occasion he more explicitly stated: 'These Tutsi have caused the extermination of their fellows ... their innocent fellows ... because of anger ... following the attacks of the *Inkotanyi*, which were unjustified.'[301]

So the effectiveness of propaganda in Rwanda rested on ideological processes familiar to societies across the world and especially salient in security politics. Many accepted the claims of hardline propaganda not out of blind obedience, but because they were epistemically dependent on state officials, military/militia units, and media output for information about the ongoing crisis.[302] RTLM gained appeal, indeed, since it was the only station to initially report on Habyarimana's assassination, due to political divisions at Radio Rwanda: 'While Radio Rwanda played classical music', one former perpetrator recalled, 'RTLM gave news about the situation.'[303] Existing ethnonationalist ideas made it relatively easy, moreover, for RTLM and hardline elites to link the invasion by a predominantly Tutsi RPF force, or President Ndadaye's assassination at the hands of Tutsi army officers in Burundi, to assertions of a terrible Tutsi threat. The emphasis perpetrators place on the transformative effect of the civil war on attitudes is also suggestive of well-researched psychological tendencies to promote in-group cohesion and out-group

---

[295] Cited in Article 19 1996, 72.
[296] Cited in ibid. 69.
[297] Cited in ibid.
[298] Cited in ibid. 68.
[299] Ibid. 71.
[300] Cited in ibid. 67–8.
[301] Ibid. 68.
[302] Ibid. 23.
[303] Cited in Li 2004, 16.

distancing in response to crisis, especially amongst those already supportive of the regime or inclined to pro-authority values.[304]

Overall, then, broad-brushed scepticism about the importance of propaganda in Rwanda is unwarranted. Obviously, propaganda alone did not cause genocide, and its power depended heavily on the context of crisis and the existing ideological infrastructure of Rwandan society. But as in other mass killings, the hardline justificatory narrative underlying the violence did not just emerge spontaneously from the population as a whole, nor did it reflect a necessary interpretation of the objective crisis facing Rwanda. It was a distinctive ideological representation of crisis and violence, disseminated by hardline authorities, in an uncertain and polarized environment. Many Rwandans, when not precommitted to more limitationist ideas or intensely sceptical of the ruling regime, found such hardline claims sufficiently plausible or pervasive to participate in genocide.

8.4 Conclusion

Faced with the appalling horror and scale of the Rwandan Genocide, it is easy to reach for an excessively monolithic image of genocidal ideology as the wellspring of violence. Even leading studies sometimes slip into strong assertions that 'Killing Tutsis was a political tradition in postcolonial Rwanda',[305] or that 'The image of the Tutsi as inherently evil and exploitative was, and still is, deeply rooted in the psyche of most Rwandans.'[306] Such claims, suggestive of a traditional-ideological perspective on mass killing, are exaggerated. Hardline ethnonationalist ideology was contested rather than monolithic, sincerely accepted to varying degrees by Rwandans, and profoundly radicalized by political crisis, rather than simply reflective of longstanding hatreds.

Yet ideology was nevertheless essential to both the occurrence and character of the genocide. No matter how fully we accept the image of Rwandan elites as cynical promoters of their own power and privilege, their self-interest did not rationally necessitate or even clearly imply genocide. Had more moderate ideological factions within Rwanda been somewhat stronger, an entirely different outcome was available. Instead, a hardline faction interpreted the crisis in Manichean terms, and embraced the potential mobilizing power of established anti-Tutsi stereotypes and authoritarian notions of state power and legitimacy as a basis for forging an extreme Hutu Power political base around a campaign of genocidal violence. In this respect, the Rwandan Genocide again demonstrates the perversity of opposing 'ideological' factors, on the one hand, to 'security', 'strategy', or 'self-interest',

---

[304] McDoom 2021, 95–111.
[305] Gourevitch 1999, 96.
[306] Uvin 1996, 13

on the other. Like Stalinist Terror, Allied area bombing, and Guatemala's mass killings, the Rwandan Genocide was perpetrated as a strategy for securing a particular political regime and community. But the allure and viability of that strategy depended on a radical ethnonationalist ideological infrastructure, and the justificatory narrative for violence that such an infrastructure enabled. In tandem with the particular crisis Rwanda faced, ideology enabled hardliners to prevail over their opponents, and underpinned the collective constellation of individual decisions that produced the genocide.

9
Conclusion

9.1 The Role of Ideology in Mass Killing

Voltaire is frequently quoted as warning that 'those who can make you believe absurdities, can make you commit atrocities'.[1] These words have been echoed by several modern scholars, who have long suggested that certain sets of beliefs and ideas, typically called 'ideologies', play a significant role in mass killings. Yet the nature and relative importance of that role has been a subject of intense debate. It is a debate troubled, moreover, by serious theoretical ambiguity. There is no consensus on what ideologies even are or how they might, in principle, be linked to violent outcomes. Profound uncertainty has therefore remained over what an assertion of ideology's importance really means and what it suggests about the mindsets, motives, and influences that guide perpetrators of mass killing.

This book has made a sustained attempt to take that debate forward and offer a more detailed and causally substantiated account of how ideologies affect the occurrence and character of mass killings. I have sought to show that much scepticism about ideology's importance rests on erroneous assumptions—as though ideologies only matter if perpetrators are zealous true believers, if more prosaic motives for violence are absent, or if the violence closely matches a preformulated ideological blueprint. Research has decisively demonstrated that ideology does not shape mass killing in this crude manner. Perpetrators are clearly guided by multiple motives and have varied levels of ideological conviction. The path to mass killings is complex and fraught with contingency, and does not simply reflect the mechanical implementation of an ideological blueprint for violence. It is, moreover, obvious that many factors other than ideology matter. These are established findings, all of which apply to Stalinist repression, Allied area bombing in World War II, Guatemala's civil war, and the Rwandan Genocide. But none of this counts against more sophisticated claims of ideology's essential role in mass killing. By strongly affirming such findings, I have tried to shift the debate towards more genuine and compelling controversies: over how ideologies and political crises interact, the extent to which ideology is rooted in broader material and social

---

[1] Voltaire's literal concern was more expansively with 'injustices'; see Torrey 1961, 277.

Ideology and Mass Killing: The Radicalized Security Politics of Genocides and Deadly Atrocities. Jonathan Leader Maynard, Oxford University Press. © Jonathan Leader Maynard (2022). DOI: 10.1093/oso/9780198776796.003.0009

developments, the kinds of ideological claims that are most critical, the dominant mechanisms of ideology's influence, and so on.

Beyond such clarificatory moves, this book has substantively argued that mass killing must be understood as a form of ideologically radicalized security politics. With very few (and partial) exceptions, mass killings are strategies to achieve familiar security-centric aims, such as defeating perceived enemies, upholding regimes, policing societies, and winning wars.[2] But the occurrence and character of such extreme strategies cannot be explained if the violence is stripped from its particular ideological context. The choices to initiate mass killing, and the targets and logics of the violence, are embedded in distinctive 'hardline' sets of ideas held by particular individuals, factions, and organizations within the societies where mass killing occurs. Those ideas represent radicalized interpretations of familiar claims of security politics—about threat, punishment, duty, warfare, and order—that allow large-scale violence against civilians to appear strategically and morally justifiable in times of crisis.

In advancing this 'neo-ideological' account, I have critiqued two key tendencies that have characterized much existing scholarship. The first is an assumption that if ideology matters in mass killing, it is through the promotion of revolutionary projects, hatreds, or values that stand in contrast to 'strategic' or 'pragmatic' rationales for violence. Such an assumption is tied to a broader tendency in much scholarship to see security policies and military action as lying 'outside' ideology—by contrast with economic, social, or welfare policy. Yet the politics of security, including war, is subject to ideological interpretation and disagreement just like any other domain of political life. Too much scholarship, typically influenced by the dominant rationalist theoretical paradigm of conflict and security studies, ignores such ideological divergence. Even if this is an acceptable simplification for some forms of political violence, it is not fit for purpose for the study of mass killings. Such extreme atrocities depend on conceptions of security and threat (and also moral legitimacy and duty) that are always highly partisan and non-obvious, rooted in broader visions of politics and society, and often substantially divorced from reality.

Yet my focus on ideologically radicalized security politics has been motivated as much by appreciation of rationalist insights about political violence as by dissatisfaction at their neglect of ideology. Rationalists are right to present mass killings as an instrumental strategy and to emphasize logics of political control and military confrontation in shaping the violence. Incentives generated by contexts of war and upheaval, the particular capacities of states, the spatial structure of conflict, and perpetrators' sources of material and political support all matter. Rationalists simply tend to overestimate the extent to which such factors create clear and

---

[2] Again, though, some of the violence may exceed such strategies, generating self-reinforcing momentum or being subverted for private ends; see Chapter 1, fn.37.

specific implications for behaviour. Contexts of crisis are interpreted in various ways, and mass killings represent strategically indeterminate and normatively extreme policies. The willingness of both elites and mass publics to implement them consequently varies, according to their existing ideological preconceptions and the ideological narratives they are exposed to.[3] We therefore need to accord distinctive ideological understandings of security politics a central focus.

My second central critique has been against the narrow way of thinking about ideology and its influence that I have referred to as the 'true believer model'. This label, I re-emphasize, denotes a rough set of analytical tendencies and assumptions— it is not a single, clearly elaborated approach. But its central features can be found in much scholarship on mass killing: namely, the assumption that ideologies are generally influential insofar as perpetrators are motivated by deep commitments to elaborate belief systems and special ideological goals. As has long been appreciated by many specialist scholars of ideology (and also of Nazi and Soviet politics), this way of thinking caricatures ideologies and their political influence. Liberalism, conservatism, fascism, communism, or nationalism are not tightly systematized programmatic visions of the ideal society, but looser bundles of concepts, narratives, values, assumptions, and preferences. They shape politics not simply through passionate belief, but through shallower adoption, instrumental manipulation, and conformity to political norms and institutions. Here, the complementarity of ideological and situationist explanations of mass killing becomes obvious, since ideologies shape social norms, institutional environments, and the resulting situational pressures individuals are exposed to. Such a synthesis, as emphasized in Chapters 2 and 3, is already welcomed by many leading situationists. To suggest that particular ideological conceptions and narratives are critical to mass killing is not, therefore, to suggest that all or even most perpetrators deeply believe in them. It is to focus on the multiple interacting psychological and social mechanisms of influence through which key components of a given ideology 'infrastructurally' shape patterns of violence.

Many who have expressed scepticism about ideology's importance in mass killing might respond by emphasizing that they have never intended to deny that ideology matters in *this* sense. This is all the better. Once ideology is disassociated from crude, monolithic, and conviction-centred portrayals, a broad consensus on its centrality to mass killing may be possible. But the argument in this book is not 'mere semantics'—just a fight over what we label ideology. If I am right, then any explanation of mass killing must attend to things that most 'sceptical' perspectives on ideology's role ignore, including the complex evolution of ideas about security held by elite decision-makers, the broader ideological infrastructures in which those elites are embedded, and the contextually specific conceptions of threat,

---

[3] Verdeja 2012; Staniland 2015; Straus 2015a.

guilt, identity, effectiveness, necessity, and valour that guide the violence. We cannot remain satisfied with explanations that, while appealing to 'strategic incentives' or 'situational pressures', fail to examine the distinctive sets of ideas that shape the actual direction, content, and force of those incentives and pressures.

Some readers may remain reluctant to broaden their conception of ideology and its political influence in the way this book recommends, preferring to keep it narrowly associated with tightly consistent belief-systems and strong commitments. Such an approach can be rendered internally coherent, but it makes it impossible to understand how things like liberalism, conservatism, communism, and fascism—phenomena that *everyone agrees are ideologies*—visibly shape real world politics. Psychological research conclusively demonstrates that most individuals who identify with and espouse such ideologies are not guided by rigid belief-systems built purely of deep commitments. We can often link the specific contents of such ideologies to policies, institutions, and patterns of collective behaviour even though the individuals who produce such outcomes are not ideological true believers. Adopting a conception of ideology that recognizes these realities enables more accurate, complex, and revealing analyses of the role of distinctive ideological frameworks in political life.

In emphasizing the multifaceted nature of ideology's influence, I do not want to be misread as claiming that 'real belief' in ideology is rare or unimportant. On the contrary, my research suggests that most perpetrators of mass killing do sincerely accept ideological justifications of violence to some significant degree, and that such internalization is often critical—especially amongst elites. Since our evidence of perpetrators' mental states is indirect, we may never *absolutely* rule out the possibility that ideological justifications are purely instrumental rationalizations. But such an interpretation should not be treated as some kind of 'default' that we fall back on unless absolute proof of sincere ideological belief can be produced, and the balance of evidence is stacked against it. Perpetrators frequently invoke ideology even when doing so does not appear to help them. Real ideological belief is often consistent with their behaviour and is declared in forms of private testimony such as diaries, personal letters, or internal government correspondence. Such internalization is often what social scientific research would lead us to expect, given the cultural context in which such individuals operated. Such ideological belief is almost always interwoven with forms of material and symbolic self-interest. But in such circumstances, assertions that ideology merely provides post-hoc rationalizations represent an article of faith, rather than an evidentially grounded conclusion.

As I have tried to emphasize throughout this book, a neo-ideological perspective on mass killing is not unprecedented, and much of my contribution represents a synthesis of key arguments and bodies of research from across the disciplines that examine mass killing. Besides drawing on rationalist and situationist scholarship, my account represents a targeted revision of traditional-ideological perspectives, not an outright rejection of them. I am not claiming that utopian ideological projects, revolutionary values, or identity-based hatreds are wholly unimportant in mass killings. There is no question, in my view, that many perpetrators come to feel hatred for their victims, rooted, in part, in ideology. In certain cases, such as the Holocaust and Khmer Rouge Cambodia, more utopian aspects of ideology loom especially large. Amongst traditional-ideological perspectives, I encourage readers to return, in particular, to Hannah Arendt's *Origins of Totalitarianism* and Eric Weitz's *A Century of Genocide: Utopias of Race and Nation*—both of which continue to offer powerful insights on the logics and mentalities that can link such radical political programmes to mass violence.

But traditional-ideological perspectives overstate the typicality and centrality of special ideological goals, values, and hatreds in mass killing and exaggerate its disconnect from more conventional strategic and moral ideas. Even in the Holocaust, Nazism's hateful, fantastical, and pseudoscientific construction of Jews was most consistently propagated through notions of threat and egregious wrongdoing.[4] Amongst the rank-and-file, Nazism had a huge influence—but primarily in its radical framing of the war effort, and its success in reconstructing mass murder as dutiful service to the fatherland.[5] The Holocaust was an outlier in the scale of its murderous ambitions and the depth of its ideological delusions about Jews and other 'subhumans'. Yet Nazism's mass appeal and political strength was still rooted in the coupling of familiar arguments about self-defence, punishment of criminals, and patriotic duty with an extremist *ideological narrative about the world*, in which Jewish-led conspiracies sought to destroy the German race. Nazi atrocities represented an ultrahardline and biopolitical form of security politics first, and a call to utopian revolution second.

My focus in this book has been on clarifying *ideology's role* in large-scale violence against civilians, and the arguments I offer do not amount to a complete causal theory of mass killing. As observed in Chapter 3, there are several ways in which other bodies of research must be plugged together with my account to generate more comprehensive explanations. This book has not, for example, attempted to explain the broader *origins of the political crises* in which mass killings tend to occur. Here, a panoply of existing historical and social scientific research on the structural, economic, institutional, and political roots of war, rebellion, coups, state collapse, security dilemmas, and other forms of insecurity must be invoked. Similarly, non-ideological *constraints* may be crucial in preventing many potential mass killings from occurring, while *some* of the violence in any mass killing will deviate from the prevailing justificatory narrative, whether through endogenous 'self-perpetuating' processes of violence, or bureaucratic, private, and

---

[4] Aronsfeld 1985; Herf 2006.
[5] Bartov 1994.

interpersonal interests that are weakly connected, if at all, to ideological content.[6] My aim has been to secure ideology's place at the heart of contemporary theories of mass killing, but also to emphasize the way it interacts with, rather than overrides, other critical causal factors.

9.2 Moving Forward

This book is intended to galvanize and empower future research. In particular, I have attempted to refocus the central debate between those scholars who are convinced that ideology's role in mass killing is crucial and those who are more sceptical. Once various confusions about ideology are cleared out of the way and crude accounts of its impact are abandoned, ideology clearly affects the occurrence and character of mass killing in important ways. Perpetrators genuinely adhere to ideologically distinctive understandings of the violence they commit and are subject to various kinds of social pressures and incentives to espouse and enact the prevailing justificatory ideological narrative for that violence. Unsurprisingly, such understandings, pressures, and incentives seem to significantly shape perpetrator behaviour. The most important remaining debate, I argue, is not over the *consequences* of hardline ideologies in mass killing—i.e. whether they affect the violence or not—but over the *causes* of such ideologies—i.e. why they become dominant and why they contain the ideas that they do.

It is, in other words, a debate about the dynamics of ideological radicalization. Some scholars will still think that the emergence and character of hardline ideologies largely reflects underlying material and social causes of mass killing, while others will stress the primacy of cultural context and ideological agency. Some will think that hardline ideologies are principally rooted in long-standing features of social structure, regime type, and national culture, while others will stress short-term activism and rapid escalation during crisis. While these positions are well staked out, they typically reflect scholars' prior theoretical commitments—to rationalism, constructivism, institutionalism, historicism, and so forth—rather than specific empirical findings. The actual substance of this controversy over ideological radicalization remains remarkably underdeveloped.[7]

I have offered three main arguments here. First, it is implausible to treat ideologies as largely reducible to broader material and social forces. Such forces are clearly key influences on ideological radicalization, but it is equally clear that radicalization depends in large part on the existing character of the ideology and broader cultural context in question. German far-right nationalism in Germany

---

[6] Many examples can be found in Gerlach 2010; Mitton 2015.
[7] See also Murray 2015.

contained a potential for radicalization towards Nazism which German liberalism lacked. Stalinism was prone to extreme self-radicalization in response to even mild social crises in a way that most other forms of revolutionary Marxism are not. Moreover, the interaction between ideological and broader material and social changes runs in both directions and is staggeringly complex. It is partly *because* of this complexity that ideology is a key variable. We typically have to take the ideologies of individuals, groups, or organizations as contingently 'given' by history because they cannot be easily predicted by other causes. Even if certain material or social conditions might lead us to anticipate *some kind* of hardline ideology emerging, the detailed content of such an ideology—which, I have shown, matters immensely—is always rooted in specific paths of cultural and political development.[8] The challenge is to better theorize how the prior character of ideologies and processes of ideological activism interact with material and social conditions to fuel or restrain radicalization, rather than reducing ideology to material and social context, or vice versa.

Second, I have sought to show that such theorization is possible, because the key dynamics of ideological radicalization towards mass killing are *largely* 'ordinary'. They rest not on unusual levels of irrationality, prejudice, or psychological abnormality, but on familiar forms of epistemic dependence and motivated reasoning, which, coupled with the right pre-existing ideological foundations and circumstances of crisis, can give even extremist claims about security politics significant social influence. Since justifications of mass killing capitalize on widely accepted strategic and moral arguments, there is no need for perpetrators to undergo some sudden, vast psychological conversion involving the abandonment of all traditional moral convictions and principles. There is no need to impute to propaganda an implausible power to simply 'inject' extremist hatreds into mass populations. One need only appeal to ubiquitous processes of perception, belief formation, norm-building, and institutionalization and understand how these can empower hardline claims about security politics in times of crisis.

Finally, I have followed other scholars (in particular Scott Straus and Omar McDoom) in emphasizing that the *ideological balance of power* matters. Ideologies are factional, and the particular ideological orientations of more hardline and more moderate factions are critical because, in principle, either faction could be dominant in a given context. Which faction is most powerful is, therefore, a key question—but this is not a reason to discount ideology. The relevance of Stalinist ideology may have depended on Stalin's power, but that power *made the particular content of Stalinist ideology* immensely consequential for Soviet politics. Moreover, ideology is itself constitutive of power: factions gain influence in part through the appeal of their ideas, they can shape outcomes partly through ideological framing and narrative building, and they can operate most effectively when they possess

---

[8] Straus 2015a, 330–2.

significant ideological cohesion and mobilization.[9] Again, power-dynamics and ideology need to be analysed in tandem.

These three claims are important, but we still need a lot more research on the precise dynamics of ideological change and radicalization in the build-up to (as well as during and after) mass killings. To what extent are certain quasi-objective material and social conditions a prerequisite for sustained ideological radicalization? How crucial is 'top-down' elite promulgation of radical ideas versus relatively 'bottom-up' processes of meaning-making, rumour circulation, and psychological bias? How far can propaganda 'at a distance' – via newspapers, radio, television, or online media – inculcate acceptance of justificatory narratives for mass killing in the absence of face-to-face mobilization or direct experience of crisis? What are the preconditions for effective moderate resistance to hardline mobilization? In what ways is radicalization self-reinforcing, and why does it sometimes reverse? While I have touched on these questions, they require a sustained, multi-method research agenda involving both within-case tracing of particular radicalization trajectories and cross-case comparisons. Ideally such research should connect with a parallel literature on radicalization in terrorism studies and with pioneering research on polarization and extremism in contemporary politics.[10]

Beyond the dynamics of ideological radicalization, establishing the broader scope of a neo-ideological approach to thinking about ideology and political violence requires further research. I have offered illustrative evidence across the book that my conclusions should extend beyond my four case studies to a range of other major mass killings, including Nazi atrocities, Lenin's Red Terror, the Indonesian anticommunist campaign of 1965–66, atrocities in the War in Yugoslavia, Cambodia under the Khmer Rouge, the Cultural Revolution in China, and colonial atrocities in the European empires and the westwards expansion of the United States.[11] This is a substantial subset of the largest cases of mass killing in modern history, but remains open to support or challenge from detailed work on the fuller universe of mass killings. I have also suggested that many arguments of the book may extend to less extreme forms of political violence, such as more limited 'one-sided violence' in civil wars, terrorism, political rebellion and more conventional levels of state repression. While ideology is especially crucial for mass killings because of their high level of strategic indeterminacy and normative extremity, no form of political violence is genuinely rendered 'necessary' by incontestable interests or self-evident strategic realities. Almost by definition, all political violence has an ideological dimension—the questions are how important that dimension is and

---

[9] Mann 1986, 22–4; Gutiérrez Sanín and Wood 2014, 217–20.
[10] E.g. Breton et al. 2002; McCauley and Moskalenko 2011; Neumann 2013; Tsintsadze-Maass and Maass 2014; Norris and Inglehart 2019.
[11] For accounts of these cases that are consistent with my main arguments see, for example, Jackson 1989b; Bartov 1994; Oberschall 2000; Semelin 2003; Madley 2004; Browning 2005; Hinton 2005; Su 2011; Madley 2016; Robinson 2017; Robinson 2018; Williams 2021.

in what ways specific kinds of ideological processes or contents affect the dynamics of violence. While the detail of the theory would be quite different, my basic account of the way ideology can shape perpetrator behaviour, and the relevance of ideology in shaping strategic calculations and situationist pressures, should be cross-applicable to these other forms of violence.

Finally, this book has not analysed 'negative' cases—where mass killings might have but did not occur—in detail, and it has only briefly discussed the role of 'limitationist' ideological doctrines in such cases.[12] There are methodological reasons for this (discussed in the Methodological Appendix), but also a more basic practical concern: that effective analysis of the role of ideology in negative cases requires independent study and theorization in its own right. Just as our most useful explanations of why planes crash are not simply the inverse of our aerodynamic theories of flight, effective explanations of how ideologies restrain extreme violence are not simply the mirror image of explanations of how they justify extreme violence. The two are linked, but ideology's role in the avoidance of mass killing requires dedicated further research, and connects most naturally with existing scholarship on the constraining power of norms on violence and the growing propagation and institutionalization of human rights and notions of civilian immunity.[13] Like much research on mass killing, however, this literature has often remained silent on the role of ideology in generating varying ideas of human rights and civilian immunity and imbuing such ideas with political power.[14] Ideology's role in the most egregious forms of violence against civilians and human rights abuses offers reason to think that this connection is more important than has been appreciated.

9.3 Prediction and Prevention

While my primary concerns in this book have been to explain why and how mass killings occur, a neo-ideological perspective has implications for efforts to *predict* and to *prevent* mass killings and other forms of 'atrocity crimes'. On the predictive side, the kinds of ideological factors I have pointed to as crucial are hard to quantify, and often very difficult to observe in real time—so it is not easy to directly 'plug in' the findings of this book into quantitative risk assessment or early warning models of mass killing. Nevertheless, the general conclusions of this book can inform qualitative, country-specific analyses by government specialists or NGOs. In particular, I suggest that the analysis of ideological risks, on the one hand, and structural or strategic factors conducive to atrocities, on the other,

---

[12] See also Straus 2012b; McLoughlin 2020.
[13] Risse-Kappen, Ropp, and Sikkink 1999; Slim 2007, 259–93; Pinker 2011; Sikkink 2011; Bellamy 2012b.
[14] Although interest is growing, see Mutua 1996; Bajaj 2011; Bellamy 2012b; David 2020.

need to be more tightly woven together. More emphasis is needed on the particular doctrines of security institutions and political and military elites in states at risk of mass atrocities, and the way these are linked to broader political narratives promoted by regimes and armed groups or their backers. This contrasts with a dominant focus in much contemporary commentary on intense identity hatreds or blatant extremism found in mass public attitudes. These do matter, and more broadly 'exclusionary' and 'revolutionary' ideologies such as extreme forms of ethnonationalism and religious fundamentalism are commonly associated with the kind of hardline ideas I have linked to mass killing. But they are not necessary for hardline radicalization to occur and should not be assumed to exhaustively encapsulate the ideological risks of atrocity. Some of the worst recent atrocities have occurred when relatively reactionary authoritarian regimes adopt intransigent, hardline responses to political challenge. The repressive mass killing of the Assad regime in the Syrian civil war, for example, has hugely outweighed the more obviously extremist and widely watched atrocities committed by ISIS.

The six justificatory mechanisms outlined in Chapter 4 provide a framework for such analysis, since the growing propagation of such justificatory mechanisms in elite and public discourse is often indicative of the escalating risk of atrocity. Applying an early version of that framework to Myanmar, for example, Matthew Walton, Matthew Schissler, and Phyu Phyu Thi identified important currents of dangerous ideological radicalization in advance of the Rohingya Crisis of 2015, which primarily revolved not around explicit intergroup hatreds or ambitions to transform Myanmar society, but assertions of 'threat and virtuous defence', as well as the destruction of alternatives.[15] The findings of this book will hopefully enable other country specialists and NGOs to better understand the risks and warning signs of such political and social radicalization, as well as the underlying sociopsychological processes that fuel them. Analysis of the six justificatory mechanisms could be adapted, moreover, for certain forms of quantitative analysis, especially new forms of big-data trawling on social media.

On the preventive side, my research has somewhat double-edged implications. Most obviously, it strongly supports those who emphasize the role of information flows, media output, propaganda, grass-roots activism, 'dangerous speech', and processes of organizational training in both the occurrence and prevention of genocides, mass atrocities, and organized violence more broadly.[16] It is true that scholars, governments, and NGOs have often asserted the importance of such 'ideational' dimensions of conflict on the basis of limited evidence, simply assuming that media output or political discourse influences audiences.[17] But the

---

[15] Schissler 2014; Schissler, Walton, and Thi 2015.
[16] Metzl 1997; Howard 2002; Price and Thompson 2002; Thompson and Price 2003; Wolfsfeld 2004; Bratić 2008; Scutari 2009; Mani and Weiss 2011; Benesch 2014a; Benesch 2014b; Petrova and Yanagizawa-Drott 2016; Gordon 2017; Benesch et al. 2021.
[17] Schoemaker and Stremlau 2014.

balance of evidence does suggest that such processes can both erode important societal restraints on mass killing over the long term and be key means of political mobilization of perpetrators and supporters in the short-term escalation of crises. They do not merely offer post-hoc rationalizations for violence, but have real effects in inflaming fear, legitimating perpetrating authorities, and promoting participation in atrocity. Consequently, as Hugo Slim argues: 'Any practical effort to encourage [an] ideology of limited warfare and its central concern for civilian protection will need to go head-to-head with these various anti-civilian ideologies and challenge the social and psychological conditioning that makes people agree to and enact them.'[18] Actions that disrupt ideological activism and mobilization or strengthen limitationist ideas and forces are not idle or 'soft' forms of prevention. They carry lifesaving potential and constitute important parts of the 'atrocity-prevention toolkit'.[19]

Yet the research presented in this book also challenges many conventional assumptions about the relevance of speech, propaganda, media, and activism in atrocity prevention. For a start, much existing preventive work again focuses on *mass public attitudes*—using peace education, media liberalization, and cultural programmes to promote positive intergroup relations and norms of peaceful political contestation. Such work is of significant value.[20] But it frequently involves the tacit assumption that the fundamental root of mass violence, or at least the most critical enabling condition, lies in 'the people'. This assumption is not well backed by available evidence: majority public support is probably unnecessary for mass killings, and in almost all cases political elites and specialized military or paramilitary organizations play the determining roles in the violence. Alternatively, preventive scholarship often focuses more generally on society-wide sources of structural resilience to the risk of mass atrocities—such as political regime type, judicial independence, or the presence of a free media—rather than the ideological orientations of key political actors.[21] This book suggests that such a focus is lopsided. Prevention efforts that seek to engage major political parties and government bureaucrats, strengthen those who oppose hardline factions, and institutionalize limitationist ideas in military and police regulations and training processes may carry more potential to counter mass killings than efforts to shift public attitudes or alter the basic political structures of a society as a whole. In most cases, indeed, local politicians, activists, and movements, rather than external states and organizations, *are themselves the best placed preventive actors*.[22] So their willingness to engage in preventive ideological activism and reject hardline ideas and narratives offers the optimal path to avoid mass killing. Even if they

---

[18] Slim 2007, 250.
[19] For more detailed discussion see Leader Maynard 2015. See also Sharma and Welsh 2015.
[20] See again, for example, the concrete impacts found in Collier and Vicente 2013.
[21] McLoughlin 2020, 1548.
[22] Straus 2015a, 326.

cannot prevent mass killings outright, past cases demonstrate that they can often obstruct and limit the violence to some degree, rescue potential victims, and save significant numbers of lives.

My research also suggests that scholars and practitioners of atrocity prevention often overfocus on the most overt forms of 'big media' propaganda—principally radio stations, newspapers, and major online output—rather than face-to-face forms of ideological activism. Existing preventive practice also fixates too centrally on countering 'hatreds', which most research suggests are (i) more of an output of hardline radicalization than a cause, and (ii) rarer than feelings of fear and duty in actually motivating or legitimating extreme violence. While countering intergroup animosities is obviously a worthy endeavour, grass-roots preventive efforts should place equal or greater emphasis on educating populations about violent mobilization, promoting sceptical attitudes and norms towards hardline narratives, and developing the ideological, and physical, infrastructure of non-violent movements and networks of civil resistance.[23] Finally, I also question the 'techno-euphoric'[24] assumption that increasing the quantity and freedom of media consistently undermines the capacity of hardliners to justify and mobilize violence. In three of my four case studies (Stalinist repression being the exception), private media and/or grass-roots groups not formally part of the state were nevertheless energetic propagators of hardline ideas. In more contemporary cases, expanded access to online social media has often empowered extremist groups (as well as state agencies) as much as moderates.[25] So media liberalization and expansion of new technologies is not a panacea to the dangers of hardline radicalization, which generally needs more proactive preventive policies organized by alliances of international and local actors. Much more research is needed, however, on the theory and practice of such efforts.

9.4 The Problem of Extremism and Atrocity

Research on genocides and deadly atrocities has often attacked a seductive but inaccurate myth about extreme political violence: that it is generally perpetrated by 'crazy' and/or wilfully amoral people, wholly unlike the rest of us, for irrational reasons incomprehensible to decent folk. This is a myth reproduced daily in films, media coverage, and a depressing amount of notionally serious intellectual output. As observed in Chapter 1, however, the much less comforting finding of most

---

[23] See also Paluck 2009; Paluck and Green 2009; Chenoweth and Cunningham 2013; Collier and Vicente 2013; Masullo 2020; Masullo 2021.
[24] As labelled by Schoemaker and Stremlau 2014, 182.
[25] Thompson 2011; Schissler 2014; Schoemaker and Stremlau 2014; Pauwels and Hardyns 2018; Neilsen 2021.

research is that the perpetrators of atrocities, up to and including the worst cases of genocide, are generally 'ordinary'.

This book has continued to press that argument. Indeed, in important ways I go further. It is not only that the *people* who perpetrate atrocities are generally fairly ordinary. They also justify the atrocities through the manipulation of *ordinary ideas and claims* that societies across the world use to think about and regulate security politics. Perpetrators appeal to self-defence, state authority, the punishment of criminals, patriotic duty, military necessity, and soldierly valour. The appeal is to highly radicalized interpretations of those ideas, and mass killings are genuinely *extremist* policies of violence. But this extremism does rely on utopian demands or the complete abandonment of conventional moral sentiments. Instead, this kind of extremism capitalizes on the selective elevation of certain ordinary moral sentiments, and couples this with hyperbolic assertions about security and crisis which most people are not in a position to authoritatively challenge. It is consequently relatively easy for societies to come under the influence of such hardline narratives when leading elites and institutions promulgate them in times of crisis. The kind of extremism typically involved in violent atrocities is rooted, in other words, in the ethical fragility of ordinary human beings as potential participants in violence—a fragility that is, in important ways, fostered and sustained by the ordinary security politics of modern societies.

There are two 'critical' implications here. First, I fear that a persistent exaggeration of the gap between the ideological foundations of mass killings and ordinary security politics weakens our capacity to regulate violence and perceive atrocity. It is easy for any individual, group, or government to think that because its aims do not seem extreme—it merely acts in self-defence—its violence cannot be akin to 'real' atrocities. We are seduced by, to use David Runciman's words, 'the politics of good intentions'.[26] Yet almost all large-scale atrocities, even those by the most radical extremists, are primarily perpetrated in pursuit of purportedly conventional, security-orientated aims. So the propriety of our *goals* is no good guide to the legitimacy of our violence. When the right narratives, forms of motivated reasoning, and relationships of epistemic dependence are in place, seemingly prosaic assertions about 'military necessity', 'responding to aggression', 'promoting national/international security', or 'ensuring victory' can allow ordinary people to see even the most appallingly destructive forms of violence as justifiable.

The second implication of this ethical fragility concerns the large body of scholarship on military ethics and the philosophy of war. A long intellectual tradition in such scholarship centred around 'just war theory' has been concerned with identifying the principles that ought to regulate violence.[27] This is a tremendously

---

[26] Runciman 2009.
[27] For outstanding examples see Walzer 2000; Orend 2006; Rodin and Shue 2008; McMahan 2009; Fabre 2012.

sophisticated and indispensable field of research, and my arguments in this book do not directly undermine any of its central ethical arguments. But I do pose an orthogonal challenge. One of the world's leading ethicists of war, Jeff McMahan, rightly observes that: 'No nation or people is genetically immune to Nazism or to similar political and ideological movements that spawn unjust wars. We as individuals are protected from becoming Nazis by *ideas*, and by the cultural and political institutions they inspire.'[28] A key question is then: which ideas? If the worst forms of *unjust* violence occur not because perpetrators adhere to ideas that are alien to just war theory and our leading philosophies of war, but precisely through the manipulation and stretching of such principles, then a critical problem is *how conventional justifications of political violence go awry*. Yet with a few important exceptions,[29] leading research on the ethics of war has focused relatively little attention on this process.[30] In other words, many leading philosophers and political theorists devote too little consideration to the way their own arguments may function when used (and abused) by actual human decision-makers.

Here, as in the many other fields of research engaged in this book, my concern is that scholars fail to appreciate the *irrevocably* ideological nature of politics: that it is always embedded in distinctive, partisan, and contestable sets of ideas that tend to have profound effects on the way people think and act in politics. As human individuals, we depend on distinctive frameworks of ideas to interpret the world around us and to make decisions in that world. As collective political subjects, we also rely on such frameworks of ideas to coordinate social behaviour, to articulate shared political norms and expectations, and to strategize our political action. This underlies our ethical fragilities in employing political violence, but it provides basis for hope as well as concern. The influence of hardline ideas is far from irresistible. Indeed, in every case of mass killing we find individuals whose doubts and preconvictions lead them to reject, dismiss, and fight against the hardline justification of violence. Self-awareness of our ideological nature, our reliance on narratives that are often weakly grounded in hard evidence, and our tendencies to conform to the views and norms to which we are most exposed, are some of the best defences we have. So by unpacking the way mass killings are ideologically justified, I have sought to contribute to the collective effort to explain such violence. Beyond this, I have tried to show how each of us can choose: to leave ourselves open to such justifications, or to understand, counter, and resist them.

---

[28] McMahan 2009, 3.
[29] E.g. Glover 1999. A key literature in critical security studies does address such concerns within a rather different paradigm—see, for example, Shaw 2002; Blakeley 2009; Sjoberg and Via 2010; Eastwood 2018; Millar 2019.
[30] See also Pauer-Studer and Velleman 2011.

APPENDIX
Methodological Appendix

Comparative Historical Analysis

While the debates seem less heated than they once were, the proper way to study human behaviour remains contentious in contemporary social science. My instincts, in such debates, are pluralist and supportive of a kind of 'analytic eclecticism'.[1] Ultimately, there are many kinds of questions that puzzle us about human society, and no methodological approach has a strong record of answering all such questions effectively, nor a legitimate warrant for claiming that the kinds of questions it addresses well are the most important ones. Methodological standards are important, but they are plural, and depend on the kind of contribution a particular research project intends to make. My methodology in this book is correspondingly eclectic—blending techniques from qualitative political science, history, discourse analysis, intellectual history, and political theory. But the overarching approach I employ in blending such techniques is a form of comparative historical analysis that systematically examines relatively detailed reconstructions of multiple cases of mass killing, primarily through my four case studies in Chapters 5–8.

Comparative historical analysis is a means of generating and substantiating causal explanations, but it does not seek to identify general 'causal laws' or measure the 'average causal effect' of a given variable across cases.[2] The focus is instead, in the oft-used jargon, on 'causes of effects' rather than 'effects of causes'.[3] In other words, rather than computing an average effect of a given cause across cases, comparative historical analysis traces the causal mechanisms (or combinations of mechanisms) that are typically necessary, or even sufficient, to bring about certain kinds of outcome.[4] Such an approach is motivated in part by an awareness of the complexity and 'causal heterogeneity' of real-world occurrences of a phenomenon such as mass killing. Since each mass killing is a highly complex and historically situated event, with multiple possible causal pathways to the final outcome ('equifinality', in the jargon), it is doubtful that law-like generalizations or average causal effects really explain much about individual occurrences of mass killing. Indeed, the very *notion* of an average causal effect is often dubious for such heterogenous phenomena.[5] Comparative historical analysis is also typically rooted in sympathies for some kind of 'scientific realism', in which genuine causal explanations need to identify the actual mechanisms that generate observed outcomes, rather than resting satisfied with identifying strongly correlated predictive associations between variables.[6]

---

[1] Sil and Katzenstein 2010.
[2] Causal laws could still underlie the processes identified by comparative historical analysis, as Carl Hempel argued, but need not—compare Hempel 1942; Skinner 1966b; Mahoney and Terrie 2008; Mahoney 2015.
[3] Mahoney and Terrie 2008; Waldner 2012, 67–71.
[4] Gerring 2007; Hedström and Ylikoski 2010. See also Norman 2021, 7–9.
[5] Norman 2021, 9 & 18–19.
[6] See, in general, Elster 2015, chs.1–2; Jackson 2016, ch.4; Kertzer 2017.

Even social scientists employing comparative historical analysis (or related forms of qualitative case study methods) often present their research methodology as conforming to a 'hypothetico-deductive' approach that roughly resembles a kind of scientific experiment.[7] In this experimental presentation, research involves three main phases: (i) 'theory-building' leads to the development of 'hypotheses' about broad causal relationships, (ii) ways to test these causal hypotheses using data from suitably varied historical cases (to create some kind of 'quasi' or 'natural' experiment) are identified, and (iii) the data are then analysed in order to sustain or falsify the hypotheses.

I do not pretend to conform to this experimental model, for two basic reasons. First, it is widely appreciated that the model is an inaccurate presentation of the way much leading research in social science actually proceeds.[8] Even setting aside some famous ambiguities about what 'falsification' involves,[9] most scholars in practice refine their own arguments and hypotheses based on how well they are supported by the data, so that the eventual theory is partially rooted in the evidence used to substantiate it. I self-consciously embrace this form of *inductive iteration*, as Sean Yom terms it: my arguments reflect empirical findings from across research on mass killing, but also evidence from my case studies themselves.[10] Such an approach is, as Peter Hays Gries puts it: 'unapologetically "applied" ... Rather than deductively testing hypotheses to build theory, it will *inductively* explore what existing and new ... data can teach us.'[11] The main concern with such an approach is with 'curve-fitting' and 'ad hoccery'—i.e. the use of contrived ad hoc modifications to make a theory fit all the available data. But this is not an automatic consequence of such an approach: accusations of ad hoccery need to show that a theory involves genuine theoretical contrivances and lacks meaningful explanatory power.[12]

Second, the appropriateness of an experimental model of research to the study of historical cases is dubious even in principle, since it simultaneously overestimates and underestimates what historical comparisons reveal. The model overestimates, because comparative historical analyses are not really experiments—or even 'quasi-experiments'—as it is rarely possible outside genuine laboratory settings to manipulate or isolate specific causal variables.[13] But the model also underestimates, because in portraying causal inference as grounded solely in experimental comparison, it obscures how much *prior theoretical and empirical knowledge* may be brought to bear on the analysis of historical cases to make reliable causal inferences. Effective comparative historical analysis goes beyond the mere observation of expected cross-case or intra-case patterns of causal variation, applying our full existing knowledge about human behaviour—such as the role of intentions and beliefs, the extent and limits of rationality, the influence of social pressure, the effects of discursive

---

[7] Elster 2015, 10. Elster is explicit that in practice, good social science will often necessarily deviate from this model. Not all hypothetico-deductive research need take a 'quasi-experimental' form.

[8] Yom 2015. See also Elster 2015, 37–8; Norman 2021, 16–17.

[9] Jackson 2016, 65–70 & 80–2.

[10] Yom 2015. On the closely related concept of 'abduction' see Douven 2011. More specifically, my neo-ideological account started from what Yom, following Michael Mann, terms 'theoretical hunches', specifically concerning the shortcomings of the 'true believer' model of ideology, and the centrality of relatively multifaceted justificatory narratives in motivating and legitimating violence; see Yom 2015, 626. Such hunches were rooted in existing general scholarship on mass killing coupled with my own examination of Nazi atrocities and Stalinist repression. I then refined the theory based on research on Allied area bombing and broader research on violence against civilians in armed conflict. I then further 'tested' the theory against the Guatemala and Rwanda cases, on the basis of which I made more minor refinements to my theoretical exegesis.

[11] Gries 2014, 14.

[12] On explanatory power see Ylikoski and Kuorikoski 2010.

[13] Skocpol and Somers 1980, 193–4; King, Keohane, and Verba 1994, 7, fn.1; Jackson 2016, 77.

communication, the processes of collective coordination, and a host of other 'nuts and bolts' of social inquiry—to analyse richly detailed real historical cases.[14] As noted in my introduction (Chapter 1), this renders causal inference in comparative historical analysis less like a laboratory experiment and more like a detective's investigation or a court case, and no less reliable for this. Comparison is still important in such an approach, since common configurations of causal mechanisms can often be identified across cases, but it is not the sole or overriding basis for causal inference.[15]

All methodological choices involve trade-offs. Comparative historical analysis tends to permit more accurate and detailed theoretical inferences of the cases examined, at the cost of somewhat reduced confidence that the argument extends to other cases (i.e. it increases internal validity at the expense of external validity). Like Yom and Gries, I think this cost worth paying, and it can be mitigated by picking diverse and difficult cases for the reiterative development and testing of theory,[16] by applying the theory to some cases that played a minimal role in the formulation of the theory,[17] and by keeping the argument in contact with existing research on the full universe of available cases. Such an approach reduces concerns that the cases used are outliers or have been 'cherry-picked' to support the theory.

Selection of my four cases—Stalinist repression, 1930–38; Allied area bombing in World War II, 1940–45; mass killing in Guatemala's civil war, 1978–83; and the Rwandan Genocide, 1990–94—was guided principally by two criteria. First, I sought to identify mass killings implemented by regimes with contrasting ideological orientations and in different contexts of occurrence, so as to provide the hardest test for claims of generalizable ideological dynamics. Thus one of my cases (Stalinist repression) considers a revolutionary communist state, the second (Allied area bombing) considers liberal/conservative democracies, the third (Guatemala) considers an authoritarian ultraconservative military regime, and the fourth (Rwanda) considers a regime with a revolutionary and authoritarian ethnonationalist political order. Similarly, one of my cases occurs outside of war (Stalinist repression), one in interstate war (Allied area bombing), and two in civil war (Guatemala and Rwanda). The second selection principle was to pick mass killings that individually provided hard cases for my neo-ideological account, and relatively easy cases for my theoretical opponents. Thus, Stalinist Repression is a classic case of 'totalitarian' or 'revolutionary' mass killing, of the kind focused on by traditional-ideological accounts. Allied area bombing and the Guatemalan civil war are cases that superficially seem likely to fit 'rationalist' theories of mass killing, which focus on strategic killing and minimize ideology's importance. Rwanda is something of a mixed case in scholarship, provoking both ideological and non-ideological explanations. But it is one in which several recent studies strongly emphasize situationist or rationalist factors.[18]

One thing I do not do, by contrast with several recent studies of mass killing, is offer a dedicated study of a case where mass killing *does not occur*. Some readers may worry that I therefore 'select on the dependent variable'—only examining cases where mass killing does occur—and thus fail to demonstrate that variation in ideology actually produces variation in outcomes. Such an accusation would be misplaced. For a start, my 'dependent variable' is not simply 'mass killing: yes/no'. My four cases provide access to considerable variation

---

[14] See also Norman 2021, 16–17. Much such foundational knowledge is discussed in Elster 2015.
[15] On the multiple uses of historical comparison see Skocpol and Somers 1980.
[16] Allied area bombing and Guatemala, in this book.
[17] Guatemala and Rwanda, in this book.
[18] Such 'most-different' research designs are somewhat unfashionable, but this preference again reflects an excessively experimental logic of inquiry in which causal inference is rooted in the capacity of case selection to 'isolate' an independent variable.

in the character of mass killing and how it unfolds, which I link to ideological variation across the cases. In addition, while my four detailed case studies only examine cases where mass killing has occurred, this is not true of the evidence base of the book as a whole. I employ a much broader array of evidence, in Chapter 3, to show that ideology is linked to the occurrence and non-occurrence of mass killing. I also discuss cases of non-occurrence in Chapter 4—providing a baseline of explanatory contrast with the observations in my case studies.

My decision to eschew a full case study where mass killing was avoided also reflects my judgement as to the principal value added by this book. As summarized in Chapter 3, several excellent recent studies—such as Scott Straus's *Making and Unmaking Nations*, Maureen Hiebert's *Constructing Genocide and Mass Violence*, Alex Bellamy's *Massacres and Morality*, or Zeynep Bulutgil's 'Ethnic Cleansing and Its Alternatives in Wartime'—already demonstrate how contrasts in ideology (or phenomena I theorize as constitutive of ideology) appear to distinguish cases and non-cases of mass killing.[19] Such work is bolstered by quantitative and large-N qualitative analyses which also find ideology a difference-maker between mass killings and their absence.[20] Once sources of scepticism about ideology's role in mass killing are dismantled—a task I spend some time on—I suspect that the vast bulk of scholars will acknowledge that ideology plays *some* role in mass killing occurrence. The real debate surrounds *how ideology makes a difference*, and to answer that question we need deeper examination of actual ideological logics and mechanisms as they manifest in specific cases of mass killing.

Studying Ideas and Violence

Even accepting the value of comparative historical analysis, some scholars worry as to whether the rigorous study of *ideology*, and ideas more generally, is really viable. Both internalized ideologies and ideological structures operate through thought processes: whether sincere understandings or expectations about social structures.[21] The methodological concern is that such thought processes are generally 'unobservable'—occurring inside the mind—and hence either impossible or extremely challenging to research scientifically.[22] In their discussion of 'culturalist' research on ethnic and nationalist violence, Rogers Brubaker and David Laitin press a moderate version of this concern effectively, noting how:

> It is difficult to know whether, when, where, to what extent, and in what manner the posited beliefs and fears were actually held. How do we know that, in India, the most 'rabid and senseless Hindu propaganda,' 'the most outrageous suggestions' about the allegedly evil, dangerous, and threatening Muslim 'other,' have come to be 'widely believed,' and to constitute 'a whole new "common sense"'? How do we know that, in

---

[19] Bellamy 2012b; Straus 2015a; Bulutgil 2017; Hiebert 2017.
[20] E.g. Harff 2003; Williams 2016; Kim 2018.
[21] Some deny this, contending that ideas fundamentally exist in intersubjective discourses or social institutions rather than minds; see, for example, Schull 1992; Laffey and Weldes 1997; Swidler 2001; Krebs and Jackson 2007; Campbell 2009, 158–9. This confuses the truth that ideas possess emergent structural properties at the collective level with an erroneous claim that ideas' effects are not rooted in individuals' mental processes. As stressed below, without assumptions about mental processes, one cannot plausibly link intersubjective phenomena like discourses or institutions to behaviour. See van Dijk 1998, 9–10 & 43–6; Mercer 2005.
[22] E.g. Swidler 2001, 83–5; Kalyvas 2006, 92; Krebs and Jackson 2007, 40. See also Morrow 2020, 32–5.

Sri Lanka in 1983, Tamils were believed to be 'superhumanly cruel and cunning and, like demons, ubiquitous' or 'agents of evil,' to be rooted out through a kind of 'gigantic exorcism'? How do we know that, in the Serb-populated borderlands of Croatia, Serbs really feared Croats as latter-day Ustashas? Lacking direct evidence (or possessing at best anecdotal evidence) of beliefs and fears, culturalist accounts often rely on nationalist propaganda tracts but are unable to gauge the extent to which or the manner in which such fearful propaganda has been internalized by its addressees.[23]

This is an important challenge to any attempt to study the impact of ideologies and ideas, and Brubaker and Laitin are right that many studies do not adequately meet it. But the challenge is not insurmountable.

For a start, claims that ideas are 'unobservable' are rhetorical exaggerations. *Direct observation* of human thinking may be impossible, but this is commonplace in scientific research, and a matter of degree.[24] For example, a scholar can never be sure that a given set of historical population statistics were collected without error, or that photographs of a now-destroyed mass grave are not forgeries, or that newspaper reports of injuries at a protest event were not exaggerated. In all these examples, our observations are *irrevocably indirect*—the bits of the world we would like to directly observe no longer exist. But this does not make the data useless. Good scholarship is not about limiting oneself to direct observations that are (purportedly) free from any chance of error, but matching the confidence of one's conclusions with the reliability of what evidence is available. Evidence on individuals' thought processes may be indirect and uncertain, but this simply makes our inferences about thinking more tentative rather than cripplingly unreliable.

Like most researchers who study ideas, I rely on examining various forms of 'discourse'— primarily in written texts or subsequent interviews—under the methodological conviction that 'we can infer psychological characteristics based upon the subject's verbal behavior'.[25] This process of inference is obviously tricky, since what individuals say is no sure-fire guarantee as to what they think. Individuals lie, forget, dissimulate, speak metaphorically, engage in rhetorical strategies, and in a wide range of other ways use discourse that may not transparently represent the thought processes we are interested in.[26] We can improve the reliability of our interpretations of discourse, however, by 'triangulating' it with other evidence, such as observable behaviour and the nature of the cultural and social context of our subjects of study.[27] This will not establish *certainty*, but it can provide a rigorous foundation for claims about ideas. Indeed, if human beings could not make fairly reliable inferences about each other's mental states in this way, most complex human interaction would be impossible, and not just discourse analysis but almost all archival, interview, and survey research methods would be useless.[28]

Crucially, while the interpretations that follow from such an inquiry are tentative, we *have to make* some interpretations of this kind in social science, since more easily observable phenomena—such as material environments—are always linked to human choices

---

[23] Brubaker and Laitin 1998, 443.
[24] Shorten 2013, 362–3.
[25] Walker & Schafer 2006: 26. See also: Owen 2010, 73; Saucier and Akers 2018, 81.
[26] Fujii 2010.
[27] On the importance and methodology of analysing discursive context see Skinner 1966a; Skinner 1969; Skinner 1974; Skinner 2002. Again, broader psychological science is invaluable in strengthening our interpretations.
[28] van Dijk 1995, 44.

and behaviour via thought processes that vary across individuals and groups.[29] As the pre-eminent theorist of 'symbolic interactionism', Herbert Blumer, wrote:

> if the scholar wishes to understand the action of people it is necessary for him to see their objects as they see them. Failure to see their objects as they see them ... is the gravest kind of error that the social scientist can commit. It leads to the setting up of a fictitious world. Simply put, people act towards things on the basis of the meaning that these things have for them, not on the basis of the meaning that these things have for the outside scholar. Yet we are confronted right and left with studies of human group life and of the behavior of people in which the scholar has made no attempt to find out how the people see what they are acting toward.[30]

Indeed, many essential explanatory concepts—knowledge, motives, beliefs, institutions, rules, norms, preferences—and a vast range of social scientific variables—such as national security, state resolve, ethnic identity, voting intentions, and consumer confidence—are fundamentally constituted by mental states. They cannot even be *specified* without making some assumptions about what people actually think. Scholars can use various more observable 'proxies', instead of actually trying to discern mental states themselves.[31] But if certain mental processes are not *assumed to be associated with that proxy*, there is no convincing way to make the explanatory link from the proxy to behaviour.

In this, I endorse the view that 'social inquiry is not just about estimating causal effects, but also providing *explanations*, "unpacking the black box of causality" by positing the processes through which effects are produced'.[32] This cannot be done in any serious way for human behaviour without making some kind of assumptions about human thinking. This does not mean that all researchers need to study ideas in detail. There are many reasonable assumptions scholars can make about mental processes. But we can only have confidence that those assumptions are reasonable if some social scientists attempt to ground them in empirical evidence rather than mere theoretical hunches. As Robert English observes, 'no quantity of evidence on what actors should have been thinking can substitute for quality of evidence on what they actually did think'.[33]

In conducting the case studies, then, I employ a form of ideological analysis involving three main elements. First, I make claims about the character of a particular ideology and its component ideas by *identifying and interpreting ideological elements that are observable in discourse*. The testimony of victims and bystanders plays a role here, but most of the relevant discourse comes from perpetrators, both through original texts produced before or during the violence and through subsequent testimony in memoirs, court cases, or interviews conducted by fieldworkers. It goes without saying that such perpetrator discourse requires careful, critical interpretation and cannot simply be taken at face value. Perpetrators often have incentives to dissimulate after the fact, and to strategically frame their actions and motives during violence itself. Problems of recall—especially for what may be

---

[29] King 2004, 453–5; Mercer 2005, 79–89.
[30] Blumer 1969, 50–1. Much scholarship draws far-reaching ontological and epistemological conclusions from arguments like these, contending that there is therefore 'nothing outside of discourse' (Campbell 1998, 4). I do not share this view. Discourses powerfully shape human experience but are not a precondition for it: basic sensory and representational capacities for the ordering of experience exist in humans, as in non-linguistic animals, prior to the acquisition of specific discourses, and humans could not acquire discourses in the first place without such prediscursive capacities. See Pinker 1994; Tomasello 1999.
[31] See also Kertzer 2017, 87–90.
[32] Ibid. 84.
[33] English 2002, 91.

traumatic events—may often trouble the data. This does not, however, render perpetrator discourse useless. Indeed, it is often the best data we have for making necessary inferences about perpetrator actions and mindsets and the broader ideological structures under which perpetrators act. Moreover, perpetrators' self-interested incentives in legal or research contexts after violence are often more likely to count *against* statements of ideological belief or compliance, by comparison with appeals to coercion or other constraints on agency, and thus understate rather that overstate the importance of ideology.

Scholars of mass killing have developed sophisticated methodologies for interpreting such perpetrator discourse, involving the kind of triangulation of statements, behaviour, and cultural and social context mentioned above.[34] I emphasize, as such, that my own interpretations of ideological discourse in my four case studies almost always follow dominant accepted readings by case-specialist historians, ethnographers, and social scientists, rather than resting on contentious readings of my own. I also draw on broader methodological contributions from discourse analysis and contextualist approaches to intellectual history, which emphasize that accurate interpretations of discourse must attend to (i) what it appears that individuals were *doing* in producing such discourse,[35] and (ii) what it is feasible to assume they were expressing given the cultural environment in which they were writing/speaking.[36] In this respect, much of the discourse available to us has grounds for being taken seriously. Discourse that precedes acts of violence is clearly of particular value, since it vitiates assertions that such discourse is merely a 'post-hoc rationalization'. More private forms of discourse—such as diaries and personal correspondence—are often more valuable than public discourse, since they reduce many incentives for dissimulation. Even when public discourse may not be a good indicator of ideological internalization, moreover, it still offers strong evidence of the prevailing ideological norms, discourses, and institutions—i.e. of ideological structures—as they were understood by those subject to them. Such discourse analysis is not 'merely descriptive'. Given broader knowledge about discourses, intentionality, and human cognition, it already takes us some way in *tentatively* substantiating causal claims about ideology's role, by revealing meanings, intentions, and expectations that would make little sense in the absence of ideology's influence. When individuals avow ideological beliefs or testify to the strength of ideological norms, and when such avowal and testimony is highly plausible contextually, this is a reasonable basis for inferring their genuine influence.

Second, I show that various *forms of behaviour in mass killing correspond with such ideological elements* while being largely inexplicable if ideology is ignored. This is, in part, a form of what is sometimes labelled *congruence-testing*.[37] I show that within-case patterns of violence fit with particular ideological categories and justifications, while cross-case contrasts in such patterns of violence can be shown to match cross-case contrasts in ideology. In addition, *process-tracing* of the concatenations of events and decisions leading to mass killing can identify ideological notions appearing to exercise important influence over the direction

---

[34] See, in particular, Straus 2006, 97–103; Fujii 2009, 23–44; Fujii 2010; Jessee 2017, 15–23 & 149–80; Schmidt 2017; Viola 2017; Viola 2018.
[35] Tully 1983, 490–4; Skinner 2002, 2–4; George and Bennett 2005, 99–108.
[36] Skinner 1974, 283 & 289. This involves, as Skinner explains, a kind of rationalist intuition, since 'what is rational to believe depends in large measure on the nature of our other beliefs', and as a result, we should 'interpret specific beliefs by placing them in the context of other beliefs [and] interpret systems of belief by placing them in wider intellectual frameworks'; see Skinner 2002, 4–5. But it also rests on attention to the forms of power manifest in the prevailing discursive and ideological structures to which individuals are exposed. See also Quine and Ullian 1978; Boudon 1989, 3–8 & 11; Howarth and Stavrakakis 2000; Fairclough 2010.
[37] George and Bennett 2005, chs.9 & 10.

chosen.[38] Again, this process involves considerable analytical judgement, and draws on both broader social scientific knowledge and detailed attention to historical context. Most particularly, quite what it means for an idea to 'correspond' to behaviour is not obvious, since specific ideas are rarely compatible with *only one* course of action.[39] But most ideas are still consistent with only a small subset of all possible actions, and once ideas are combined into sets and understood in their specific cultural context, the range of behaviour that is plausibly consistent with such ideas narrows considerably. By contrast with some comparative studies, I therefore emphasize, here, detailed attention to *ideological content*. Merely making reference to broad ideological labels—such as communism or ethnonationalism—or highly general ideas within such ideologies—such as class struggle or ethnic community—is unlikely to provide much explanatory leverage, since ideas at this level of breadth or generality lack clear behavioural implications. Only by examining ideological worldviews in detail, and tracing how ideological claims manifest in specific beliefs, norms, institutions, policy-making processes, and areas of practice, can we generally link ideologies to behaviour in a revealing way.

This form of theoretical inference depends on continuously considering alternative explanations of mass killing that either do not accord ideology a central role, or accord it a very different role to my own account. As set up in Chapter 1, my comparative focus, here, is primarily on traditional-ideological accounts, rationalist ideology-sceptics, and situationist ideology-sceptics. Each of these categories encompasses a number of specific theories and explanations, so my precise consideration of alternatives is somewhat dependent on the specific context of each of my case studies. The process of comparatively assessing explanations is, moreover, a complex interpretive and abductive activity, and I resist efforts to lay it out in overly mechanical terms orientated around lists of (purportedly) specific hypotheses that are then straightforwardly falsified, or not, against the world.[40]

Nevertheless, although somewhat schematic, it is worth being explicit about the very broad implications I take each of these three explanatory approaches to have, by comparison with my own argument.[41] If *traditional-ideological accounts* are correct, we should generally expect mass killings to occur in contexts of large-scale social re-engineering projects; to be organized around targeting categories, logics, and local variables linked to radical political goals; and to be justified by perpetrators with reference to strong pre-existing beliefs in such radical political goals and/or radical denigration of conventional moral principles and norms. If, by contrast, *rationalist-sceptical accounts* are correct, we would generally expect mass killings to occur when strong rational incentives for such violence can be identified; to be organized around targeting categories, logics, and local variables clearly linked to physical threat, political control, and/or material acquisition; and to be justified primarily through arguments about strategic security needs and self-interest, with more overtly 'ideological' claims emerging only as a late-stage or post-hoc rationalization for such underlying rationales. If *situationist-sceptical accounts* are correct, finally, we would generally expect mass killings to occur in contexts of sustained institutional radicalization largely driven by bureaucratic organizational interests; to be organized around logics of organizational routine or expediency and processes of escalation driven by intragroup social pressure; and

---

[38] Waldner 2012; Mahoney 2015.
[39] Kalyvas 1999, 247–51; Fearon and Laitin 2000, 863–4.
[40] For some philosophical issues that arise with such a depiction of theory-testing see Jackson 2016, 65–82. My more central concern is that this mechanical presentation oversimplifies the kind of analytical and interpretive judgements involved in deeming explanations unconvincing or falsified.
[41] This is a slight expansion of the summary given in Chapter 1. The logic behind these implications is explained across Chapters 3 and 4.

to be justified primarily through appeals to authority, coercion, and peer-pressure with, again, ideological justifications largely emerging as post-hoc rationalizations. Both kinds of sceptical account also suggest that either there should be a limited match between ideology and specific patterns of violence (barring those reconstructed in post-hoc ideological rationalizations), or that such ideological patterns should be predictably linked to (and thus reducible to) the causal factors emphasized by rationalists and situationists. They should also expect little psychological evidence of real causal effects associated with ideological justifications of violence.

Third, I use *psychological research* (primarily summarized in Chapters 2 and 4) to substantiate the claim that ideologies do indeed (contra the sceptics) have the specific *causal powers* to shape behaviour, and encourage violence, that I impute to them—specifically via the four forms of ideological influence identified in Chapter 2 and the six justificatory mechanisms explicated in Chapter 4. This reflects, in part, a scientific realist contention that good explanations should be able to point to mechanisms that have such apparent causal powers to shape the outcomes we seek to explain.[42] For example, I show not only that framing civilians as guilty criminals *corresponds* with violence enacted against them in historical cases, but also that process-tracing can highlight such framing as occupying a key place in the deliberations, decisions, and activities leading to violence. I also point to psychological evidence that such assertions of guilt are typically linked to the kinds of revenge motives and legitimating frameworks that *can produce* such behaviour. Similarly, I argue that we should infer that ideological constructions of threat tend to enhance the willingness of individuals to initiate, implement, and support violence against civilians, not just because this is a central and recurring feature of historical cases of mass killing, but because threat-priming has been shown, experimentally, to increase such willingness, even in the absence of 'objective' threats. I do not claim that psychological science therefore definitively supports my account and rules out any other explanations of mass killing. But the fact that the causal processes I identify as linking ideology to violence are consistent with findings from more genuinely experimental settings in psychology, coupled with the apparent centrality of such processes in historical cases to explain the path to and nature of violence, offers strong evidence for my account.

Conclusion

The theoretical and methodological approach you find most convincing depends on the kinds of knowledge you most value. As I have suggested, my advocacy of the detailed study of ideologies and their effects reflects a preference for social scientific theories that embrace complexity, richness, and realism.[43] This stands in contrast to approaches that prioritize parsimony and predictive specificity, typically by identifying key causal variables that generate important predictive regularities in behaviour.[44] Parsimonious theories are useful, and they may reveal, to borrow Kenneth Waltz's famous phrase, 'a small number of big and important things'.[45] But many scholars now accept that, to the degree that highly parsimonious theories rest on implausible characterizations of underlying causal processes, they remain at best problematically incomplete and at worst false. The fact that they might be easily

---

[42] Jackson 2016, 103–4.
[43] Within scholarship on mass killing see Verdeja 2012; Owens, Su, and Snow 2013; Williams and Pfeiffer 2017. See also Pierson 2004; Tilly and Goodin 2006; Hedström and Ylikoski 2010.
[44] See also Smith, Booth, and Zalewski 1996; Hedström and Ylikoski 2010.
[45] Waltz 1986, 329.

modelled or conveniently utilized to generate predictions 'as if' they were true does not change this.[46]

Consequently, we must complement parsimonious theories with rich descriptive histories and ethnographies, models of complex systems and interactions, and theoretical emphasis of context-specificity, subjectivity, and agency.[47] Such methodologies make our causal *explanations* of mass killing more powerful and realistic, in part by allowing more plausible *interpretations* of the mindsets of perpetrators of such violence and the subjective meanings of their actions. Against the common suggestion that 'explanatory' and 'interpretive' approaches to social science are incompatible, I therefore see them as profoundly complementary.[48] But this is not to make interpretation merely the servant to explanation. Making some effort to cross the 'hermeneutic gap' to perpetrators—studying ideas not just to identify their causal impact but to render perpetrators' actions intelligible, even if still inexcusable—is also essential if we are to meaningfully evaluate such acts and reflect on their broader implications for human life.

Examining ideas and ideologies is not easy and, as Francisco Gutiérrez Sanín and Elisabeth Jean Wood emphasize, 'a non-rigorous approach can use ideas as a wild card to explain anything.'[49] Too much social scientific research remains essentially anecdotal—working off very limited data and making bold inferences about organizations or whole societies grounded in little more than a few key speeches or some indirect proxies for ideology.[50] But no research method is so reliable that it cannot be misused, and, again, social science cannot eschew the study of ideas entirely. As Robert Jervis argued in his famous study of *Perception and Misperception in International Politics*, 'it is often impossible to explain crucial decisions and policies without reference to the decision-makers' beliefs about the world and their images of others … these cognitions are part of the proximate cause of the relevant behaviour and other levels of analysis cannot immediately tell us what they will be.'[51] In recognition of this fact, and the broader 'socially constructed' character of political contexts and political action, this book is devoted to the project of making the *best explanatory and interpretive inferences we can* about the character and role of ideology in mass killing.

---

[46] George and Bennett 2005, 208; Shapiro 2005; Hedström and Ylikoski 2010, 60–1; Ylikoski 2012, 41–2.
[47] Browder 2003, 493–6; King 2004, 439–47; Mitchell 2004; Fujii 2009; Owens, Su, and Snow 2013.
[48] Contra, for example, Hollis and Smith 1990; Bevir and Blakely 2018.
[49] Gutiérrez Sanín and Wood 2014, 214.
[50] Ibid. 215–17.
[51] Jervis 1976, 28.

Bibliography

Abrajano, Marisa, and Zoltan Hajnal. *White Backlash: Immigration, Race, and American Politics*. Princeton: Princeton University Press, 2015.

Ackerman, Gary, and Michael Burnham. "Towards a Definition of Terrorist Ideology." Terrorism and Political Violence, Early Release Online Version (2019).

Adams, Richard N. "Ethnic Images and Strategies in 1944." In *Guatemalan Indians and the State, 1540–1988*, edited by Carol A. Smith, 141–162. Austin: University of Texas Press, 1990.

Adamson, Fiona B. "Global Liberalism Versus Political Islam: Competing Ideological Frameworks in International Politics." *International Studies Review* 7, no. 4 (2005): 547–569.

Adena, Maja, Ruben Enikolopov, Maria Petrova, Veronica Santarosa, and Ekaterina Zhuravskaya. "Radio and the Rise of the Nazis in Prewar Germany." *The Quarterly Journal of Economics* 130, no. 4 (2015): 1885–1939.

Adler, R.N., C.E. Loyle, J. Globerman, and E.B. Larson. "Transforming Men into Killers: Attitudes Leading to Hands-on Violence During the 1994 Rwandan Genocide." *Global Public Health* 3, no. 3 (2008): 291–307.

Afflitto, Frank M. "The Homogenizing Effects of State-Sponsored Terrorism: The Case of Guatemala." In *Death Squad: The Anthropology of State Terror*, edited by Jeffrey A. Sluka, 114–126. Philadelphia: University of Pennsylvania Press, 2000.

Agamben, Georgio. *State of Exception*. Chicago: University of Chicago Press, 2005.

Aguilar, Pilar, Silvina Brussino, and José-Miguel Fernández-Dols. "Psychological Distance Increases Uncompromising Consequentialism." *Journal of Experimental Social Psychology* 49 (2013): 449–452.

Allen, Michael Thad. *The Business of Genocide: The S.S., Slave Labour, and the Concentration Camps*. Chapel Hill: University of North Carolina Press, 2002.

Alperovitz, Gar, Robert L. Messer, and Barton J. Bernstein. "Correspondence: Marshall, Truman, and the Decision to Drop the Bomb." *International Security* 16, 3 (1991/1992): 204–221.

Althusser, Louis. "Ideology and Ideological State Apparatuses (Notes Towards an Investigation)." In *Lenin and Philosophy and Other Essays*, 85–126. New York: Monthly Review Press, 1971.

Alvarez, Alex. "Destructive Beliefs: Genocide and the Role of Ideology." In *Supranational Criminology: Towards a Criminology of International Crimes*, edited by A. Smeulers and R. Haveman, 213–232. Antwerpen: Intersentia, 2008.

Alvarez, Alex. *Genocidal Crimes*. Abingdon: Routledge, 2010.

Alvarez, Alex. "Making Enemies: The Uses and Abuses of Tainted Identities." *Crosscurrents* 65, no. 3 (2015): 311–320.

Aly, Götz. *Hitler's Beneficiaries: Plunder, Racial War, and the Nazi Welfare State*. New York: Henry Holt and Company, 2008.

Anderson, Benedict. *Imagined Communities: Reflections on the Origin and Spread of Nationalism*. London: Verso, 1983/2006.

Anderson, Kjell. *Perpetrating Genocide: A Criminological Account.* Abingdon: Routledge, 2017a.
Anderson, Kjell. "Who Was I to Stop the Killing?: Moral Neutralization Among Rwandan Genocide Perpetrators." *Journal of Perpetrator Research* 1, no. 1 (2017b): 39–63.
Anderton, Charles H. "Choosing Genocide: Economic Perspectives on the Disturbing Rationality of Race Murder." *Defence and Peace Economics* 21, no. 5–6 (2010): 459–486.
Anderton, Charles H. "Datasets and Trends of Genocides, Mass Killings, and Other Civilian Atrocities." In *Economic Aspects of Genocides, Other Mass Atrocities, and Their Prevention*, edited by Charles H. Anderton and Jurgen Brauer, Chapter 3. Oxford: Oxford University Press, 2016.
Anderton, Charles H., and Jurgen Brauer, eds. *Economic Aspects of Genocides, Other Mass Atrocities, and Their Prevention.* Oxford: Oxford University Press, 2016a.
Anderton, Charles H., and Jurgen Brauer, eds. "On the Economics of Genocides, Other Mass Atrocities, and Their Prevention." In *Economic Aspects of Genocides, Other Mass Atrocities, and Their Prevention*, edited by Charles H. Anderton and Jurgen Brauer, Chapter 1. Oxford: Oxford University Press, 2016b.
Arendt, Hannah. *The Origins of Totalitarianism.* Orlando, FL: Harcourt Books, 1951/1976.
Arendt, Hannah. *Eichmann in Jerusalem: A Report on the Banality of Evil.* London: Penguin Books, 1963/2006.
Arendt, Hannah. *On Violence.* Orlando, FL: Harcourt Books, 1970.
Arias, Arturo. "Changing Indian Identity: Guatemala's Violent Transition to Modernity." In *Guatemalan Indians and the State, 1540–1988*, edited by Carol A. Smith, 230–255. Austin: University of Texas Press, 1990.
Ariely, Dan. *Predictably Irrational: The Hidden Forces That Shape Our Decisions.* Revised and Expanded ed. London: HarperCollins, 2009.
Aronsfeld, C.C. *The Text of the Holocaust: A Study of the Nazis' Extermination Propaganda, from 1919–1945.* Marblehead, MA: Micah Publications, 1985.
Aronson, Ronald. "Societal Madness: Impotence, Power and Control." In *Toward the Understanding and Prevention of Genocide: Proceedings of the International Conference on the Holocaust and Genocide*, edited by Israel Charny, Chapter 11. Boulder, CO: Westview Press, 1984.
Arreguín-Toft, Ivan. "How the Weak Win Wars: A Theory of Asymmetric Conflict." *International Security* 26, no. 1 (2001): 93–128.
Article 19. *Broadcasting Genocide: Censorship, Propaganda, and State-Sponsored Violence in Rwanda, 1990–1994.* London: Article 19, 1996.
Asal, Victor, and R. Karl Rethemeyer. "The Nature of the Beast: Organizational Structures and the Lethality of Terrorist Attacks." *The Journal of Politics* 70, no. 2 (2008): 437–449.
Asch, Solomon E. "Studies of Independence and Conformity: I. A Minority of One Against a Unanimous Majority." *Psychological Monographs: General and Applied* 70, no. 9 (1956): 1–70.
Atran, Scott, and Robert Axelrod. "Reframing Sacred Values." *Negotiation Journal* 24, no. 3 (2008): 221–246.
Atran, Scott, and Jeremy Ginges. "Religious and Sacred Imperatives in Human Conflict." *Science* 336 (2012): 855–857.
Baberowski, Jörg, and Anselm Doering-Manteuffel. "The Quest for Order and the Pursuit of Terror: National Socialist Germany and the Stalinist Soviet Union as Multiethnic Empires." Translated by Barry Haneberg. In *Beyond Totalitarianism: Stalinism and Nazism Compared*, edited by Michael Geyer and Sheila Fitzpatrick, 180–228. Cambridge: Cambridge University Press, 2009.

Backer, David. "Reconstructing the Rwandan Genocide: In Search of Local Dynamics." *Taiwan Journal of Democracy* 4, no. 1 (2008): 179–185.

Bajaj, Monisha. "Human Rights Education: Ideology, Location and Approaches." *Human Rights Quarterly* 33, no. 2 (2011): 481–508.

Balcells, Laia. "Rivalry and Revenge: Violence Against Civilians in Conventional Civil Wars." *International Studies Quarterly* 54 (2010): 291–313.

Balcells, Laia. *Rivalry and Revenge: The Politics of Violence During Civil War*. Cambridge: Cambridge University Press, 2017.

Balcells, Laia, and Stathis N. Kalyvas. Revolutionary Rebels and the Marxist Paradox. 2015. Unpublished Manuscript.

Balkin, J.M. *Cultural Software: A Theory of Ideology*. New Haven: Yale University Press, 1998.

Ball, Patrick, Paul Kobrak, and Herbert F. Spirer. *State Violence in Guatemala, 1960–1996: A Quantitative Reflection*. Washington, DC: American Association for the Advancement of Science, 1999.

Bandura, Albert. "Moral Disengagement in the Perpetration of Inhumanities." *Personality and Social Psychology Review* 3, no. 3 (1999): 193–209.

Bandura, Albert, Bill Underwood, and Michael E. Fromson. "Disinhibition of Aggression Through Diffusion of Responsibility and Dehumanization of Victims." *Journal of Research in Personality* 9 (1975): 253–269.

Bar-Tal, Daniel. "Sociopsychological Foundations of Intractable Conflicts." *American Behavioural Scientist* 50, no. 11 (2007): 1430–1453.

Barber, Benjamin, and Charles Miller. "Propaganda and Combat Motivation: Radio Broadcasts and German Soldiers' Performance in World War II." *World Politics* 71, no. 3 (2019): 457–502.

Barkan, Rachel, Shahar Ayal, and Dan Ariely. "Ethical Dissonance, Justifications, and Moral Behavior." *Current Opinion in Psychology* 6 (2015): 157–161.

Barnes, Nicholas. "Criminal Politics: An Integrated Approach to the Study of Organized Crime, Politics and Violence." *Perspectives on Politics* 15, no. 4 (2017): 967–987.

Barnes, Samuel H. "Ideology and the Organization of Conflict: On the Relationship Between Political Thought and Behavior." *The Journal of Politics* 28, no. 3 (1966): 513–530.

Bartov, Omer. *Hitler's Army: Soldiers, Nazis and War in the Third Reich*. Oxford: Oxford Paperbacks, 1994.

Bartov, Omer. "'Fields of Glory': War, Genocide, and the Glorification of Violence." In *Catastrophe and Meaning: The Holocaust and the Twentieth Century*, edited by Moishe Postone and Eric Santner, Chapter 6. Chicago: University of Chicago Press, 2003.

Bateson, Regina. "The Socialization of Civilians and Militia Members: Evidence from Guatemala." *Journal of Peace Research* 54, no. 5 (2017): 634–647.

Baugher, Amy R., and Julie A. Gazmararian. "Masculine Gender Role Stress and Violence: A Literature Review and Future Directions." *Aggression and Violent Behavior* 24 (2015): 107–112.

Baum, Steven K. *The Psychology of Genocide*. Cambridge: Cambridge University Press, 2008.

Bauman, Zygmunt. *Modernity and the Holocaust*. Cambridge: Polity Press, 1989.

Baurmann, Michael. "Rational Fundamentalism? An Explanatory Model of Fundamentalist Beliefs." *Episteme* 4, no. 2 (2007): 150–166.

Beech, Hannah. "Denial and Loathing in Myanmar." *The New York Times: International Edition* (New York), 25 October 2017, 1 & 4.

Bell, P.M.H. *The Origins of the Second World War in Europe*. Harlow: Longman, 1986.
Bellamy, Alex J. "The Ethics of Terror Bombing: Beyond Supreme Emergency." *Journal of Military Ethics* 7, no. 1 (2008): 41–65.
Bellamy, Alex J. "Mass Atrocities and Armed Conflict: Links, Distinctions, and Implications for the Responsibility to Prevent." *Stanley Foundation Policy Analysis Brief* (2011).
Bellamy, Alex J. "Mass Killing and the Politics of Legitimacy: Empire and the Ideology of Selective Extermination." *Australian Journal of Politics and History* 58 (2012a): 159–180.
Bellamy, Alex J. *Massacres and Morality: Mass Atrocities in an Age of Civilian Immunity*. Oxford: Oxford University Press, 2012b.
Bénabou, Roland, and Jean Tirole. "Belief in a Just World and Redistributive Politics." *The Quarterly Journal of Economics* 121, no. 2 (2006): 699–746.
Benesch, Susan. "The Ghost of Causation in International Speech Crime Cases." In *Propaganda, War Crimes Trials and International Law: From Speakers' Corner to War Crimes*, edited by Predrag Dojčinović, 254–268. Abingdon: Routledge, 2012a.
Benesch, Susan. "Words as Weapons." *World Policy Journal* 29 (2012b): 7–12.
Benesch, Susan. Dangerous Speech: A Proposal to Prevent Group Violence. http://www.worldpolicy.org/susan-benesch, 2012c (accessed 15 October 2012).
Benesch, Susan. Countering Dangerous Speech to Prevent Mass Violence in Kenya's 2013 Elections. https://dangerousspeech.org/countering-dangerous-speech-kenya-2013/, 2014a.
Benesch, Susan. Countering Dangerous Speech: New Ideas for Genocide Prevention. Working Paper. Washington, DC: United States Holocaust Memorial Museum, 2014b.
Benesch, Susan, Cathy Buerger, Tonei Glavinic, Sean Manion, and Dan Bateyko. *Dangerous Speech: A Practical Guide*. The Dangerous Speech Project, 2021.
Berger, Peter L., and Thomas Luckmann. *The Social Construction of Reality: A Treatise in the Sociology of Knowledge*. New York: Anchor Books, 1967.
Berlin, Isaiah. "Historical Inevitability." In *Liberty: Isaiah Berlin*, edited by Henry Hardy, 94–165. Oxford: Oxford University Press, 1954/2002.
Berlin, Isaiah. "The Pursuit of the Ideal." In *The Crooked Timber of Humanity: Chapters in the History of Ideas*, edited by Henry Hardy, Chapter 1. London: Pimlico, 2013.
Bernstein, Seth. "Introduction to the English-Language Edition." In *Agents of Terror: Ordinary Men and Extraordinary Violence in Stalin's Secret Police*, edited by Alexander Vatlin, xix–xxxii. Madison: University of Wisconsin Press, 2016.
Berrington, Hugh. "When Does Personality Make a Difference? Lord Cherwell and the Area Bombing of Germany." *International Political Science Review* 10, no. 1 (1989): 9–34.
Bevir, Mark, and Jason Blakely. *Interpretive Socal Science: An Anti-Naturalist Approach*. Oxford: Oxford University Press, 2018.
Bhavnani, Ravi. "Ethnic Norms and Interethnic Violence: Accounting for Mass Participation in the Rwandan Genocide." *Journal of Peace Research* 43, no. 6 (2006): 651–669.
Bilinsky, Yaroslav. "Was the Ukrainian Famine of 1932–1933 Genocide?" *Journal of Genocide Research* 1, no. 2 (1999): 147–156.
Blair, Tony. Religious Difference, Not Ideology, Will Fuel This Century's Epic Battles. *The Observer*, 25 January 2014. http://tinyurl.com/l2uh3w3.
Blakeley, Ruth. *State Terrorism and Neoliberalism: The North in the South*. Abingdon: Routledge, 2009.
Block, Jack, and Jeanne H. Block. "Nursery School Personality and Political Orientation Two Decades Later." *Journal of Research in Personality* 40 (2006): 734–749.
Bloxham, Donald. "The Armenian Genocide of 1915–16: Cumulative Radicalization and the Development of a Destruction Policy." *Past and Present* 181 (2003): 141–191.

Bloxham, Donald. *The Great Game of Genocide: Imperialism, Nationalism and the Destruction of the Ottoman Armenians*. Oxford: Oxford University Press, 2005.

Bloxham, Donald. "Organized Mass Murder: Structure, Participation, and Motivation in Comparative Perspective." *Holocaust and Genocide Studies* 22, no. 2 (2008): 203-245.

Bloxham, Donald, and A. Dirk Moses. "Editors' Introduction: Changing Themes in the Study of Genocide." In *The Oxford Handbook of Genocide Studies*, edited by Donald Bloxham and A. Dirk Moses, 1-16. Oxford: Oxford University Press, 2010.

Blumer, Herbert. *Symbolic Interactionism: Perspective and Method*. Eaglewood Cliffs, NJ: Prentice-Hall, 1969.

Bodley, John H. "Victims of Progress." In *Genocide: An Anthropological Reader*, edited by Alexander Laban Hinton, Chapter 7. Malden, MA: Blackwell Publishing, 2002.

Bosi, Lorenzo, and Donatella Della Porta. "Micro-Mobilization into Armed Groups: Ideological, Instrumental and Solidaristic Paths." *Qualitative Sociology* 35 (2012): 361-383.

Boudon, Raymond. *The Analysis of Ideology*. Cambridge: Polity Press, 1989.

Bradshaw, Peter. The Look of Silence: Act of Killing Director's Second Film Is as Horrifically Gripping as First—Venice Film Festival Review. *The Guardian (London)*, 28 August 2014. http://www.theguardian.com/film/2014/aug/27/the-look-of-silence-review-act-of-killing-venice-film-festival.

Brady, Anne-Marie. "Mass Persuasion as a Means of Legitimation and China's Popular Authoritarianism." *American Behavioral Scientist* 53, no. 3 (2009): 434-457.

Brandenberger, David. *Propaganda State in Crisis: Soviet Ideology, Indoctrination, and Terror under Stalin, 1927-1941*. New Haven: Yale University Press, 2011.

Brass, Paul. *The Production of Hindu-Muslim Violence in Contemporary India*. Seattle: University of Washington Press, 2005.

Bratić, Vladimir. "Examining Peace-Oriented Media in Areas of Violent Conflict." *The International Communication Gazette* 70, no. 6 (2008): 487-503.

Breton, Albert, Gianluigi Galeotti, Pierre Salmon, and Ronald Wintrope, eds. *Political Extremism and Rationality*. Cambridge: Cambridge University Press, 2002.

Brett, Roddy. *The Origins and Dynamics of Genocide: Political Violence in Guatemala*. Houndmills: Palgrave Macmillan, 2016.

Brooks, Jeffrey. "Socialist Realism in *Pravda*: Read All About It!" *Slavic Review* 53, no. 4 (1994): 973-991.

Brooks, Stephen G., and William C. Wohlforth. "Power, Globalization, and the End of the Cold War: Reevaluating a Landmark Case for Ideas." *International Security* 25, no. 3 (2000/2001): 5-53.

Browder, George C. "Perpetrator Character and Motivation: An Emerging Consensus?" *Holocaust and Genocide Studies* 17, no. 3 (2003): 480-497.

Browning, Christopher R. "The Government Experts." In *The Holocaust: Ideology, Bureaucracy, and Genocide: The San José Papers*, edited by Henry Friedlander and Sybil Milton, 183-197. Millwood, NY: Kraus International, 1980.

Browning, Christopher R. *Ordinary Men: Reserve Police Battalion 101 and the Final Solution in Poland*. London: Penguin Books, 1992/2001.

Browning, Christopher R. *The Origins of the Final Solution: The Evolution of Nazi Jewish Policy 1939-1942*. London: Arrow Books, 2005.

Browning, Christopher R. "Foreword." In *Becoming Evil: How Ordinary People Commit Genocide and Mass Killing*, edited by James Waller, vii-ix. Oxford: Oxford University Press, 2007.

Browning, Christopher R., and Lewis H. Siegelbaum. "Frameworks for Social Engineering: Stalinist Schema of Identification and the Nazi Volksgemeinschaft." Translated by

Barry Haneberg. In *Beyond Totalitarianism: Stalinism and Nazism Compared*, edited by Michael Geyer and Sheila Fitzpatrick, Chapter 6. Cambridge: Cambridge University Press, 2009.

Brubaker, Rogers, and David D. Laitin. "Ethnic and Nationalist Violence." *Annual Review Sociology* 24 (1998): 423–452.

Brzezinski, Zbigniew. "Totalitarianism and Rationality." *American Political Science Review* 50, no. 3 (1956): 751–763.

Bubandt, Nils. "Rumors, Pamphlets and the Politics of Paranoia in Indonesia." *The Journal of Asian Studies* 67, no. 3 (2008): 789–817.

Bulutgil, H. Zeynep. *The Roots of Ethnic Cleansing in Europe*. Cambridge: Cambridge University Press, 2016.

Bulutgil, H. Zeynep. "Ethnic Cleansing and Its Alternatives in Wartime: A Comparison of the Austro-Hungarian, Ottoman and Russian Empires." *International Security* 41, no. 4 (2017): 169–201.

Burleigh, Michael. *The Third Reich: A New History*. London: Pan Books, 2001.

Butcher, Charles, Benjamin E. Goldsmith, Sascha Nanlohy, Arcot Sowmya, and David Muchlinski. "Introducing the Targeted Mass Killing Data Set for the Study and Forecasting of Mass Atrocities." Journal of Conflict Resolution, Early Release Online Version (2020).

Buzan, Barry. *People, States and Fear: An Agenda for International Security Studies in the Post-Cold War Era*. Colchester: ECPR Press, 2007.

Buzan, Barry, Ole Wæver, and Jaap de Wilde. *Security: A New Framework for Analysis*. Boulder, CO: Lynne Riener Publishers, 1998.

Byman, Daniel. "Understanding the Islamic State—a Review Essay." *International Security* 40, no. 4 (2016): 127–165.

Campbell, Bradley. "Genocide as Social Control." *Sociological Theory* 27, no. 2 (2009): 150–172.

Campbell, David. *Writing Security: United States Foreign Policy and the Politics of Identity*. Revised ed. Minneapolis: University of Minnesota Press, 1998.

Cangemi, Michael. "Ambassador Frank Ortiz and Guatemala's 'Killer President,' 1976–1980." *Diplomatic History* 42, no. 4 (2018): 613–639.

Canh, Nguyen Van. *Vietnam under Communism, 1975–1982*. Stanford, CA: Hoover Institution Press, 1983.

Capoccia, Giovanni, and R. Daniel Kelemen. "The Study of Critical Junctures: Theory, Narrative and Counterfactuals in Historical Institutionalism." *World Politics* 59, no. 3 (2007): 341–369.

Caprara, Gian Vittorio, and Philip G. Zimbardo. "Personalizing Politics: A Congruency Model of Political Preference." *American Psychologist* (2004): 581–594.

Caprara, Gianvittorio, Donata Francescato, Minour Mebane, Roberta Sorace, and Michele Vecchione. "Personality Foundations of Ideological Divide: A Comparison of Women Members of Parliament and Women Voters in Italy." *Political Psychology* 31, no. 5 (2010): 739–762.

Card, Claudia. *The Atrocity Paradigm: A Theory of Evil*. Oxford: Oxford University Press, 2002.

Carlsmith, Kevin M., John M. Darley, and Paul H. Robinson. "Why Do We Punish? Deterrence and Just Deserts as Motives for Punishment." *Journal of Personality and Social Psychology* 83, no. 2 (2002): 284–299.

Casier, Tom. "The Shattered Horizon: How Ideology Mattered to Soviet Politics." *Studies in Eastern European Thought* 51 (1999): 35–59.

Cass, Lewis. "Removal of the Indians." *North American Review* 30, no. 64 (1830): 69–71.
Cassels, Alan. *Ideology and International Relations in the Modern World*. London: Routledge, 1996.
Cavanaugh, William T. *The Myth of Religious Violence: Secular Ideology and the Roots of Modern Conflict*. Oxford: Oxford University Press, 2009.
Ceadel, Martin. *Pacifism in Britain, 1914–1945: The Defining of a Faith*. Oxford: Clarendon Press, 1980.
Ceadel, Martin. *Thinking About Peace and War*. Oxford: Oxford University Press, 1987.
Chalk, Frank, and Kurt Jonassohn. *The History and Sociology of Genocide: Analyses and Case Studies* New Haven: Yale University Press, 1990.
Chandler, David. *Voices from S-21: Terror and History in Pol Pot's Secret Prison*. Chiang Mai: Silkworm Books, 2000.
Checkel, Jeffrey T. *Ideas and International Political Change: Soviet/Russian Behavior and the End of the Cold War*. New Haven: Yale University Press, 1997.
Checkel, Jeffrey T. "Socialization and Violence: Introduction and Framework." *Journal of Peace Research* 54, no. 5 (2017): 592–605.
Chenoweth, Erica, and Kathleen Gallagher Cunningham. "Understanding Nonviolent Resistance: An Introduction." *Journal of Peace Research* 50, no. 3 (2013): 271–276.
Chenoweth, Erica, and Pauline Moore. *The Politics of Terror*. Oxford: Oxford University Press, 2018.
Chirot, Daniel, and Clark McCauley. *Why Not Kill Them All?: The Logic and Prevention of Mass Political Murder*. Princeton: Princeton University Press, 2006.
Clodfelter, Mark. *The Limits of Airpower: The American Bombing of North Vietnam*. New York: Free Press, 1989.
Coady, C.A.J. "Terrorism, Just War and Supreme Emergency." In *Terrorism and Justice: Moral Argument in a Threatened World*, edited by C.A.J. Coady and Michael O'Keefe, 105–125. Melbourne: Melbourne University Press, 2002.
Cohen-Almagor, Raphael. "Foundations of Violence, Terror and War in the Writings of Marx, Engels and Lenin." *Terrorism and Political Violence* 3, no. 2 (1991): 1–24.
Cohen, Geoffrey L. "Party Over Policy: The Dominating Impact of Group Influence on Political Beliefs." *Journal of Personality and Social Psychology* 85, no. 5 (2003): 808–822.
Cohrs, J. Christopher. "Ideological Bases of Violent Conflict." In *Oxford Handbook of Intergroup Conflict*, edited by L. R. Tropp, 53–71. New York: Oxford University Press, 2012.
Coleman, Heather J., Alan Barenberg, Wendy Z. Goldman, and Tanja Penter. "A Roundtable on Lynne Viola's *Stalinist Perpetrators on Trial: Scenes from the Great Terror in Soviet Ukraine*." *Canadian Slavonic Papers/Revue Canadienne des Slavistes* 61, no. 2 (2019): 225–243.
Collier, Paul, and Pedro C. Vicente. "Votes and Violence: Evidence from a Field Experiment in Nigeria." *The Economic Journal* 124 (2013): F327–F355.
Collins, Randall. *Violence: A Microsociological Approach*. Princeton: Princeton University Press, 2008.
Collins, Randall. "Micro and Macro Causes of Violence." *International Journal of Conflict and Violence* 3, no. 1 (2009): 9–22.
Collins, Randall. "Micro and Macro Sociological Causes of Violent Atrocities." *Sociologia, Problemas e Práticas* 71 (2013): 9–22.
Connelly, Mark. "The British People, the Press and the Strategic Air Campaign Against Germany, 1939–45." *Contemporary British History* 16, no. 2 (2002): 39–58.

Converse, Philip Ernest. "The Nature of Belief Systems in Mass Publics." In *Ideology and Discontent*, edited by David Apter, Chapter 6. London: Free Press of Glencoe, 1964.
Costalli, Stefano, and Andrea Ruggeri. "Indignation, Ideologies, and Armed Mobilization: Civil War in Italy, 1943–45." *International Security* 40, no. 2 (2015): 119–157.
Costalli, Stefano, and Andrea Ruggeri. "Emotions, Ideologies, and Violent Political Mobilization." *PS: Political Science and Politics* 50, no. 4 (2017): 923–927.
Courtois, Stéphane, Nicolas Werth, Jean-Louis Panné, Andrzej Paczkowski, Karel Bartošek, and Jean-Louis Margolin. *The Black Book of Communism: Crimes, Terror, Repression*. Translated by Jonathan Murphy and Mark Kramer. Cambridge, MA: Harvard University Press, 1999.
Covington, Coline. "Hannah Arendt, Evil and the Eradication of Thought." *International Journal of Psychoanalysis* 93, no. 5 (2012): 1215–1236.
Covington, Coline. *Everyday Evils*. Abingdon: Routledge, 2017.
Covington, Coline, Paul Williams, Jean Arundale, and Jean Knox, eds. *Terrorism and War: Unconscious Dynamics of Political Violence*. London: Karnac Books, 2006.
Crawford, Neta. C. *Argument and Change in World Politics: Ethics, Decolonization and Humanitarian Intervention*. Cambridge: Cambridge University Press, 2002.
Crelinsten, Ronald D. "The World of Torture: A Constructed Reality." *Theoretical Criminology* 7, no. 3 (2003): 293–318.
Cromartie, Alan, ed. *Liberal Wars: Anglo-American Strategy, Ideology and Practice*. Abingdon: Routledge, 2015.
Dallaire, Roméo. *Shake Hands with the Devil: The Failure of Humanity in Rwanda*. London: Arrow Books, 2003.
Danning, Gordon. "Did Radio Rtlm Really Contribute Meaningfully to the Rwandan Genocide?: Using Qualitative Information to Improve Causal Inference from Measures of Media Availability." *Civil Wars* 20, no. 4 (2018): 529–554.
Danning, Gordon. "'It Ain't So Much the Things We Don't Know That Get Us in Trouble. It's the Things We Know That Ain't So': The Dubious Intellectual Foundations of the Claim That 'Hate Speech' Causes Political Violence." *Pepperdine Law Review* 98 (2019): 98–124.
Das, Veena. "Official Narratives, Rumour, and the Social Production of Hate." *Social Identities* 4, no. 1 (1998): 109–130.
Davenport, Christian. "State Repression and Political Order." *Annual Review of Political Science* 10, no. 1 (2007): 1–23.
David, Lea. "Human Rights as an Ideology? Obstacles and Benefits." *Critical Sociology* 46, no. 1 (2020): 37–50.
Davies, Lawrence William, Oleg V. Khlevniuk, and E. Arfon Rees, eds. *The Stalin–Kaganovich Correspondence, 1931–36*. New Haven: Yale University Press, 2003.
Davies, Sarah. *Popular Opinion in Stalin's Russia*. Cambridge: Cambridge University Press, 1997.
Davis Biddle, Tami. *Rhetoric and Reality in Air Warfare: The Evolution of British and American Ideas About Strategic Bombing, 1914–1945*. Princeton: Princeton University Press, 2002.
Davis, Shelton H. "Introduction: Showing the Seeds of Violence." In *Harvest of Violence: The Maya Indians and the Guatemala Crisis*, edited by Robert M. Carmack, 3–36. Norman: University of Oklahoma Press, 1988.
Dawidowicz, Lucy S. *The War Against the Jews 1933–45*. London: Penguin Books, 1987.

de Figueiredo, Rui J. P., and Barry R. Weingast. "The Rationality of Fear: Political Opportunism and Ethnic Conflict." In *Civil Wars, Insecurity, and Intervention*, edited by Barbara F. Walter and Jack Snyder, 261–302. New York: Columbia University Press, 1999.

Der Derian, James. "The Terrorist Discourse: Signs, States, and Systems of Global Political Violence." In *Critical Practices in International Theory*, edited by James Der Derian, Chapter 5. Abingdon: Routledge, 2009.

Des Forges, Alison. "The Ideology of Genocide." *Issue: A Journal of Opinion* 23, no. 2 (1995): 44–47.

Dieng, Adama, and Jennifer Welsh. "Assessing the Risk of Atrocity Crimes." *Genocide Studies and Prevention* 9, no. 3 (2016): 4–12.

Dill, Janina. *Legitimate Targets?: Social Construction, International Law and U.S. Bombing*. Cambridge: Cambridge University Press, 2014.

Dillon, Michael, and Julian Reid. *The Liberal Way of War: Killing to Make Life Live*. Abingdon: Routledge, 2009.

Doris, John M., and Dominic Murphy. "From My Lai to Abu Ghraib: The Moral Psychology of Atrocity." *Midwest Studies in Philosophy* 31 (2007): 25–55.

Douhet, Giulio. *The Command of the Air*. Translated by Dino Ferrari. Washington, DC: Office of Air Force History, 1921/1983.

Douven, Igor. "Abduction." Stanford Encyclopedia of Philosophy. https://plato.stanford.edu/entries/abduction/, 2011.

Downes, Alexander B. "Desperate Times, Desperate Measures: The Causes of Civilian Victimization in War." *International Security* 30, no. 4 (2006): 152–195.

Downes, Alexander B. *Targeting Civilians in War*. Ithaca, NY: Cornell University Press, 2008.

Downes, Alexander B., and Kathryn McNabb Cochran. "Targeting Civilians to Win? Assessing the Military Effectiveness of Civilian Victimization in Interstate War." In *Rethinking Violence: States and Non-State Actors in Conflict*, edited by Erica Chenoweth and Adria Lawrence, Chapter 2. Cambridge, MA: MIT Press, 2010.

Dragojević, Mila. *Amoral Communities: Collective Crimes in Times of War*. Ithaca, NY: Cornell University Press, 2019.

Drake, Charles J.M. "The Role of Ideology in Terrorists' Target Selection." *Terrorism and Political Violence* 10, no. 2 (1998): 53–85.

du Preez, Peter. *Genocide: The Psychology of Mass Murder*. London: Bowerdean and Boyars, 1994.

Dukalskis, Alexander, and Johannes Gerschewski. "Adapting or Freezing? Ideological Reactions of Communist Regimes to a Post-Communist World." Government and Opposition, Early Release Online Version (2018): 1–22.

Dumitru, Diana, and Carter Johnson. "Constructing Interethnic Conflict and Cooperation: Why Some People Harmed Jews and Others Helped Them During the Holocaust in Romania." *World Politics* 63, no. 1 (2011): 1–42.

Dutton, Donald G. *The Psychology of Genocide, Massacres and Extreme Violence: Why 'Normal' People Come to Commit Atrocities*. Westport, CT: Praeger Security International, 2007.

Dutton, Donald G., Ehor O. Boyanowsky, and Michael Harris Bond. "Extreme Mass Homicide: From Military Massacre to Genocide." *Aggression and Violent Behavior* 10 (2005): 437–473.

Eagleton, Terry. *Ideology: An Introduction*. London: Verso, 1991.

Earl, Jennifer. "Political Repression: Iron Fists, Velvet Gloves, and Diffuse Control." *Annual Review of Sociology* 37 (2011): 261–284.

Eastwood, James. "Rethinking Militarism as Ideology: The Critique of Violence After Security." *Security Dialogue* 49, no. 1-2 (2018): 44-56.

Eck, Kristine, and Lisa Hultman. "One-Sided Violence Against Civilians in War: Insights from New Fatality Data." *Journal of Peace Research* 44, no. 2 (2007): 233-246.

Edel, Mirjam, and Maria Josua. "How Authoritarian Rulers Seek to Legitimize Repression: Framing Mass Killings in Egypt and Uzbekistan." *Democratization* 25, no. 5 (2018): 882-900.

Edelman, Murray. *Political Language: Words That Succeed and Policies That Fail*. New York: Academic Press, 1977.

Edgerton, David. *England and the Aeroplane: Militarism, Modernity and Machines*. London: Penguin Books, 1991/2013.

Ehrenreich, Barbara. *Blood Rites: Origins and History of the Passions of War*. London: Virago Press, 1997.

Elshtain, Jean Bethke. *Women and War*. New York: Basic Books, 1987.

Elster, Jon. "Belief, Bias and Ideology." In *Rationality and Relativism*, edited by Martin Hollis and Steven Lukes, 123-148. Cambridge, MA: The MIT Press, 1982.

Elster, Jon. "Introduction." In *Rational Choice*, edited by Jon Elster, 1-33. New York: New York University Press, 1986.

Elster, Jon. "Norms." In *The Oxford Handbook of Analytical Sociology*, edited by Peter Hadström and Peter Bearman, Chapter 9. Oxford: Oxford University Press, 2009.

Elster, Jon. *Explaining Social Behavior: More Nuts and Bolts for the Social Sciences*. Revised ed. Cambridge: Cambridge University Press, 2015.

English, Robert D. "Power, Ideas, and New Evidence on the Cold War's End: A Reply to Brooks and Wohlforth." *International Security* 26, no. 4 (2002): 70-92.

Esteban, Joan, Massimo Morelli, and Dominic Rohner. "Strategic Mass Killings." *Journal of Political Economy* 123, no. 5 (2015): 1087-1132.

Fabre, Cécile. *Cosmopolitan War*. Oxford: Oxford University Press, 2012.

Fairclough, Norman. *Critical Discourse Analysis: The Critical Study of Language*. Harlow: Pearson, 2010.

Falla, Ricardo. *Massacres in the Jungle: Ixcán, Guatemala, 1975-1982*. Translated by Julia Howland. Boulder, CO: Westview Press, 1994.

Farr, Martin. "The Labour Party and Strategic Bombing in the Second World War." *Labour History Review* 77, no. 1 (2012): 133-153.

Fearon, James D., and David D. Laitin. "Violence and the Social Construction of Ethnic Identity." *International Organization* 54, no. 4 (2000): 845-877.

Fearon, James D., and David D. Laitin. "Ethnicity, Insurgency, and Civil War." *The American Political Science Review* 97, no. 1 (2003): 75-90.

Fein, Helen. *Accounting for Genocide: National Responses and Jewish Victimization During the Holocaust*. New York: Free Press, 1979.

Fein, Helen. *Genocide: A Sociological Perspective*. London: Sage Publications, 1990.

Fein, Helen. "Accounting for Genocide After 1945: Theories and Some Findings." *International Journal on Group Rights* 1 (1993): 79-106.

Fiala, Andrew, ed. *The Routledge Handbook of Pacifism and Nonviolence*. Abingdon: Routledge, 2018.

Figes, Orlando. *A People's Tragedy: The Russian Revolution, 1891-1924*. London: Pimlico, 1996.

Figes, Orlando. *The Whisperers: Private Life in Stalin's Russia*. London: Penguin Books, 2002.

Finer, Samuel. *The Man on Horseback: The Role of the Military in Politics*. New York: Routledge, 1962/2017.
Finnemore, Martha, and Kathryn Sikkink. "International Norm Dynamics and Political Change." *International Organization* 52, no. 4 (1998): 887–917.
Fischer, Lars. "The Meanings of Genocide." *The Political Quarterly* 77, no. 2 (2006): 295–299.
Fiske, Alan Page, and Tage Shakti Rai. *Virtuous Violence: Hurting and Killing to Create, Sustain, End, and Honor Social Relationships*. Cambridge: Cambridge University Press, 2014.
Fitzpatrick, Sheila. *Everyday Stalinism: Ordinary Life in Extraordinary Times: Soviet Russia in the 1930s*. Oxford: Oxford University Press, 1999.
Fitzpatrick, Sheila. "Ascribing Class: The Construction of Social Identity in Soviet Russia." In *Stalinism: New Directions*, edited by Sheila Fitzpatrick, 20–46. London: Routledge, 2000.
Fjelde, Hanne, and Lisa Hultman. "Weakening the Enemy: A Disaggregated Study of Violence Against Civilians in Africa." *Journal of Conflict Resolution* 58, no. 7 (2014): 1230–1257.
Fleming, Marie. "Genocide and the Body Politic in the Time of Modernity." In *The Specter of Genocide: Mass Murder in Historical Perspective*, edited by Robert Gellately and Ben Kiernan, 97–114. Cambridge: Cambridge University Press, 2003.
Forges, Alison Des. *"Leave None to Tell the Story": Genocide in Rwanda*. New York: Human Rights Watch, 1999.
Forges, Alison Des. "Call to Genocide: Radio in Rwanda, 1994." In *The Media and the Rwandan Genocide*, edited by Allan Thompson, 41–54. London: Pluto Press, 2007.
Foucault, Michel. *Discipline and Punish: The Birth of the Prison*. London: Penguin Books, 1977.
Foucault, Michel. "The Subject and Power." *Critical Inquiry* 8, no. 4 (1982): 777–795.
Fox, Nicole, and Nyseth Brehm Hollie. "'I Decided to Save Them': Factors That Shaped Participation in Rescue Efforts During Genocide in Rwanda." *Social Forces* 96, no. 4 (2018): 1625–1648.
Fox, Nicole, Nyseth Brehm Hollie, and John Gasana Gasasira. "The Impact of Religious Beliefs, Practices, and Social Networks on Rwandan Rescue Efforts During Genocide." *Genocide Studies and Prevention* 15, no. 1 (2021): 97–114.
Frazer, Elizabeth, and Kimberly Hutchings. *Can Political Violence Ever Be Justified?* Cambridge: Polity, 2019.
Frazer, Elizabeth, and Kimberly Hutchings. *Violence and Political Theory*. Cambridge: Polity Press, 2020.
Freeden, Michael. *Ideologies and Political Theory: A Conceptual Approach*. Oxford: Oxford University Press, 1996.
Freeden, Michael. "Confronting the Chimera of a 'Post-Ideological' Age." *Critical Review of International Social and Political Philosophy* 8, no. 2 (2005): 247–262.
Freeden, Michael. "Thinking Politically and Thinking About Politics: Language, Interpretation, and Ideology." In *Political Theory: Methods and Approaches*, edited by David Leopold and Marc Stears, 196–215. Oxford: Oxford University Press, 2008.
Freeden, Michael, Lyman Tower Sargent, and Marc Stears, eds. *The Oxford Handbook of Political Ideologies*. Oxford: Oxford University Press, 2013.
Fricker, Miranda. *Epistemic Injustice: Power & the Ethics of Knowing*. Oxford: Oxford University Press, 2009.
Fujii, Lee Ann. "Transforming the Moral Landscape: The Diffusion of a Genocidal Norm in Rwanda." *Journal of Genocide Research* 6, no. 1 (2004): 99–114.

Fujii, Lee Ann. "The Power of Local Ties: Popular Participation in the Rwandan Genocide." *Security Studies* 17, no. 3 (2008): 568–597.

Fujii, Lee Ann. *Killing Neighbors: Webs of Violence in Rwanda*. Ithaca, NY: Cornell University Press, 2009.

Fujii, Lee Ann. "Shades of Truth and Lies: Interpreting Testimonies of Peace and War." *Journal of Peace Research* 47, no. 2 (2010): 231–241.

Fujii, Lee Ann. "The Puzzle of 'Extra-Lethal Violence.'" *Perspectives on Politics* 11, no. 2 (2013): 410–426.

Fulbrook, Mary. *Dissonant Lives: Generations and Violence Through the German Dictatorships*. Oxford: Oxford University Press, 2011.

Furnham, Adrian. "Belief in a Just World: Research Progress Over the Past Decade." *Personality and Individual Differences* 34, no. 5 (2003): 795–817.

Gade, Emily Kalah, Mohammed M. Hafez, and Michael Gabbay. "Fratricide in Rebel Movements: A Network Analysis of Syrian Militant Infighting." Journal of Peace Research, Early View Online Version (2019): 1–15.

Gagnon, V.P. "Ethnic Nationalism and International Conflict: The Case of Serbia." *International Security* 19, no. 3 (1994/1995): 130–166.

Gagnon, Valère Philip. *The Myth of Ethnic War: Serbia and Croatia in the 1990s*. Ithaca, NY: Cornell University Press, 2004.

Garrard-Burnett, Virginia. *Terror in the Land of the Holy Spirit: Guatemala Under General Efrain Rios Montt, 1982–1983*. Oxford: Oxford University Press, 2009.

Gartzke, Erik, and Kristian Skrede Gleditsch. "Identity and Conflict: Ties That Bind and Differences That Divide." *European Journal of International Relations* 12, no. 1 (2006): 53–87.

Gault, William Barry. "Some Remarks on Slaughter." *American Journal of Psychiatry* 128, no. 4 (1971): 450–454.

Gause, F. Gregory. "Balancing What? Threat Perception and Alliance Choice in the Gulf." *Security Studies* 13, no. 2 (2003): 273–305.

Gawronski, Vincent T., and Richard Stuart Olson. "Disasters as Crisis Triggers for Critical Junctures? The 1976 Guatemala Case." *Latin American Politics and Society* 55, no. 2 (2013): 133–149.

Geertz, Clifford. "Ideology as a Cultural System." In *Ideology and Discontent*, edited by David Apter, 47–76. London: Free Press of Glencoe, 1964.

Gellately, Robert. "The Third Reich, the Holocaust, and Visions of Serial Genocide." In *The Specter of Genocide: Mass Murder in Historical Perspective*, edited by Robert Gellately and Ben Kiernan, 241–263. Cambridge: Cambridge University Press, 2003.

Gellately, Robert, and Ben Kiernan. "The Study of Mass Murder and Genocide." In *The Specter of Genocide: Mass Murder in Historical Perspective*, edited by Robert Gellately and Ben Kiernan, 3–26. Cambridge: Cambridge University Press, 2003.

Gentile, Gian Peri. "Advocacy or Assessment? The United States Strategic Bombing Survey of Germany and Japan." *Pacific Historical Review* 66, no. 1 (1997): 53–79.

George, Alexander L., and Andrew Bennett. *Case Studies and Theory Development in the Social Sciences*. Cambridge, MA: MIT Press, 2005.

Gerges, Fawad. *The Far Enemy: Why Jihad Went Global*. Cambridge: Cambridge University Press, 2009.

Gerlach, Christian. *Extremely Violent Societies: Mass Violence in the Twentieth-Century World*. Cambridge: Cambridge University Press, 2010.

Gerlach, Christian, and Nicolas Werth. "State Violence—Violent Societies." In *Beyond Totalitarianism: Stalinism and Nazism Compared*, edited by Michael Geyer and Sheila Fitzpatrick, Chapter 4. Cambridge: Cambridge University Press, 2009.

Gerring, John. "Ideology: A Definitional Analysis." *Political Research Quarterly* 50, no. 4 (1997): 957–994.

Gerring, John. "Review Article: The Mechanismic Worldview: Thinking Inside the Box." *British Journal of Political Science* 38 (2007): 161–179.

Gert, Bernard, and Joshua Gert. "The Definition of Morality." *Stanford Encyclopedia of Philosophy*. https://plato.stanford.edu/entries/morality-definition/, 2016.

Getachew, Adom. *Worldmaking After Empire: The Rise and Fall of Self-Determination*. Princeton: Princeton University Press, 2019.

Getty, J. Arch. *The Origins of the Great Purges: The Soviet Communist Party Reconsidered, 1933-1938*. New ed. Cambridge: Cambridge University Press, 1985/2008.

Getty, J. Arch. "Excesses Are Not Permitted': Mass Terror and Stalinist Governance in the Late 1930s." *The Russian Review* 61 (2002): 113–138.

Getty, John Arch, and Oleg V. Naumov. *The Road to Terror: Stalin and the Self-Destruction of the Bolsheviks, 1932-1939*. New Haven: Yale University Press, 1999.

Ghiglieri, Michael P. *The Dark Side of Man: Tracing the Origins of Male Violence*. New York: Helix Books/Basic Books, 1999.

Giddens, Anthony. *The Constitution of Society*. Oakland, CA: University of California Press, 1984.

Ginges, Jeremy, and Scott Atran. "War as a Moral Imperative (Not Just Practical Politics by Other Means)." *Proceedings of the Royal Society B* 278 (2011): 2930–2938.

Glazer, Amihai, and Bernard Grofman. "Why Representatives Are Ideologists Though Voters Are Not." *Public Choice* 61 (1989): 29–39.

Glover, Jonathan. *Humanity: A Moral History of the Twentieth Century*. New Haven: Yale University Press, 1999.

Goldhagen, Daniel. *Hitler's Willing Executioners: Ordinary Germans and the Holocaust*. London: Abacus, 1996.

Goldhagen, Daniel. *Worse Than War: Genocide, Eliminationism and the Ongoing Assault on Humanity*. London: Abacus, 2010.

Goldman, Wendy Z. *Inventing the Enemy: Denunciation and Terror in Stalin's Russia*. Cambridge: Cambridge University Press, 2011.

Goldstein, Jeffrey. "Emergence as a Construct: History and Issues." *Emergence* 1, no. 1 (1999): 49–72.

Gong, Geret. *The Standard of 'Civilization' in International Society*. Oxford: Oxford University Press, 1984.

Goodwin, Jeff. "'The Struggle Made Me a Nonracialist': Why There Was So Little Terrorism in the Anti-Apartheid Struggle." *Mobilization: An International Quarterly Review* 12, no. 2 (2007): 193–203.

Gordon, Gregory S. *Atrocity Speech Law: Foundation, Fragmentation, Fruition*. New York: Oxford University Press, 2017.

Gordy, Eric. *The Culture of Power in Serbia: Nationalism and the Destruction of Alternatives*. University Park: Pennsylvania State University Press, 1999.

Gould-Davies, Nigel. "Rethinking the Role of Ideology in International Politics During the Cold War." *Journal of Cold War Studies* 1, no. 1 (1999): 90–109.

Gourevitch, Philip. *We Wish to Inform You That Tomorrow We Will Be Killed with Our Families*. London: Picador, 1999.

Graham, Jesse, Jonathan Haidt, and Brian A. Nosek. "Liberals and Conservatives Rely on Different Sets of Moral Foundations." *Journal of Personality and Social Psychology* 96, no. 5 (2009): 1029–1046.
Grandin, Greg. *The Last Colonial Massacre: Latin America in the Cold War*. Chicago: The University of Chicago Press, 2011.
Grandin, Greg. "Politics by Other Means: Guatemala's Quiet Genocide." In *Quiet Genocide: Guatemala 1981–1983*, edited by Etelle Higonnet, 1–14. Abingdon: Routledge, 2017.
Granovetter, Mark. "Threshold Models of Collective Behavior." *American Journal of Sociology* 83, no. 6 (1978): 1420–1443.
Graziosi, Andrea. "Political Famines in the Ussr and China: A Comparative Analysis." *Journal of Cold War Studies* 19, no. 3 (2017): 42–103.
Gries, Peter Hays. *The Politics of American Foreign Policy: How Ideology Divides Liberals and Conservatives Over Foreign Affairs*. Stanford, CA: Stanford University Press, 2014.
Grossman, Dave. *On Killing: The Psychological Cost of Learning to Kill in War and Society*. New York: Back Bay Books, 2009.
Guichaoua, André. *From War to Genocide: Criminal Politics in Rwanda 1990–1994*. Translated by Don E. Webster. Madison: The University of Wisconsin Press, 2015.
Gupta, Davashree. "The Limits of Radicalization: Escalation and Restraint in the South African Liberation Movement." In *Dynamics of Political Violence: A Process-Oriented Perspective on Radicalization and the Escalation of Political Conflict*, edited by Lorenzo Bosi, Chares Demetriou, and Stefan Malthaner, Chapter 7. Farnham: Ashgate, 2014.
Gutiérrez Sanín, Francisco, and Elisabeth Jean Wood. "Ideology in Civil War: Instrumental Adoption and Beyond." *Journal of Peace Research* 51, no. 2 (2014): 213–226.
Haas, François. "German Science and Black Racism—Roots of the Nazi Holocaust." *The FASEB Journal* 22 (2008): 332–337.
Haas, Mark L. *The Ideological Origins of Great Power Politics, 1789–1989*. Ithaca, NY: Cornell University Press, 2005.
Haas, Mark L. *The Clash of Ideologies: Middle Eastern Politics and American Security*. Oxford: Oxford University Press, 2012.
Haas, Mark L. "Ideological Polarity and Balancing in Great Power Politics." *Security Studies* 23, no. 4 (2014): 715–753.
Hafez, Mohammed M. "The Curse of Cain: Why Fratricidal Jihadis Fail to Learn from Their Mistakes." *CTC Sentinel* (2017): 1–7.
Hafez, Mohammed M., Emily Kalah Gade, and Michael Gabbay. "Ideology in Civil Wars." In *Routledge Handbook of Ideology and International Relations*, edited by Mark L. Haas and Jonathan Leader Maynard. Abingdon: Routledge, 2022 (forthcoming).
Hafner-Burton, Emilie M., Stephen Haggard, David A. Lake, and David G. Victor. "The Behavioural Revolution and International Relations." *International Organization* 71, no. Supplement (2017): S1–S31.
Hagan, John, and Wenona Rymond-Richmond. "The Collective Dynamics of Racial Dehumanization and Genocidal Victimization in Darfur." *American Sociological Review* 73 (2008): 875–902.
Hagenloh, Paul. "'Socially Harmful Elements' and the Great Terror." In *Stalinism: New Directions*, edited by Sheila Fitzpatrick, 286–308. London: Routledge, 2000.
Hagenloh, Paul. *Stalin's Police: Public Order and Mass Repression in the USSR, 1926–1941*. Baltimore: Johns Hopkins University Press, 2009.
Haidt, Jonathan. *The Righteous Mind: Why Good People Are Divided by Religion and Politics*. London: Allen Lane, 2012.

Haidt, Jonathan. Nationalism Rising: When and Why Nationalism Beats Globalism. *The American Interest*, 7 October 2016. https://www.the-american-interest.com/2016/07/10/when-and-why-nationalism-beats-globalism/.

Haidt, Jonathan, Jesse Graham, and Craig Joseph. "Above and Below Left-Right: Ideological Narratives and Moral Foundations." *Psychological Inquiry* 20, no. 2-3 (2009): 110-119.

Halfin, Igal. *Stalinist Confessions: Messianism and Terror at the Leningrad Communist University*. Pittsburgh: University of Pittsburgh Press, 2009.

Hall, Norris F., Zechariah Chafee Jr., and Manley O. Hudson. *The Next War: Three Addresses Delivered at a Symposium at Harvard University*. Cambridge, MA: The Harvard Alumni Bulletin Press, 1925.

Hall, Stuart. "The Problem of Ideology: Marxism Without Guarantees." In *Stuart Hall: Critical Dialogues in Cultural Studies*, edited by David Morley and Kuan-Hsing Chen, 24-45. London: Routledge, 1996.

Hall, Todd H., and Andrew A.G. Ross. "Affective Politics After 9/11." *International Organization* 69, no. 4 (2015): 847-879.

Haller, John J., and Michael A. Hogg. "All Power to Our Great Leader: Political Leadership Under Uncertainty." In *Power, Politics and Paranoia: Why People Are Suspicious of Their Leaders*, edited by Jan-Willem van Prooijen and Paul Al. M. van Lange, 130-149. Cambridge: Cambridge University Press, 2014.

Hamilton, Malcolm B. "The Elements of the Concept of Ideology." *Political Studies* 35 (1987): 18-38.

Hammack, Philip L. "Narrative and the Cultural Psychology of Identity." *Personality and Social Psychology Review* 12, no. 3 (2008): 222-247.

Hardin, Russell. "The Crippled Epistemology of Extremism." In *Political Extremism and Rationality*, edited by Albert Breton, Gianluigi Galeotti, Pierre Salmon and Ronald Wintrope, 3-22. Cambridge: Cambridge University Press, 2002.

Hardwig, John. "Epistemic Dependence." *Journal of Philosophy* 82, no. 7 (1985): 335-349.

Harff, Barbara. "No Lessons Learned from the Holocaust? Assessing Risks of Genocide and Political Mass Murder Since 1955." *American Political Science Review* 97, no. 1 (2003): 57-73.

Harmer, Tanya. "The "Cuban Question" and the Cold War in Latin America, 1959-1964." *Journal of Cold War Studies* 21, no. 3 (2019): 114-151.

Harmon, Christopher C. *"Are We Beasts?": Churchill and the Moral Question of World War II "Area Bombing"*. The Newport Papers. Newport, RI: Naval War College, 1991.

Harris, James. *The Great Fear: Stalin's Terror of the 1930s*. Oxford: Oxford University Press, 2016.

Haslam, Nick. "Dehumanization: An Integrative Review." *Personality and Social Psychology Review* 10, no. 3 (2006): 252-264.

Hastings, Max. *Bomber Command*. London: Pan Macmillan, 2010.

Hatzfeld, Jean. *A Time for Machetes—the Rwandan Genocide: The Killers Speak*. Translated by Linda Coverdale. London: Serpent's Tail, 2005.

Hedin, Astrid. "Stalinism as Civilization: New Perspectives on Communist Regimes." *Political Studies Review* 2 (2004): 166-184.

Hedström, Peter, and Petri Ylikoski. "Causal Mechansims in the Social Sciences." *Annual Review of Sociology* 36 (2010): 49-67.

Hempel, Carl G. "The Function of General Laws in History." *The Journal of Philosophy* 39, no. 2 (1942): 35-48.

Herf, Jeffrey. *The Jewish Enemy: Nazi Propaganda During World War II and the Holocaust*. Cambridge, MA: Harvard University Press, 2006.

Hiebert, Maureen S. "The Three 'Switches' of Identity Construction in Genocide: The Nazi Final Solution and the Cambodian Killing Fields." *Genocide Studies and Prevention* 3, no. 1 (2008): 5–29.
Hiebert, Maureen S. *Constructing Genocide and Mass Violence: Society, Crisis, Identity*. Abingdon: Routledge, 2017.
Hilberg, Raul. *The Destruction of the European Jews*. Student ed. New York: Holmes & Meier, 1985.
Hintjens, Helen M. "Explaining the 1994 Genocide in Rwanda." *The Journal of Modern African Studies* 37, no. 2 (1999): 241–286.
Hinton, Alexander Laban. "Why Did You Kill?: The Cambodian Genocide and the Dark Side of Face and Honour." *The Journal of Asian Studies* 57 (1998), 93–122.
Hinton, Alexander Laban. "Introduction: Genocide and Anthropology." In *Genocide: An Anthropological Reader*, edited by Alexander Laban Hinton, 1–22. Malden, MA: Blackwell Publishers, 2002.
Hinton, Alexander Laban. *Why Did They Kill? Cambodia in the Shadow of Genocide*. Berkeley: University of California Press, 2005.
Hirschberger, Gilad, Tom Pyszczynski, and Tsachi Ein-Dor. "Why Does Existential Threat Promote Intergroup Violence? Examining the Role of Retributive Justice and Cost–Benefit Utility Motivations." *Frontiers in Psychology* 6 (2015): 1–9.
Hochschild, Jennifer L. "Where You Stand Depends on What You See: Connections Among Values, Perceptions of Fact, and Political Prescriptions." In *Citizens and Politics: Perspectives from Political Psychology*, edited by James H. Kuklinski, 313–340. Cambridge: Cambridge University Press, 2001.
Hodson, Gordon, Sarah M. Hogg, and Cara C. MacInnis. "The Role of 'Dark Personalities' (Narcissism, Machiavellianism, Psychopathy), Big Five Personality Factors, and Ideology in Explaining Prejudice." *Journal of Research in Personality* 43 (2009): 686–690.
Hodson, Randy, Dusko Sekulic, and Garth Massey. "National Tolerance in the Former Yugoslavia." *American Journal of Sociology* 99, no. 6 (1994): 1534–1558.
Hoess, Rudolf. *Commandant of Auschwitz*. London: Weidenfeld & Nicholson, 1959.
Hoffman, David L. *Stalinist Values: The Cultural Norms of Soviet Modernity, 1917–1941*. Ithaca, NY: Cornell University Press, 2003.
Hogg, Michael A. "From Uncertainty to Extremism: Social Categorization and Identity Processes." *Current Directions in Psychological Science* 23, no. 5 (2014): 338–342.
Holbrook, Donald, and John Horgan. "Terrorism and Ideology: Cracking the Nut." *Perspectives on Terrorism* 13, no. 6 (2019): 2–15.
Hollis, Martin, and Steve Smith. *Explaining and Understanding International Relations*. Oxford: Clarendon Press, 1990.
Holquist, Peter. *Making War, Forging Revolutions: Russia's Continuum of Crisis, 1914–1921*. Cambridge, MA: Harvard University Press, 2002.
Holquist, Peter. "Violent Russia, Deadly Marxism? Russia in the Epoch of Violence, 1905–21." *Kritika: Explorations in Russian and Eurasian History* 4, no. 3 (2003): 627–652.
Hoover Green, Amelia. "The Commander's Dilemma: Creating and Controlling Armed Group Violence." *Journal of Peace Research* 53, no. 5 (2016): 619–632.
Howard, Michael. *War and the Liberal Conscience*. Oxford: Oxford University Press, 1978.
Howard, Ross. *An Operational Framework for Media and Peacebuilding*. Vancouver, BC: Institute for Media, Policy and Civil Society, 2002.
Howarth, David, and Yannis Stavrakakis. "Introducing Discourse Theory and Political Analysis." In *Discourse Theory and Political Analysis: Identities, Hegemonies and Social*

Change, edited by David Howarth, Aletta Norval, and Yannis Stavrakakis, Chapter 1. Manchester: Manchester University Press, 2000.

Howarth, David, Aletta Norval, and Yannis Stavrakakis, eds. *Discourse Theory and Political Analysis: Identities, Hegemonies and Social Change*. Manchester: Manchester University Press, 2000.

Huddy, Leonie. "From Social to Political Identity: A Critical Examination of Social Identity Theory." *Political Psychology* 22, no. 1 (2001): 127–156.

Hudson, Valerie. "Foreign Policy Analysis: Actor-Specific Theory and the Ground of International Relations." *Foreign Policy Analysis* 1, no. 1 (2005): 1–30.

Hull, Isabel V. "Military Culture and the Production of 'Final Solutions' in the Colonies—the Example of Wilhelminian Germany." In *The Specter of Genocide: Mass Murder in Historical Perspective*, edited by Robert Gellately and Ben Kiernan, Chapter 7. Cambridge: Cambridge University Press, 2003.

Humphreys, Macartan, and Jeremy Weinstein. "Handling and Manhandling Civilians in Civil War." *American Political Science Review* 100, no. 3 (2006): 429–447.

Huntington, Samuel P. *The Soldier and the State: The Theory and Politics of Civil-Military Relations*. Cambridge, MA: Harvard University Press, 1957.

Huntington, Samuel P. *The Clash of Civilizations and the Remaking of World Order*. New York: Free Press, 2002.

Inkeles, Alex, and Raymond A. Bauer. *The Soviet Citizen*. Cambridge, MA: Harvard University Press, 1959.

International Panel of Eminent Personalities. *Rwanda: The Preventable Genocide*. Addis Ababa: African Union. https://www.refworld.org/docid/4d1da8752.html, 2000 (accessed 24 October 2020).

Jabri, Vivienne. *Discourses on Violence: Conflict Analysis Reconsidered*. Manchester: Manchester University Press, 1996.

Jackson, Karl D., ed. *Cambodia 1975–1978: Rendezvous with Death*. Princeton: Princeton University Press, 1989a.

Jackson, Karl D., ed. "The Ideology of Total Revolution." In *Cambodia 1975–1978: Rendezvous with Death*, edited by Karl D. Jackson, 37–78. Princeton: Princeton University Press, 1989b.

Jackson, Patrick Thaddeus. *The Conduct of Inquiry in International Relations: Philosophy of Science and Its Implications for the Study of World Politics*. Abingdon: Routledge, 2016.

Janis, Irving. *Groupthink*. Boston: Houghton Mifflin Company, 1982.

Jervis, Robert. *Perception and Misperception in International Politics*. Princeton: Princeton University Press, 1976.

Jervis, Robert. *System Effects: Complexity in Political and Social Life*. Princeton: Princeton University Press, 1997.

Jessee, Erin. *Negotiating Genocide in Rwanda: The Politics of History*. Cham: Palgrave Macmillan, 2017.

Johnson, Dominic. *Overconfidence and War: The Havoc and Glory of Positive Illusions*. Cambridge, MA: Harvard University Press, 2004.

Johnston, Alastair Iain. "Thinking About Strategic Culture." *International Security* 19, no. 4 (1995): 32–64.

Jost, John T. "The End of the End of Ideology." *American Psychologist* 61 (2006): 651–670.

Jost, John T. "Ideological Asymmetries and the Essence of Political Psychology." *Political Psychology* 38, no. 2 (2017): 167–207.

Jost, John T., and Orsolya Hunyady. "Antecedents and Consequences of System-Justifying Ideologies." *Current Directions in Psychological Science* 14, no. 5 (2005): 260–265.

Jost, John T., and Brenda Major, eds. *The Psychology of Legitimacy: Emerging Perspectives on Ideology, Justice and Intergroup Relations*. Cambridge: Cambridge University Press, 2001.

Jost, John T., Jack Glaser, Arie W. Kruglanski, and Frank J. Sulloway. "Political Conservatism as Motivated Social Cognition." *Psychological Bulletin* 129 (2003): 339–375.

Jost, John T., Christopher M. Federico, and Jaime Napier. "Political Ideology: Its Structure, Functions and Elective Affinities." *Annual Review of Psychology* 60 (2009): 307–337.

Kahneman, Daniel. *Thinking, Fast and Slow*. London: Penguin Books, 2012.

Kahneman, Daniel, and Amos Tversky. "Choices, Values, and Frames." *American Psychologist* 39, no. 4 (1984): 341–350.

Kaldor, Mary. *New and Old Wars: Organized Violence in a Global Era*. 3rd ed. Stanford: Stanford University Press, 2012.

Kallis, Aristotle A. *Genocide and Fascism: The Eliminationist Drive in Fascist Europe*. New York: Routledge, 2009.

Kallis, Aristotle A. "Race, 'Value' and the Hierarchy of Human Life: Ideological and Structural Determinants of National Socialist Policy-Making." *Journal of Genocide Research* 7, no. 1 (2005): 5–30.

Kalyvas, Stathis N. "Wanton and Senseless?: The Logic of Massacres in Algeria." *Rationality and Society* 11, no. 3 (1999): 243–285.

Kalyvas, Stathis N. "The Ontology of 'Political Violence': Action and Identity in Civil Wars." *Perspectives on Politics* 1, no. 3 (2003): 475–494.

Kalyvas, Stathis N. *The Logic of Violence in Civil War*. Cambridge: Cambridge University Press, 2006.

Kalyvas, Stathis N. "Conflict." In *Oxford Handbook of Analytical Sociology*, edited by Peter Hedström and Peter Bearman, 592–616. Oxford: Oxford University Press, 2009.

Kalyvas, Stathis N. "Micro-Level Studies of Violence in Civil War: Refining and Extending the Control-Collaboration Model." *Terrorism and Political Violence* 24 (2012): 658–668.

Kalyvas, Stathis N. "Is Isis a Revolutionary Group and If Yes, What Are the Implications?" *Perspectives on Terrorism* 9, no. 4 (2015): 42–47.

Kalyvas, Stathis N. "Jihadi Rebels in Civil War." *Daedalus* 147, no. 1 (2018): 36–47.

Kalyvas, Stathis N., and Laia Balcells. "International System and Technologies of Rebellion: How the End of the Cold War Shaped Internal Conflict." *American Political Science Review* 104, no. 3 (2010): 415–429.

Kaplan, Robert D. *Balkan Ghosts: A Journey Through History*. New York: Picador, 2005.

Karagiannis, Emmanuel, and Clark McCauley. "Hizb Ut-Tahrir Al-Islami: Evaluating the Threat Posed by a Radical Islamic Group That Remains Nonviolent." *Terrorism and Political Violence* 18, no. 2 (2006): 315–334.

Karstedt, Susanne. "Contextualizing Mass Atrocity Crimes: The Dynamics of 'Extremely Violent Societies.'" *European Journal of Criminology* 9, no. 5 (2012): 499–513.

Karstedt, Susanne, Nyseth Brehm Hollie, and Laura C. Frizzell. "Genocide, Mass Atrocity, and Theories of Crime: Unlocking Criminology's Potential." *Anual Review of Sociology* 4 (2021): 10.1–10.23.

Katzenstein, Peter, ed. *The Culture of National Security: Norms and Identity in World Politics*. Edited by Peter J. Katzenstein. New York: Columbia University Press, 1996a.

Katzenstein, Peter, ed. "Introduction: Alternative Perspectives on National Security." In *The Culture of National Security: Norms and Identity in World Politics*, edited by Peter J. Katzenstein, Chapter 1. New York: Columbia University Press, 1996b.

Kaufman, Stuart J. *Modern Hatreds: The Symbolic Politics of Ethnic War*. Ithaca, NY: Cornell University Press, 2001.

Kaufman, Stuart J. "Symbolic Politics or Rational Choice: Testing Theories of Extreme Ethnic Violence." *International Security* 30, no. 4 (2006): 45–86.

Kaufman, Stuart J. *Nationalist Passions*. Ithaca, NY: Cornell University Press, 2015.

Keen, Sam. *Faces of the Enemy: Reflections of the Hostile Imagination*. New York: Harper & Row, 1986.

Kellow, Christine L., and H. Leslie Steeves. "The Role of Radio in the Rwandan Genocide." *Journal of Communication* 48, no. 3 (1998): 107–128.

Kelman, Herbert C. "Compliance, Identification, and Internalization: Three Processes of Attitude Change." *Journal of Conflict Resolution* 2, no. 1 (1958): 51–60.

Kelman, Herbert C. "Violence Without Moral Restraint: Reflections on the Dehumanization of Victims and Victimizers." *Journal of Social Issues* 29, no. 4 (1973): 25–61.

Kelman, Herbert C. "Social-Psychological Dimensions of International Conflict." In *Peacemaking in International Conflict: Methods and Techniques*, edited by I. William Zartman, 61–107. Washington, DC: United States Institute of Peace, 2007.

Kelman, Herbert C., and V. Lee Hamilton. *Crimes of Obedience: Toward a Social Psychology of Authority and Responsibility*. New Haven: Yale University Press, 1989.

Kershaw, Ian. "'Working Towards the Führer.' Reflections on the Nature of the Hitler Dictatorship." *Contemporary European History* 2, no. 2 (1993): 103–118.

Kertzer, Joshua D. "Microfoundations in International Relations." *Conflict Management and Peace Science* 34, no. 1 (2017): 81–97.

Kiernan, Ben. "Twentieth-Century Genocides: Underlying Ideological Themes from Armenia to East Timor." In *The Specter of Genocide: Mass Murder in Historical Perspective*, edited by Robert Gellately and Ben Kiernan, 29–52. Cambridge: Cambridge University Press, 2003.

Kim, Dongsuk. "What Makes State Leaders Brutal? Examining Grievances and Mass Killing During Civil War." *Civil Wars* 12, no. 3 (2010): 237–260.

Kim, Nam Kyu. "Revolutionary Leaders and Mass Killing." *Journal of Conflict Resolution* 62, no. 2 (2018): 289–317.

Kimonyo, Jean-Paul. *Rwanda's Popular Genocide: A Perfect Storm*. Translated by Wandia Njoya. Boulder, CO: Lynne Rienner Publishers, 2016.

Kinder, Donald R., and Nathan P. Kalmoe. *Neither Liberal Nor Conservative: Ideological Innocence in the American Public*. Chicago: University of Chicago Press, 2017.

King, Charles. "The Micropolitics of Social Violence." *World Politics* 56 (2004): 431–455.

King, Charles. "Can There Be a Political Science of the Holocaust?" *Perspectives on Politics* 10, no. 2 (2012): 323–341.

King, Gary, Robert O. Keohane, and Sidney Verba. *Designing Social Inquiry: Scientific Inference in Qualitative Research*. Princeton: Princeton University Press, 1994.

Kirkpatrick, Jean. "Dictatorships and Double Standards." *Commentary* 68, no. 5 (1979): 34–45.

Klusemann, Stefan. "Micro-Situational Antecedents of Violent Atrocity." *Sociological Forum* 25, no. 2 (2010): 272–295.

Klusemann, Stefan. "Massacres as Process: A Micro-Sociological Theory of Internal Patterns of Mass Atrocities." *European Journal of Criminology* 9, no. 5 (2012): 468–480.

Knott, Kim, and Benjamin J. Lee. "Ideological Transmission in Extremist Contexts: Towards a Framework of How Ideas Are Shared." *Politics, Religion & Ideology* 21, no. 1 (2020): 1–23.

Kołakowski, Leszek. *Main Currents of Marxism*. New York: W.W. Norton & Company, 1978/2008.

Koonings, Kees, and Dirk Kruijt, eds. *Political Armies: The Military and Nation Building in the Age of Democracy*. London: Zed Books, 2002.

Kotkin, Stephen. *Magnetic Mountain: Stalinism as Civilization*. Berkeley, CA: University of California Press, 1995.

Kowert, Paul, and Jeffrey Legro. "Norms, Identity and Their Limits: A Theoretical Reprise." In *The Culture of National Security: Norms and Identity in World Politics*, edited by Peter J. Katzenstein, Chapter 12. New York: Columbia University Press, 1996.

Krain, Matthew. "State-Sponsored Mass Murder: The Onset and Severity of Genocides and Politicides." *Journal of Conflict Resolution* 41, no. 3 (1997): 331–360.

Kramer, Mark. "Ideology and the Cold War." *Review of International Studies* 25 (1999): 539–576.

Krcmaric, Daniel. "Varieties of Civil War and Mass Killing: Reassessing the Relationship Between Guerilla Warfare and Civilian Victimization." *Journal of Peace Research* 55, no. 1 (2018): 18–31.

Krebs, Ronald R. *Narrative and the Making of US National Security*. Cambridge: Cambridge University Press, 2015.

Krebs, Ronald R., and Patrick Thaddeus Jackson. "Twisting Tongues and Twisting Arms: The Power of Political Rhetoric." *European Journal of International Relations* 13, no. 1 (2007): 35–66.

Kressel, Neil J. *Mass Hate: The Global Rise of Genoicde and Terror*. Cambridge, MA: Westview Press, 2002.

Kruglanski, Arie W., and Edwad Orehek. "The Role of the Quest for Personal Significance in Motivating Terrorism." In *The Psychology of Social Conflict and Aggression*, edited by Joseph P. Forgas, Arie W. Kruglanski, and Kipling D. Williams, 153–164. New York: Psychology Press, 2011.

Kruit, Dirk. "Guatemala's Political Transitions, 1960s–1990s." *International Journal of Political Economy* 30, no. 1 (2000): 9–35.

Kubota, Yuichi. "Explaining State Violence in the Guatemalan Civil War: Rebel Threat and Counterinsurgency." *Latin American Politics and Society* 59, no. 3 (2017): 48–71.

Kühne, Thomas. "Male Bonding and Shame Culture: Hitler's Soldiers and the Moral Basis of Genocidal Warfare." In *Ordinary Peple as Mass Murderers: Perpetrators in Comparative Perspective*, edited by Olaf Jensen and Claus-Christian W. Szejnmann, 55–77. Houndmills: Palgrave Macmillan, 2008.

Kühne, Thomas. "Great Men and Large Numbers: Undertheorising a History of Mass Killing." *Contemporary European History* 21, no. 2 (2012): 133–143.

Kuper, Leo. *Genocide: Its Political Use in the Twentieth Century*. New Haven: Yale University Press, 1981.

Kuperman, Alan J. "Provoking Genocide: A Revised History of the Rwandan Patriotic Front." *Journal of Genocide Research* 6, no. 1 (2004): 61–84.

Kuran, Timur. "Sparks and Prairie Fires: A Theory of Unanticipated Political Revolution." *Public Choice* 61 (1989): 41–74.

Kuran, Timur. "Now Out of Never: The Element of Surprise in the East European Revolution of 1989." *World Politics* 44, no. 1 (1991): 7–48.

Laffey, Mark, and Jutta Weldes. "Beyond Belief: Ideas and Symbolic Technologies in the Study of International Relations." *European Journal of International Relations* 3, no. 2 (1997): 193–237.

Lakoff, George. *Moral Politics: How Liberals and Conservatives Think*. 2nd ed. Chicago: The University of Chicago Press, 2002.

Landau, Mark J., Sheldon Solomon, Jeff Greenberg, Florette Cohen, Tom Pyszczynski, Jamie Arndt, Claude H. Miller, Daniel M. Ogilvie, and Alison Cook. "Deliver Us from Evil: The Effects of Mortality Salience and Reminders of 9/11 on Support for President George W. Bush." *Personality and Social Psychology Bulletin* 30, no. 9 (2004): 1136–1150.

Lane, Robert E. *Political Ideology: Why the American Common Man Believes What He Does*. New York: The Free Press of Glencoe, 1962.

Lang, Johannes. "Questioning Dehumanization: Intersubjective Dimensions of Violence in the Nazi Concentration and Death Camps." *Holocaust and Genocide Studies* 24, no. 2 (2010): 225–246.

Lang, Johannes. "The Limited Importance of Dehumanization in Collective Violence." *Current Opinion in Psychology* 35 (2020): 17–20.

Lankford, Adam. *Human Killing Machines: Systematic Indoctrination in Irwan, Nazi Germany, Al Qaeda, and Abu Ghraib*. Lanham, MD: Lexington Books, 2009.

Larrain, Jorge. *The Concept of Ideology*. London: Hutchinson & Co., 1979.

Leader Maynard, Jonathan. "Rethinking the Role of Ideology in Mass Atrocities." *Terrorism and Political Violence* 26, no. 5 (2014): 821–841.

Leader Maynard, Jonathan. "Preventing Mass Atrocities: Ideological Strategies and Interventions." *Politics and Governance* 3, no. 3 (2015): 67–84.

Leader Maynard, Jonathan. "Ideologies, Identities and Speech in Atrocities." In *The Oxford Handbook of Atrocity Crimes*, edited by Barbara Hola, Nyseth Brehm Hollie, and Maartje Weerdesteijn, Chapter 9. Oxford: Oxford University Press, 2022.

Lee, Ronan. *Myanmar's Rohingya Genocide: Identity, History and Hate Speech*. London: I.B. Tauris, 2021.

Lefèvre, Raphaël. *Jihad in the City: Militant Islam and Contentious Politics in Tripoli*. Cambridge: Cambridge University Press, 2021.

Legro, Jeffrey W. *Cooperation Under Fire: Anglo-German Resrraint During World War II*. Ithaca, NY: Cornell University Press, 1995.

Legro, Jeffrey W. *Rethinking the World: Great Power Strategies and International Order*. Ithaca, NY: Cornell University Press, 2005.

Lemarchand, René. *Rwanda and Burundi*. London: Pall Mall Press, 1970.

Lemarchand, René. "Disconnecting the Threads: Rwanda and the Holocaust Reconsidered." *Journal of Genocide Research* 4, no. 4 (2002): 499–518.

Levene, Mark. *Genocide in the Age of the Nation State I: The Meaning of Genocide*. London: I.B. Tauris & Co., 2008.

Lewandowsky, Stephan, Ullrich K.H. Ecker, Colleen M. Seifert, Norbert Schwarz, and John Cook. "Misinformation and Its Correction: Continued Influence and Successful Debiasing." *Psychological Science in the Public Interest* 13, no. 3 (2012): 106–131.

Lewis, Bruce. *Aircrew: The Story of the Men Who Flew the Bombers*. London: Cassell, 1991.

Li, Darryl. "Echoes of Violence: Considerations on Radio and Genocide in Rwanda." *Journal of Genocide Research* 6, no. 1 (2004): 9–27.

Lickel, Brian. "Retribution and Revenge." In *The Oxford Handbook of Intergroup Conflict*, edited by Linda R. Tropp, 89–102. Oxford: Oxford University Press, 2012.

Lickel, Brian, Normal Miller, Douglas M. Stenrstrom, Thomas F. Denson, and Toni Schmader. "Vicarious Retribution: The Role of Collective Blame in Intergroup Aggression." *Personality and Social Psychology Review* 10, no. 4 (2006): 372–390.

Lih, Lars T., Oleg V. Naumov, and Oleg V. Khlevniuk, eds. *Stalin's Letters to Molotov: 1925–1936*. New Haven: Yale University Press, 1995.

Linz, Juan J. *Totalitarian and Authoritarian Regimes*. Boulder, CO: Lynne Rienner Publishers, 1975/2000.
Lipton, Peter. *Inference to the Best Explanation*. London: Routledge, 2004.
Littman, Rebecca, and Elizabeth Levy Paluck. "The Cycle of Violence: Understanding Individual Participation in Collective Violence." *Advances in Political Psychology* 36, 1 (2015): 79–99.
Long, Austin. *The Soul of Armies: Counterinsurgency Doctrine and Military Culture in the US and UK*. Ithaca, NY: Cornell University Press, 2016.
Longman, Timothy. "Placing Genocide in Context: Research Priorities for the Rwandan Genocide." *Journal of Genocide Research* 6, no. 1 (2004): 29–45.
Lukes, Steven. *Marxism and Morality*. Oxford: Clarendon Press, 1985.
Lukes, Steven. "On the Moral Blindness of Communism." *Human Rights Review* 2, no. 2 (2001): 113–124.
Lyons, Robert, and Scott Straus. *Intimate Enemy: Images and Voices of the Rwandan Genocide*. New York: Zone Books, 2006.
Maat, Eelco van der. "Genocidal Consolidation: Final Solution to Elite Rivalry." *International Organization* 74, no. 4 (2020): 773–809.
Mackie, J. L. "Causes and Conditions." *American Philosophical Quarterly* 2, no. 4 (1965): 245–264.
Madley, Benjamin. "Patterns of Frontier Genocide 1803–1910: The Aboriginal Tasmanians, the Yuki of California and the Herero of Namibia." *Journal of Genocide Research* 6, no. 2 (2004): 167–192.
Madley, Benjamin. "From Africa to Auschwitz." *European History Quarterly* 35 (2005): 429–464.
Madley, Benjamin. *An American Genocide: The United States and the California Indian Catastrophe*. New Haven: Yale University Press, 2016.
Mahoney, James. "Path Dependence in Historical Sociology." *Theory and Society* 29 (2000): 507–548.
Mahoney, James. "Process Tracing and Historical Explanation." *Security Studies* 24, no. 2 (2015): 200–218.
Mahoney, James, and Larkin Terrie. "Comparative-Historical Analysis in Contemporary Political Science." In *The Oxford Handbook of Political Methodology*, edited by Janet M. Box-Steffensmeier, Henry E. Brady, and David Collier, 3–38. Oxford: Oxford University Press, 2008.
Malešević, Siniša. *Identity as Ideology: Understanding Ethnicity and Nationalism*. Houndmills: Palgrave Macmillan, 2006.
Malešević, Siniša. *The Sociology of War and Violence*. Cambridge: Cambridge University Press, 2010.
Malia, Martin. "Foreword: The Uses of Atrocity." In *The Black Book of Communism: Crimes, Terorr, Repression*, edited by Stéphane Courtois, Nicolas Werth, Jean-Louis Panné, Andrzej Paczkowski, Karel Bartošek, and Jean-Louis Margolin, ix–xx. Cambridge, MA: Harvard University Press, 1999.
Mamdani, Mahmood. *When Victims Become Killers: Colonialism, Nativism, and the Genocide in Rwanda*. Kampala: Fountain Publishers, 2001.
Mani, Rama, and Thomas G. Weiss. "Can Culture Prevent Massacres?" *Global Governance* 17 (2011): 417–428.
Mann, Michael. "The Autonomous Power of the State: Its Origins, Mechanisms and Results." *European Journal of Sociology* 25, no. 2 (1984): 185–213.

Mann, Michael. *The Sources of Social Power*. Vol. 1. Cambridge: Cambridge University Press, 1986.
Mann, Michael. "Were the Perpetrators of Genocide 'Ordinary Men' or 'Real Nazis'? Results from Fifteen Hundred Biographies." *Holocaust and Genocide Studies* 14, no. 3 (2000): 331–366.
Mann, Michael. *Fascists*. Cambridge: Cambridge University Press, 2004.
Mann, Michael. *The Dark Side of Democracy: Explaining Ethnic Cleansing*. Cambridge: Cambridge University Press, 2005.
Manz, Beatriz. "The Transformation of La Esperanza, an Ixcán Village." In *Harvest of Violence: The Maya Indians and the Guatemala Crisis*, edited by Robert M. Carmack, 70–89. Norman: University of Oklahoma Press, 1988.
March, James G., and Johan P. Olsen. "The Logic of Appropriateness." In *The Oxford Handbook of Public Policy*, edited by Robert E. Goodin, Michael Moran, and Martin Rein, 689–708. Oxford: Oxford University Press, 2008.
Marchak, Patricia. *Reigns of Terror*. Montreal & Kingston: McGill-Queen's University Press, 2003.
Markusen, Eric, and David Kopf. *The Holocaust and Strategic Bombing: Genocide and Total War in the Twentieth Century*. Oxford: Westview Press, 1995.
Martin, Terry. "The Origins of Soviet Ethnic Cleansing." *The Journal of Modern History* 70 (1998): 813–861.
Mason, T. David, and Dale A. Krane. "The Political Economy of Death Squads: Toward a Theory of the Impact of State-Sanctioned Terror." *International Studies Quarterly* 33 (1989): 175–198.
Masullo, Juan. "Civilian Contention in Civil War: How Ideational Factors Shape Community Responses to Armed Groups." Comparative Political Studies, Online First Version (2020): 1–36.
Masullo, Juan. "Refusing to Cooperate with Armed Groups: Civilian Agency and Civilian Noncooperation in Armed Conflicts." International Studies Review, Online First Version (2021): 1–27.
Matthäus, Jürgen. "Controlled Escalation: Himmler's Men in the Summer of 1941 and the Holocaust in the Occupied Soviet Territories." *Holocaust and Genocide Studies* 21, no. 2 (2007): 218–242.
Mayer, Robert. "Strategies of Justification in Authoritarian Ideology." *Journal of Political Ideologies* 6, no. 2 (2001): 147–168.
McCarthy, Helen. "Democratizing British Foreign Policy: Rethinking the Peace Ballot, 1934–1935." *Journal of British Studies* 49, no. 2 (2010): 358–387.
McCauley, Clark, and Sophia Moskalenko. *Friction: How Radicalization Happens to Them and Us*. Oxford: Oxford University Press, 2011.
McCullough, Michael, Robert Kurzban, and Benjamin A. Tabak. "Evolved Mechanisms for Revenge and Forgiveness." In *Human Aggression and Violence: Causes, Manifestations, and Consequences*, edited by Philip R. Shaver and Mario Mikulincer, 221–239. Washington, DC: American Psychological Association., 2011.
McDermott, Rose. "The Feeling of Rationality: The Meaning of Neuroscientific Advances for Political Science." *Perspectives on Politics* 2, no. 4 (2004): 691–706.
McDoom, Omar Shahabudin. "The Psychology of Threat in Intergroup Conflict: Emotions, Rationality, and Opportunity in the Rwandan Genocide." *International Security* 37, no. 2 (2012): 119–155.

McDoom, Omar Shahabudin. "Who Killed in Rwanda's Genocide? Micro-Space, Social Influence and Individual Participation in Intergroup Violence." *Journal of Peace Research* 50, no. 4 (2013): 453–467.

McDoom, Omar Shahabudin. "Contested Counting: Toward a Rigorous Estimate of the Death Toll in the Rwandan Genocide." *Journal of Genocide Research* 22, no. 1 (2020a): 83–93.

McDoom, Omar Shahabudin. "Radicalization as Cause and Consequence of Violence in Genocide and Mass Killings." Violence: An International Journal 1, no.1 (2020b): 123-143

McDoom, Omar Shahabudin. *The Path to Genocide in Rwanda: Security, Opportunity, and Authority in an Ethnocratic State*. Cambridge: Cambridge University Press, 2021.

McKinney, Tiffany. "Radio Jamming: The Disarmament of Radio Propaganda." *Small Wars and Insurgencies* 13, no. 3 (2002): 111–144.

McLellan, David. *Ideology*. Buckingham: Open University Press, 1995.

McLellan, David, ed. *Karl Marx: Selected Writings*. Oxford: Oxford Universty Press, 2000.

McLoughlin, Barry, and Kevin McDermott. "Rethinking Stalinist Terror." In *Stalin's Terror: High Politics and Mass Repression in the Soviet Union*, edited by Barry McLoughlin and Kevin McDermott, 1–18. Houndmills: Palgrave Macmillan, 2003.

McLoughlin, Stephen. "The Role of Political Leaders in Mitigating the Risk of Mass Atrocities: An Analysis of Khama, Kaunda and Nyerere." *International Affairs* 96, no. 6 (2020): 1547–1564.

McMahan, Jeff. *Killing in War*. Oxford: Oxford University Press, 2009.

McMahon, Patrice. *Taming Ethnic Hatred: Ethnic Cooperation and Transnational Networks in Eastern Europe*. Syracuse: Syracuse University Press, 2007.

Meierhenrich, Jens. "How Many Victims Were There in the Rwandan Genocide? A Statistical Debate." *Journal of Genocide Research* 22, no. 1 (2020): 72–82.

Melson, Robert. *Revolution and Genocide: On the Origins of the Armenian Genocide and the Holocaust*. Chicago: University of Chicago Press, 1992.

Melvern, Linda. *Conspiracy to Murder: The Rwandan Genocide*. London: Verso, 2004.

Mercer, Jonathan. "Rationality and Psychology in International Politics." *International Organization* 59 (2005): 77–106.

Mercer, Jonathan. "Emotional Beliefs." *International Organization* 64 (2010): 1–31.

Messer, Robert L. "New Evidence on Truman's Decision." In *A History of Our Time: Readings on Postwar America*, edited by William H. Chafe and Harvard Sitkoff, 50–56. New York: Oxford University Press, 1995.

Messer, Robert L. "'Accidental Judgments, Casual Slaughters': Hiroshima, Nagasaki, and Total War." In *A World at Total War: Global Conflict and the Politics of Destruction 1937-1945*, edited by Roger Chickering, Stig Förster, and Bernd Greiner, Chapter 16. Cambridge: Cambridge University Press, 2005.

Metzl, Jamie Frederic. "Information Intervention: When Switching Channels Isn't Enough." *Foreign Affairs* 76, no. 6 (1997): 15–20.

Midlarsky, Manus. *The Killing Trap: Genocide in the Twentieth Century*. Cambridge: Cambridge University Press, 2005.

Midlarsky, Manus. "Territoriality and the Onset of Mass Violence: The Political Extremism of Joseph Stalin." *Journal of Genocide Research* 11, no. 2-3 (2009): 265–283.

Midlarsky, Manus. *Origins of Political Extremism: Mass Violence in the Twentieth Century and Beyond*. Cambridge: Cambridge University Press, 2011.

Milgram, Stanley. *Obedience to Authority: An Experimental View*. London: Pinter & Martin, 1974/2010.

Millar, Katharine M. "The Plural of Soldier Is Not Troops: The Politics of Groups in Legitimating Militaristic Violence." *Security Dialogue* 50, no. 3 (2019): 201–219.
Miller, Alice. *For Your Own Good: The Roots of Violence in Child-Rearing*. London: Virago Press, 1983.
Mills, Charles W. *Blackness Visible: Essays on Philosophy and Race*. Ithaca, NY: Cornell University Press, 2015.
Mirilovic, Nikola, and Myunghee Kim. "Ideology and Threat Perceptions: American Public Opinion Toward China and Iran." *Political Studies* 65, no. 1 (2017): 179–198.
Mironko, Charles. "*Igitero*: Means and Motive in the Rwandan Genocide." *Journal of Genocide Research* 6, no. 1 (2004): 47–60.
Mironko, Charles. "The Effect of RTLM's Rhetoric of Ethnic Hatred in Rural Rwanda." In *The Media and the Rwandan Genocide*, edited by Allan Thompson, 125–135. London: Pluto Press, 2007.
Mitchell, Neil J. *Agents of Atrocity: Leaders, Followers, and the Violation of Human Rights in Civil War*. New York: Palgrave Macmillan, 2004.
Mitton, Kieran. *Rebels in a Rotten State: Understanding Atrocity in Sierra Leone*. Oxford: Oxford University Press, 2015.
Mommsen, Hans. "Changing Historical Perspectives on the Nazi Dictatorship." *European Review* 17, no. 1 (2009): 73–80.
Monroe, Kristen Renwick. *Ethics in an Age of Terror and Genocide: Identity and Moral Choice*. Princeton: Princeton University Press, 2011.
Montalvo, Jose G., and Marta Reynal-Querol. "Discrete Polarisation with an Application to the Determinants of Genocide." *The Economic Journal* 118 (2008): 1835–1865.
Morgan, G. Scott, and Daniel C. Wisneski. "The Structure of Political Ideology Varies Between and Within People: Implications for Theories About Ideology's Causes." *Social Cognition* 35, no. 4 (2017): 395–414.
Morrow, Paul. "The Thesis of Norm Transformation in the Theory of Mass Atrocity." *Genocide Studies and Prevention* 9, no. 1 (2015): 66–82.
Morrow, Paul. *Unconscionable Crimes: How Norms Explain and Constrain Mass Atrocities*. Cambridge, MA: The MIT Press, 2020.
Moses, A. Dirk. "Structure and Agency in the Holocaust: Daniel J. Goldhagen and His Critics." *History and Theory* 37, no. 2 (1998): 194–219.
Moses, A. Dirk. *The Problem of Genocide: Permanent Security and the Language of Transgression*. Cambridge: Cambridge University Press, 2021.
Moshman, Daniel. "Us and Them: Identity and Genocide." *Identity* 7, no. 2 (2007): 115–135.
Mozur, Paul. A Genocide Incited on Facebook, with Posts from Myanmar's Military. *The New York Times*, 15 October 2018. https://www.nytimes.com/2018/10/15/technology/myanmar-facebook-genocide.html?nl=top-stories&nlid=21067275ries&ref=cta.
Mueller, John. "The Banality of 'Ethnic War.'" *International Security* 25, no. 1 (2000): 42–70.
Mugesera, Léon. Speech Made by Léon Mugesera at a Meeting of the MRND. 1992. https://nsarchive2.gwu.edu/NSAEBB/NSAEBB508/docs/Transcript%20Annex%20II%20Documents%20Referenced.pdf
Mullins, Willard. "On the Concept of Ideology in Political Science." *American Political Science Review* 66, no. 2 (1972): 498–510.
Murray, Elisabeth Hope. "Re-Evaluating Otherness in Genocidal Ideology." *Nations and Nationalism* 20, no. 1 (2014): 37–55.
Murray, Elisabeth Hope. *Disrupting Pathways to Genocide: The Process of Ideological Radicalization*. Houndmills: Palgrave Macmillan, 2015.

Mutua, Makau wa. "The Ideology of Human Rights." *Virginia Journal of International Law* 36, no. 3 (1996): 589–658.
Nabulsi, Karma. *Traditions of War: Occupation, Resistance, and the Law.* Oxford: Oxford University Press, 1999.
Naimark, Norman N. *Fires of Hatred: Ethnic Cleansing in Twntieth-Century Europe.* Cambridge, MA: Harvard University Press, 2001.
Naimark, Norman. *Stalin's Genocides.* Princeton: Princeton University Press, 2010.
National Foreign Assessment Center. *Guatemala: The Climate for Insurgency.* Langley: Central Intelligence Agency, 1981.
Ndlovu-Gatsheni, Sabelo J. *Coloniality of Power in Postcolonial Africa: Myths of Decolonization.* Dakar: CODESRIA, 2013.
Neilsen, Rhiannon S. "'Toxification' as a More Precise Early Warning Sign for Genocide Than Dehumanization? An Emerging Research Agenda." *Genocide Studies and Prevention* 9, no. 1 (2015): 83–95.
Neilsen, Rhiannon. "Cyber-Humanitarian Interventions: The Viability and Ethics of Using Cyber-Operations to Disrupt Perpetrators. Means and Motivations for Atrocities in the Digital Age." Doctor of Philosophy, University of New South Wales, 2021.
Nelson, Diane M. *A Finger in the Wound: Body Politics in Quincentennial Guatemala.* Berkeley, CA: University of California Press, 1999.
Neumann, Peter R. "The Trouble with Radicalization." *International Affairs* 89, no. 4 (2013): 873–893.
Newbury, Catherine. *The Cohesion of Oppression: Clientship and Ethicity in Rwanda, 1860–1960.* New York: Columbia University Press, 1988.
Newman, Leonard S. "What Is a 'Social-Psychological' Account of Perpetrator Behavior? The Person Versus the Situation in Goldhagen's Hitler's Willing Executioners." In *Understanding Genocide: The Social Psychology of the Holocaust* edited by Leonard S. Newman and Ralph Erber, 43–67. New York: Oxford University Press, 2002.
Newman, Leonard S., and Ralph Erber, eds. *Understanding Genocide: The Social Psychology of the Holocaust.* New York: Oxford University Press, 2002.
Noakes, Jeremy, and Geoffrey Pridham. *Nazism 1919–1945: A Documentary Reader.* 2nd ed. Vol. 3, Liverpool: Liverpool University Press, 2001.
Noelle-Neumann, Elisabeth. "The Spiral of Silence: A Theory of Public Opinion." *Journal of Communication* 24, no. 2 (1974): 43–51.
Norman, Ludvig. "Rethinking Causal Explanation in Interpretive International Studies." European Journal of International Relations, Online First Version (2021): 1–24.
Norris, Pippa, and Ronald Inglehart. *Cultural Backlash: Trump, Brexit and Authoritarian Populism.* Cambridge: Cambridge University Press, 2019.
Norval, Aletta. "The Things We Do with Words—Contemporary Approaches to the Analysis of Ideology." *British Journal of Political Science* 30, no. 2 (2000): 313–346.
Nove, Alec. "Victims of Stalinism: How Many?" In *Stalinist Terrror: New Perspectives,* edited by John Arch Getty and Roberta T. Manning, Chapter 13. Cambridge: Cambridge University Press, 1993.
Nyhan, Brendan, and Jason Reifler. "When Corrections Fail: The Persistence of Political Misperceptions." *Political Behavior* 32, no. 2 (2010): 303–330.
Nyseth Brehm, Holly. "State Context and Exclusionary Ideologies." *American Behavioral Scientist* 60, no. 2 (2016): 131–149.
Nyseth Brehm, Hollie. "Subnational Determinants of Killing in Rwanda." *Criminology* 55, no. 1 (2017): 5–31.

Nyseth Brehm, Hollie, Jared F. Edgerton, and Laura C. Frizzell. "Analyzing Participants in the 1994 Genocide in Rwanda." *Journal of Peace Research* (2022) (forthcoming).

Oberschall, Anthony. "The Manipulation of Ethnicity: From Ethnic Cooperation to Violence and War in Yugoslavia." *Ethnic and Racial Studies* 23, no. 6 (2000): 982–1001.

ODHAG. *Guatemala: Never Again! Recovery of the Historical Memory Project, the Official Report of the Human Rights Office, Archdiocese of Guatemala*. Maryknoll, NY: Orbis Books, 1999.

Oglesby, Elizabeth, and Diane M. Nelson. "Guatemala's Genocide Trial and the Nexus of Racism and Counterinsurgency." *Journal of Genocide Research* 18, no. 2-3 (2016): 133–142.

Oliver, Pamela E., and Hank Johnston. "What a Good Idea! Ideologies and Frames in Social Movement Research." *Mobilization: An International Quarterly Review* 4, no. 1 (2000): 37–54.

Olusanya, Olaoluwa. *Emotions, Decision-Making and Mass Atrocities: Through the Lens of the Macro-Micro Integrated Theoretical Model*. Farnham: Ashgate, 2014.

Opotow, Susan. "Moral Exclusion and Injustice: An Introduction." *Journal of Social Issues* 46, no. 1 (1990): 1–20.

Oppenheim, Ben, and Michael Weintraub. "Doctrine and Violence: The Impact of Combatant Training on Civilian Killings." *Terrorism and Political Violence* 29, no. 6 (2017): 1126–1148.

Orange, Richard. "Anders Behring Breivik Was Insane Five Years Ago, Mother Says." *The Telegraph*, 30 November 2011.

Orend, Brian. *The Morality of War*. Peterborough, ON: Broadview Press, 2006.

Osborn, Michelle. "Fuelling the Flames: Rumour and Politics in Kibera." *Journal of Eastern African Studies* 2, no. 2 (2008): 315–327.

O'Shaughnessy, Nicholas Jackson. *Politics and Propaganda*. Manchester: Manchester University Press, 2004.

Overy, Richard J. *The Dictators*. London: Allen Lane, 2004.

Overy, Richard J. *The Air War 1939–1945*. Washington, DC: Potomac Books, 2005a.

Overy, Richard J. "Allied Bombing and the Destruction of German Cities." In *A World at Total War: Global Conflict and the Politics of Destruction 1937–1945*, edited by Roger Chickering, Stig Förster, and Bernd Greiner, Chapter 15. Cambridge: Cambridge University Press, 2005b.

Overy, Richard J. "'The Weak Link'? The Perception of the German Working Class by Raf Bomber Command, 1940–1945." *Labour History Review* 77, no. 1 (2012): 11–33.

Overy, Richard J. "Pacifism and the Blitz, 1940–1941." *Past and Present* 219, no. 1 (2013): 201–236.

Overy, Richard J. *The Bombing War: Europe 1939–1945*. London: Penguin Books, 2014.

Owen, John M. *Liberal Peace, Liberal War: American Politics and International Security*. Ithaca, NY: Cornell University Press, 1997.

Owen, John M. *The Clash of Ideas in World Politics: Transnational Networks, States and Regime Change, 1510–2010*. Princeton, NJ: Princeton University Press, 2010.

Owen, Leah. "'A Terrible War of Defence': Examining the Role of Dehumanisation in Genocidal Mobilisation." DPhil in International Relations, University of Oxford, 2021.

Owens, Peter B., Yang Su, and David A. Snow. "Social Scientific Inquiry into Genocide and Mass Killing: From Unitary Outcome to Complex Processes." *Annual Review of Sociology* 39 (2013): 69–84.

Paluck, Elizabeth Levy. "Reducing Intergroup Prejudice and Conflict Using the Media: A Field Experiment in Rwanda." *Journal of Personality and Social Psychology* 96, no. 3 (2009): 574–587.
Paluck, Elizabeth Levy, and Donald P. Green. "Deference, Dissent, and Dispute Resolution: An Experimental Intervention Using Mass Media to Change Norms and Behavior in Rwanda." *American Political Science Review* 103, no. 4 (2009): 622–644.
Pape, Robert A. *Bombing to Win: Air Power and Coercion in War*. Ithaca, NY: Cornell University Press, 1996.
Parkinson, Sarah E. "Practical Ideology in Militant Organizations." *World Politics* 73, no. 1 (2021): 52–81.
Patterson, Molly, and Kristen Renwick Monroe. "Narrative in Political Science." *Annual Review of Political Science* 1 (1998): 315–331.
Pauer-Studer, Herlinde, and J. David Velleman. "Distortions of Normativity." *Ethical Theory and Moral Practice* 14, no. 3 (2011): 329–356.
Pauwels, Lieven J.R., and Wim Hardyns. "Endorsement for Extremism, Exposure to Extremism Via Social Media and Self-Reported Political/Religious Aggression." *International Journal of Developmental Science* 12, no. 1–2 (2018): 51–69.
Percival, Val, and Thomas Homer-Dixon. "Environmental Scarcity and Violent Conflict: The Case of Rwanda." *Journal of Environment & Development* 5, no. 3 (1996): 270–291.
Petersen, Roger D. *Understanding Ethnic Violence: Fear, Hatred, and Resentment in Twentieth-Century Eastern Europe*. Cambridge: Cambridge University Press, 2002.
Peterson, V. Spike. "Gendered Identities, Ideologies, and Practices in the Context of War and Militarism." In *Gender, War, and Militarism: Feminist Perspectives*, edited by Laura Sjoberg and Sandra Via, Chapter 1. Santa Barbara: Praeger, 2010.
Petrinovich, Lewis, and Patricia O'Neill. "Influence of Wording and Framing Effects on Moral Intuitions." *Ethology and Sociobiology* 17, no. 145–171 (1996).
Petrova, Maria, and David Yanagizawa-Drott. "Media Persuasion, Ethnic Hatred and Mass Violence." In *Economic Aspects of Genocides, Other Mass Atrocities, and Their Prevention*, edited by Charles H. Anderton and Jurgen Brauer, Chapter 12. Oxford: Oxford University Press, 2016.
Pierson, Paul. *Politics in Time: History, Institutions, and Social Analysis*. Princeton: Princeton University Press, 2004.
Pinker, Steven. *The Language Instinct*. London: Penguin Books, 1994.
Pinker, Steven. *The Better Angels of Our Nature: The Decline of Violence in History and Its Causes*. London: Allen Lane, 2011.
Pion-Berlin, David. "The National Security Doctrine, Military Threat Perception, and the 'Dirty War' in Argentina." *Comparative Political Studies* 21, no. 3 (1988): 382–407.
Plant, Roger. *Guatemala: Unnatural Disaster*. London: The Latin American Bureau, 1978.
Popper, Karl. *The Open Society and Its Enemies Vol. 2: Hegel and Marx*. Abingdon: Routledge, 1945/2003.
Popper, Karl. *Conjectures and Refutations: The Growth of Scientific Knowledge*. Abingdon: Routledge, 1963/2002.
Powell, Christopher J. *Barbaric Civilization: A Critical Sociology of Genocide*. Montreal: McGill-Queen's University Press, 2011.
Powell, Christopher J. "Genocidal Moralities: A Critique." In *New Directions in Genocide Research*, edited by Adam Jones, 37–54. Abingdon: Routledge, 2012.
Pratto, Felicia, Jim Sidanius, and Shana Levin. "Social Dominance Theory and the Dynamics of Intergroup Relations: Taking Stock and Looking Forward." *European Review of Social Psychology* 17 (2006): 271–320.

Price, Monroe E., and Mark Thompson, eds. *Forging Peace: Intervention, Human Rights and the Management of Media Space*. Bloomington: Indiana University Press, 2002.

Priestland, David. *Stalinism and the Politics of Mobilization: Ideas, Power, and Terror in Inter-War Russia*. Oxford: Oxford University Press, 2007.

Prunier, Gérard. *The Rwanda Crisis: History of a Genocide*. London: Hurst & Company, 1995.

Querido, Cyanda M. "State-Sponsored Mass Killing in African Wars—Greed or Grievance?" *International Advances in Economic Research* 15 (2009): 351–361.

Quine, W.V., and J.S. Ullian. *The Web of Belief*. New York: McGraw-Hill, 1978.

Rai, Tage S., Piercalo Valdesolo, and Jesse Graham. "Dehumanization Increases Instrumental Violence, But Not Moral Violence." *Proceedings of the National Academy of Sciences of the United States of America* 114, no. 32 (2017): 8511–8516.

Ramet, Sabrina P. "Review Article: Stalin Revisited—II." *Totalitarian Movements and Political Religions* 7, no. 4 (2006): 511–513.

Rathbun, Brian C., and Rachel Stein. "Greater Goods: Morality and Attitudes Toward the Use of Nuclear Weapons." *Journal of Conflict Resolution* 64, no. 5 (2020): 787–816.

Reicher, Stephen, S. Alexander Haslam, and Rakshi Rath. "Making a Virtue of Evil: A Five-Step Social Identity Model of the Development of Collective Hate." *Social and Personality Psychology Compass* 2, no. 3 (2008): 1313–1344.

Revkin, Mara, and Elisabeth Jean Wood. "The Islamic State's Pattern of Sexual Violence: Ideology and Institutions, Policies and Practices." Journal of Global Strategic Studies, Early Release Online (2020): 1–20.

Richter, Elihu D., Dror Kris Markus, and Casey Tait. "Incitement, Genocide, Genocidal Terror, and the Upstream Role of Indoctrination: Can Epidemiologic Models Predict and Prevent?" *Public Health Reviews* 39, no. 30 (2018): 1–22.

Rigterink, Anouk S., and Mareike Schomerus. "The Fear Factor Is a Main Thing: How Radio Influences Anxiety and Politial Attitudes." *The Journal of Development Studies* 53, no. 8 (2017): 1123–1146.

Risse-Kappen, Thomas, Stephen C. Ropp, and Kathryn Sikkink, eds. *The Power of Human Rights: International Norms and Domestic Change*. Cambridge: Cambridge University Press, 1999.

Rittersporn, Gábor Tamás. *Stalinist Simplifications and Soviet Complications: Social Tensions and Political Conflicts in the USSR, 1933–1953*. Chur, NY: Harwood Academic Publishers, 1991.

Rittersporn, Gábor T. "The Omnipresent Conspiracy: On Soviet Imagery of Politics and Social Relations in the 1930s." In *Stalinist Terrror: New Perspectives*, edited by John Arch Getty and Roberta T. Manning, 99–115. Cambridge: Cambridge University Press, 1993.

Robinson, Geoffrey. *"If You Leave Us Here, We Will Die": How Genocide Was Stopped in East Timor*. Princeton: Princeton University Press, 2010.

Robinson, Geoffrey. "'Down to the Very Roots': The Indonesian Army's Role in the Mass Killings of 1965–66." *Journal of Genocide Research* 19, no. 4 (2017): 465–486.

Robinson, Geoffrey. *The Killing Season: A History of the Indonesian Massacres, 1965–66*. Princeton: Princeton University Press, 2018.

Robinson, Neil. *Ideology and the Collapse of the Soviet System: A Critical History of Soviet Ideological Discourse*. Aldershot: Edward Elgar, 1995.

Rochat, François, and Andre Modigliani. "The Ordinary Quality of Resistance: From Milgram's Laboratory to the Village of Le Chambon." *Journal of Social Issues* 51, no. 3 (1995): 195–210.

Rodin, David, and Henry Shue. *Just and Unjust Warriors: The Moral and Legal Status of Soldiers*. Oxford: Oxford University Press, 2008.

Roemer, John E. "Rationalizing Revolutionary Ideology." *Econometrica* 53, no. 1 (1985): 85–108.

Roniger, Luis. "US Hemispheric Hegemony and the Descent into Genocidal Practices in Latin America." In *State Violence and Genocide in Latin America: The Cold War Years*, edited by Marcia Esparza, Henry R. Huttenbach, and Daniel Feierstein, Chapter 1. London: Routledge, 2010.

Roseman, Mark. "Beyond Conviction? Perpetrators, Ideas, and Action in the Holocaust in Historiographical Perspective." In *Conflict, Catastrophe, and Continuity: Essays on Modern German History*, edited by Frank Biess, Mark Roseman, and Hanna Schissler, Chapter 4. New York: Berghahn Books, 2007.

Rosen, Michael. "*On Voluntary Servitude* and the Theory of Ideology." *Constellations* 7, no. 3 (2000): 393–407.

Ross, Andrew A.G. *Mixed Emotions: Beyond Fear and Hatred in International Conflict*. Chicago: University of Chicago Press, 2014.

Roth, Paul A. "Hearts of Darkness: 'Perpetrator History' and Why There Is No Why." *History of the Human Sciences* 17, no. 2/3 (2004): 211–251.

Roth, Paul A. "Social Psychology and Genocide." In *The Oxford Handbook of Genocide Studies*, edited by Donald Bloxham and A. Dirk Moses, 198–216. Oxford: Oxford University Press, 2005.

Rothenberg, Daniel, ed. *Memory of Silence: The Guatemalan Truth Commission Report*. New York: Palgrave Macmillan, 2012.

Rousseau, David L. *Identifying Threats and Threatening Identities: The Social Construction of Realism and Liberalism*. Stanford, CA: Stanford University Press, 2006.

Rummel, Rudolph J. *Death by Government*. New Brunswick, NJ: Transactions Publishers, 1994.

Rummel, Rudolph J. "Democracy, Power, Genocide, and Mass Murder." *Journal of Conflict Resolution* 39, no. 1 (1995): 3–26.

Runciman, David. *The Politics of Good Intentions: History, Fear and Hypocrisy in the New World Order*. Princeton: Princeton University Press, 2009.

Russell, Edmund P. "'Speaking of Annihilation': Mobilizing for War Against Human and Insect Enemies, 1914–1945." *The Journal of American History* 82, no. 4 (1996): 1505–1529.

Russell, Luke. *Evil: A Philosophical Investigation*. Oxford: Oxford University Press, 2014.

Ryan, James. *Lenin's Terror: The Ideological Origins of Early Soviet State Violence*. Abingdon: Routledge, 2012.

Rydgren, Jens. "Beliefs." In *The Oxford Handbook of Analytical Sociology*, edited by Peter Hadström and Peter Bearman, 72–93. Oxford: Oxford University Press, 2009.

Saab, Rim, Russell Spears, Nicole Tausch, and Julia Sasse. "Predicting Aggressive Collective Action Based on the Efficacy of Peaceful and Aggressive Actions." *European Journal of Social Psychology* 46 (2016): 529–543.

Sagan, Scott D., and Benjamin A. Valentino. "Revisiting Hiroshima in Iran: What Americans Really Think About Using Nuclear Weapons and Killing Noncombatants." *International Security* 42, no. 1 (2017): 41–79.

Salazar, Egla Martínez. *Global Coloniality of Power in Guatemala: Racism, Genocide, Citizenship*. Lanhan: Lexington Books, 2012.

Salehyan, Idean, David Siroky, and Reed M. Wood. "External Rebel Sponsorship and Civilian Abuse: A Principal-Agent Analysis of Wartime Atrocities." *International Organization* 68, no. 3 (2014): 633–661.

Saucier, Gerard. "Isms Dimensions: Toward a More Comprehensive and Integrative Model of Belief-System Components." *Journal of Personality and Social Psychology* 104, no. 5 (2013): 921–929.

Saucier, Gerard, and Laura Akers. "Democidal Thinking: Patterns in the Mindset Behind Organized Mass Killing." *Genocide Studies and Prevention* 12, no. 1 (2018): 80–97.

Savage, Rowan. "Modern Genocidal Dehumanization: A New Model." *Patterns of Prejudice* 47, no. 2 (2013): 139–161.

Scanlon, Sandra. *The Pro-War Movement: Domestic Support for the Vietnam War and the Making of Modern American Conservatism*. Amherst: University of Massachusetts Press, 2013.

Schaff, Adam. "Marxist Theory on Revolution and Violence." *Journal of the History of Ideas* 34, no. 2 (1973): 263–270.

Schaffer, Ronald. *Wings of Judgement*. Oxford: Oxford University Press, 1988.

Schaffer, Ronald. "The Bombing Campaigns in World War II: The European Theatre." In *Bombing Civilians: A Twentieth-Century History*, edited by Yuki Tanaka and Marilyn B. Young, 30–45. New York: The New Press, 2009.

Scharpf, Adam. "Ideology and State Terror: How Officer Beliefs Shaped Repression During Argentina's 'Dirty War.'" *Journal of Peace Research* 55, no. 2 (2018): 206–221.

Scheffer, David. "Genocide and Atrocity Crimes." *Genocide Studies and Prevention* 1, no. 3 (2006): 229–250.

Schimmelfennig, Frank. "The Community Trap: Liberal Norms, Rhetorical Action, and the Eastern Enlargement of the European Union." *International Organization* 55, 1 (2001): 47–80.

Schirmer, Jennifer. *The Guatemalan Miliatry Project: A Violence Called Democracy*. Philadelphia: University of Pennsylvania Press, 1998.

Schissler, Matt. "Echo Chambers in Myanmar: Social Media and the Ideological Justifications for Mass Violence." Communal Conflict in Myanmar: Characteristics, Causes, Consequences, Yangon, Myanmar, 17–18 March 2014.

Schissler, Matthew, Matthew J. Walton, and Phyu Phyu Thi. *Threat and Virtuous Defence: Listening to Narratives of Religious Conflict in Six Myanmar Cities*. Oxford: Myanmar Media and Society Project (2015).

Schmidt, Sibylle. "Perpetrators' Knowledge: What and How Can We Learn from Perpetrator Testimony?" *Journal of Perpetrator Research* 1, no. 1 (2017): 85–104.

Schoemaker, Emrys, and Nicole Stremlau. "Media and Conflict: An Assessment of the Evidence." *Progress in Development Studies* 14, no. 2 (2014): 181–195.

Schubiger, Livia Isabella, and Matthew Zelina. "Ideology in Armed Groups." *PS: Political Science and Politics* 50, no. 4 (2017): 948–951.

Schull, Joseph. "What Is Ideology? Theoretical Problems and Lessons from Soviet-Type Societies." *Political Studies* 40 (1992): 728–741.

Schulzke, Marcus. *Pursuing Moral Warfare: Ethics in American, British, and Israeli Counterinsurgency*. Washington, DC: Georgetown University Press, 2019.

Schuurman, Bart, and Max Taylor. "Reconsidering Radicalization: Fanaticism and the Link Between Ideas and Violence." *Perspectives on Terrorism* 12, no. 1 (2018): 3–22.

Schwartz, Rachel A., and Scott Straus. "What Drives Violence Agianst Civilians in Civil War? Evidence from Guatemala's Conflict Archives." *Journal of Peace Research* 55, no. 2 (2018): 222–235.

Scott, James C. *Seeing Like a State: How Certain Schemes to Improve the Human Condition Have Failed*. New Haven: Yale University Press, 1998.

Scull, Nicholas C., Christophe D. Mbonyingabo, and Mayrian Kotb. "Transforming Ordinary People into Killers: A Psychosocial Examination of Hutu Participation in the Tutsi Genocide." *Peace and Conflict: Journal of Peace Psychology* 22, no. 4 (2016): 334–344.

Scutari, Jacqueline. *Hate Speech and Group-Targeted Violence: The Role of Speech in Violent Conflicts*. Washington, DC: United States Holocaust Memorial Museum, 2009.

Searle, Thomas R. "'It Made a Lot of Sense to Kill Skilled Workers': The Firebombing of Tokyo in March 1945." *The Journal of Military History* 66 (2002): 103–134.

Sebag Montefiore, Simon. *Stalin: The Court of the Red Tsar*. London: Phoenix, 2004.

Segal, Raz. "The Modern State, the Question of Genocide, and Holocaust Scholarship." *Journal of Genocide Research* 20, no. 1 (2018): 108–133.

Selden, Mark. "A Forgotten Holocaust: U.S. Bombing Strategy, the Destruction of Japanese Cities, and the American Way of War from the Pacific War to Iraq." In *Bombing Civilians: A Twentieth-Century History*, edited by Yuki Tanaka and Marilyn B. Young, Chapter 4. New York: The New Press, 2009.

Semelin, Jacques. *Unarmed Against Hitler: Civilian Resistance in Europe, 1939–1943*. Westport, CN: Praeger, 1993.

Semelin, Jacques. "Analysis of a Mass Crime: Ethnic Cleansing in the Former Yugoslavia, 1991–1999." In *The Specter of Genocide: Mass Murder in Historical Perspective*, edited by Robert Gellately and Ben Kiernan, 353–370. Cambridge: Cambridge University Press, 2003.

Semelin, Jacques. *Purify and Destroy: The Political Uses of Massacre and Genocide*. London: Hurst & Company, 2007.

Sewell Jr., William H. "The Concept(s) of Culture." In *Practicing History: New Directions in Historical Writing after the Linguistic Turn*, edited by Gabrielle M. Spiegel, 76–95. New York: Routledge, 2005.

Shapiro, Ian. *The Flight from Reality in the Human Sciences*. Princeton: Princeton University Press, 2005.

Sharma, Serena K., and Jennifer Welsh, eds. *The Responsibility to Prevent: Overcoming the Challenges of Atrocity Prevention*. Oxford: Oxford University Press, 2015.

Shaw, Martin. "Risk-Transfer Militarism, Small Massacres and the Historic Legitimacy of War." *International Relations* 16, no. 3 (2002): 343–359.

Shaw, Martin. *War and Genocide*. Cambridge: Polity Press, 2003.

Shaw, Martin. *What Is Genocide?* Cambridge: Polity Press, 2007.

Shearer, David R. *Policing Stalin's Socialism: Repression and Social Order in the Soviet Union, 1924–1953*. New Haven: Yale University Press, 2009.

Shearer, David R., and Vladimir Khaustov. *Stalin and the Lubianka: A Documentary History of the Political Police and Security Organs in the Soviet Union, 1922–1953*. New Haven: Yale University Press, 2015.

Sherry, Michael S. *The Rise of American Air Power: The Creation of Armageddon*. New Haven: Yale University Press, 1987.

Shorten, Richard. *Modernism and Totalitarianism*. Houndmills: Palgrave Macmillan, 2012.

Shorten, Richard. "How to Study Ideas in Politics and 'Influence': A Typology." *Contemporary Politics* 19, no. 4 (2013): 361–378.

Sibomana, André. *Hope for Rwanda: Conversations with Laure Guilbert and Hervé Deguine*. Translated by Carina Tertsakian. London: Pluto Press, 1999.

Sidanius, Jim, and Felicia Pratto. *Social Dominance: An Intergroup Theory of Hierarchy and Oppression*. Cambridge: Cambridge University Press, 1999.

Sikkink, Kathryn. *The Justice Cascade: How Human Rights Prosecutions Are Changing World Politics*. The Norton Series in World Politics. New York: WW Norton & Company, 2011.

Sil, Rudra, and Peter J. Katzenstein. "Analytic Eclecticism in the Study of World Politics: Reconfiguring Problems and Mechanisms Across Research Traditions." *Perspectives on Politics* 8, no. 2 (2010): 411–431.

Silber, Laura, and Allan Little. *Yugoslavia: Death of a Nation*. London: Penguin Books, 1997.

Simonds, A. P. "Ideological Domination and the Political Information Market." *Theory and Society* 18, no. 2 (1989): 181–211.

Sjoberg, Laura. "Gendering the Empire's Soldiers: Gender Ideologies, the U.S. Military and the 'War on Terror.'" In *Gender, War, and Militarism: Feminist Perspectives*, edited by Laura Sjoberg and Sandra Via, 209–218. Santa Barbara: Praeger, 2010.

Sjoberg, Laura. *Gendering Global Conflict: Toward a Feminist Theory of War*. New York: Columbia University Press, 2013.

Sjoberg, Laura, and Sandra Via. *Gender, War, and Militarism: Feminist Perspectives*. Santa Barbara: Praeger, 2010.

Skinner, Quentin. "History and Ideology in the English Revolution." *The Historical Journal* 8, no. 2 (1965): 151–178.

Skinner, Quentin. "The Ideological Context of Hobbes's Political Thought." *The Historical Journal* 9, no. 3 (1966a): 286–318.

Skinner, Quentin. "The Limits of Historical Explanations." *Philosophy* 41, no. 157 (1966b): 199–215.

Skinner, Quentin. "Meaning and Understanding in the History of Ideas." *History and Theory* 8, no. 1 (1969): 3–53.

Skinner, Quentin. "Some Problems in the Analysis of Political Thought and Action." *Political Theory* 2, no. 3 (1974): 277–303.

Skinner, Quentin. *The Foundations of Modern Political Thought*. Vol. 1. Cambridge: Cambridge University Press, 1978.

Skinner, Quentin. *Visions of Politics*. Vol. 1, Cambridge: Cambridge University Press, 2002.

Skocpol, Theda, and Margaret Somers. "The Uses of Comparative History in Macrosocial Inquiry." *Comparative Studies in Society and History* 22, no. 2 (1980): 174–197.

Slessor, John. *Air Power and Armies*. Tuscaloosa: University of Alabama Press, 2009.

Slim, Hugo. *Killing Civilians: Method, Madness and Morality in War*. London: Hurst & Company, 2007.

Slovic, Paul, David Zionts, Andrew K. Woods, Ryan Goodman, and Derek Jinks. "Psychic Numbing and Mass Atrocity." In *The Behavioural Foundations of Public Policy*, edited by Eldar Shafir, 126–142. Princeton: Princeton University Press, 2013.

Smeulers, Alette. "What Transforms Ordinary People Into Gross Human Rights Violators?" In *Understanding Human Rights Violations: New Systematic Studies*, edited by Sabine C. Carey and Steven C. Poe, 239–253. Aldershot: Ashgate Publishing, 2004.

Smeulers, Alette. "Perpetrators of International Crimes: Towards a Typology." In *Supranational Criminology: Towards a Criminology of International Crimes*, edited by A. Smeulers and R. Haveman, Chapter 10. Antwerpen: Intersentia, 2008.

Smith, Carol A. "Introduction: Social Relations in Guatemala Over Time and Space." In *Guatemalan Indians and the State, 1540–1988*, edited by Carol A. Smith, 1–30. Austin: University of Texas Press, 1990.

Smith, Christian. *Moral, Believing Animals*. Oxford: Oxford University Press, 2009.

Smith, David Livingstone. *Less Than Human: Why We Demean, Enslave and Exterminate Others*. New York: St. Martin's Griffin, 2011.

Smith, Steve. "The Contested Concept of Security." In *Critical Security Studies and World Politics*, edited by Ken Booth, Chapter 2. Boulder: Lynne Rienner, 2004.

Smith, Steve, Ken Booth, and Marysia Zalewski, eds. *International Theory: Positivism & Beyond*. Cambridge: Cambridge University Press, 1996.

Snow, David A. "Framing Processes, Ideology, and Discursive Fields." In *The Blackwell Companion to Social Movements*, edited by David A. Snow, Sarah A. Soule, and Hanspeter Kriesi, 380–412. Oxford: Blackwell Publishing, 2004.

Snow, David A., and Robert D. Benford. "Ideology, Frame Resonance, and Participant Mobilization." *International Social Movement Research* 1 (1988): 197–217.

Snow, David A., and Scott C. Byrd. "Ideology, Framing Processes, and Islamic Terrorist Movements." *Mobilization: An International Quarterly Review* 12, no. 1 (2007): 119–136.

Snyder, Jack. *From Voting to Violence: Democratization and Nationalist Conflict*. New York: W.W. Norton and Company, 2000.

Sofsky, Wolfgang. *Violence: Terrorism, Genocide, War*. London: Granta Books, 2002.

Solonari, Vladimir. *Purifying the Nation: Population Exchange and Ethnic Cleansing in Nazi-Allied Romania*. Washington, DC: Woodrow Wilson Center Press, 2010.

Souleimanov, Emil Aslan, and Huseyn Aliyev. "Blood Revenge and Violent Mobilization: Evidence from the Chechen Wars." *International Security* 40, no. 2 (2015): 158–180.

Stackelberg, Roderick, and Sally A. Winkle. *The Nazi Germany Sourcebook: An Anthology of Texts*. London: Routledge, 2002.

Staniland, Paul. "Militias, Ideology, and the State." *Journal of Conflict Resolution* 59, no. 5 (2015): 770–793.

Stanley, Jason. *How Propaganda Works*. Princeton: Princeton University Press, 2015.

Stanton, Gregory H. The Ten Stages of Genocide. http://genocidewatch.net/genocide-2/8-stages-of-genocide/, 2016 (accessed 8 July 2021).

Stanton, Jessica. *Violence and Restraint in Civil War: Civilian Targeting in the Shadow of International Law*. Cambridge: Cambridge University Press, 2016.

Star, Susan Leigh. "The Ethnography of Infrastructure." *American Behavioral Scientist* 43, no. 3 (1999): 377–391.

Starkov, Boris A. "Narkom Ezhov." In *Stalinist Terrror: New Perspectives*, edited by John Arch Getty and Roberta T. Manning, Chapter 1. Cambridge: Cambridge University Press, 1993.

Statiev, Alexander. "Soviet Ethnic Deportations: Intent Versus Outcome." *Journal of Genocide Research* 11, no. 2–3 (2009): 243–264.

Staub, Ervin. *The Roots of Evil*. Cambridge: Cambridge University Press, 1989.

Staub, Ervin. "The Roots of Evil: Social Conditions, Culture, Personality, and Basic Human Needs." *Personality and Social Psychology Review* 3, no. 3 (1999): 179–192.

Staub, Ervin. "The Psychology of Bystanders, Perpetrators, and Heroic Helpers." In *Understanding Genocide: The Social Psychology of the Holocaust*, edited by Leonard S. Newman and Ralph Erber, 291–324. New York: Oxford University Press, 2002.

Steele, Brent J. "Revenge, Affect, and Just War." In *Just War: Authority, Tradition, and Practice*, edited by Anthony F. Lang, Cian O'Driscoll, and John Williams, Chapter 11. Washington, DC: Georgetown University Press, 2013.

Stefano, Paul Di. "Understanding Rescuing During the Rwandan Genocide." *Peace Review: A Journal of Social Justice* 28, no. 2 (2016): 195–202.

Steger, Manfred B. "Political Ideologies in the Age of Globalization." In *The Oxford Handbook of Political Ideologies*, edited by Michael Freeden, Lyman Tower Sargent, and Marc Stears, 214–231. Oxford: Oxford University Press, 2013.

Steizinger, Johannes. "The Significance of Dehumanization: Nazi Ideology and Its Psychological Consequences." *Politics, Religion & Ideology* 19, no. 2 (2018): 139–157.

Stepanova, Ekaterina. *Terrorism in Asymmetrical Conflict: Ideological and Structural Aspects*. Sipri Research Reports. Oxford: Oxford University Press, 2008.

Stephan, Maria J., and Erica Chenoweth. "Why Civil Resistance Works: The Strategic Logic of Nonviolent Conflict." *International Security* 33, no. 1 (2008): 7–44.

Stoll, David. "Evangelicals, Guerrillas, and the Army: The Ixil Triangle Under Ríos Montt." In *Harvest of Violence: The Maya Indians and the Guatemala Crisis*, edited by Robert M. Carmack, 90–116. Norman: University of Oklahoma Press, 1988.

Stoll, David. *Between Two Armies in the Ixil Towns of Guatemala*. New York: Columbia University Press, 1993.

Stoltzfus, Nathan. *Resistance of the Heart: Intermarriage and the Rosenstrasse Protest in Nazi Germany*. New Brunswick, NJ: Rutgers University Press, 2001.

Stouffer, Samuel A., Arthur A. Lumsdaine, Marion Harper Lumsdaine, Jr., Robin M. Williams, Brewster Smith, Irving L. Janis, Shirley A. Star, and Leonard S. Cottrell Jr. *The American Soldier: Combat and Its Aftermath*. Studies in Social Psychology in World War II. Vol. 2. Princeton: Princeton University Press, 1949.

Straus, Scott. "How Many Perpetrators Were There in the Rwandan Genocide? An Estimate." *Journal of Genocide Research* 6, no. 1 (2004): 85–98.

Straus, Scott. *The Order of Genocide: Race, Power and War in Rwanda*. Ithaca, NY: Cornell University Press, 2006.

Straus, Scott. "Second-Generation Comparative Research on Genocide." *World Politics* 59, no. 3 (2007a): 476–501.

Straus, Scott. "What Is the Relationship Between Hate Radio and Violence? Rethinking Rwanda's 'Radio Machete.'" *Politics and Society* 35 (2007b): 609–637.

Straus, Scott. "'Destroy Them to Save Us': Theories of Genocide and the Logics of Political Violence." *Terrorism and Political Violence* 24, no. 4 (2012a): 544–560.

Straus, Scott. "Retreating from the Brink: Theorizing Mass Violence and the Dynamics of Restraint." *Perspectives on Politics* 10 (2012b): 342–362.

Straus, Scott. *Making and Unmaking Nations: War, Leadership and Genocide in Modern Africa*. Ithaca, NY: Cornell University Press, 2015a.

Straus, Scott. "Triggers of Mass Atrocities." *Politics and Governance* 3, no. 3 (2015b): 5–15.

Straus, Scott. *Fundamentals of Genocide and Mass Atrocity Prevention*. Washington, DC: United States Holocaust Memorial Museum, 2016.

Straus, Scott. "The Limits of a Genocide Lens: Violence Against Rwandans in the 1990s." *Journal of Genocide Research* 21, no. 4 (2019): 504–524.

Su, Yang. *Collective Killings in Rural China During the Cultural Revolution*. Cambridge: Cambridge University Press, 2011.

Sullivan, Christopher Michael. "Blood in the Village: A Local-Level Investigation of State Massacres." *Conflict Management and Peace Science* 29, 4 (2012): 373–396.

Sunstein, Cass R. "Cognition and Cost–Benefit Analysis." *Journal of Legal Studies* 29 (2000): 1059–1103.

Suny, Ronald Grigory. "Why We Hate You: The Passions of National Identity and Ethnic Violence." *Berkeley Program in Societ and Post-Soviet Studies Working Paper Series*. Berkeley: University of California, 2004.

Svolik, Milan W. *The Politics of Authoritarian Rule*. Cambridge: Cambridge University Press, 2012.

Swain, Geoffrey. *The Origins of the Russian Civil War*. London: Longman, 1996.

Swidler, Ann. "What Anchors Cultural Practices." In *The Practice Turn in Contemporary Theory*, edited by Theordore R. Schatzki, Karin Knorr Cetina, and Eike von Savigny, 74–92. London: Routledge, 2001.

Swire, Briony, Adam J. Berinsky, Stephan Lewandowsky, and Ullrich K.H. Ecker. "Processing Political Misinformation: Comprehending the Trump Phenomenon." *Royal Society Open Science* 4, no. 3 (2017): 1–21.

Szejnmann, Claus-Christian W. "Perpetrators of the Holocaust: A Historiography." In *Ordinary Peple as Mass Murderers: Perpetrators in Comparative Perspective*, edited by Olaf Jensen and Claus-Christian W. Szejnmann, 25–54. Houndmills: Palgrave Macmillan, 2008.

Tanaka, Yuki. "Introduction." In *Bombing Civilians: A Twentieth-Century History*, edited by Yuki Tanaka and Marilyn B. Young, 1–7. New York: The New Press, 2009a.

Tanaka, Yuki. "British 'Humane Bombing' in Iraq During the Interwar Era." In *Bombing Civilians: A Twentieth-Century History*, edited by Yuki Tanaka and Marilyn B. Young, Chapter 1. New York: The New Press, 2009b.

Tannenwald, Nina. "The Nuclear Taboo: The United States and the Normative Basis of Nuclear Non-Use." *International Organization* 53, no. 3 (1999): 433–468.

Tannenwald, Nina. "Ideas and Explanations: Advancing the Theoretical Agenda." *Journal of Cold War Studies* 7, no. 2 (2005): 13–42.

Tausch, Nicole, Julia C. Becker, Russell Spears, Oliver Christ, Rim Saab, Purnima Singh, and Roomana N. Siddiqui. "Explaining Radical Group Behavior: Developing Emotion and Efficacy Routes to Normative and Nonnormative Collective Action." *Journal of Personality and Social Psychology* 101, no. 1 (2011): 129–148.

Taylor, Christopher C. *Sacrifice as Terror: The Rwandan Genocide of 1994*. Oxford: Berg, 1999.

Taylor, Kathleen E. "Intergroup Atrocities in War: A Neuroscientific Perspective." *Medicine, Conflict and Survival* 22, no. 3 (2006): 230–244.

Terkel, Studs. *"The Good War": An Oral History of World War Two*. New York: Ballantine Books, 1984.

Tesón, Fernando R. "The Liberal Case for Humanitarian Intervention." In *Humanitarian Intervention: Ethical, Legal and Political Dilemmas*, edited by J.L. Holzgrefe and Robert O. Keohane, 93–129. Cambridge: Cambridge University Press, 2003.

Tetlock, Philip E., and A.S.R. Manstead. "Impression Management Versus Intrapsychic Explanations in Social Psychology: A Useful Dichotomy?" *Psychological Review* 92, no. 1 (1985): 59–77.

Tezcür, Güneş Murat. "Ordinary People, Extraordinary Risks: Participation in an Ethnic Rebellion." *American Political Science Review* 110, no. 2 (2016): 247–264.

Thagard, Paul. "The Cognitive-Affective Structure of Political Ideologies." In *Emotion in Group Decision and Negotiation*, edited by Bilyana Martinovski, 51–72. Berlin: Springer, 2014.

Thaler, Kai M. "Ideology and Violence in Civil Wars: Theory and Evidence from Mozambique and Angola." *Civil Wars* 14, no. 4 (2012): 546–567.

"The Lighter Side." *Nottingham Evening Post*, 2 May 1941.

Theweleit, Klaus. *Male Fantasies*. Vol. 1. Cambridge: Polity Press, 1989.

Thomas, Ward. *The Ethics of Destruction: Norms and Force in International Relations*. Ithaca, NY: Cornell University Press, 2001.

Thompson, Allan, ed. *The Media and the Rwandan Genocide*. London: Pluto Press, 2007.

Thompson, John. *Studies in the Theory of Ideology*. Cambridge: Polity Press, 1984.

Thompson, Mark, and Monroe E. Price. "Intervention, Media and Human Rights." *Survival* 45, no. 1 (2003): 183–202.
Thompson, Robin L. "Radicalization and the Use of Social Media." *Journal of Strategic Security* 4, no. 4 (2011): 167–190.
Thorisdottir, Hulda, John T. Jost, and Aaron C. Kay. "On the Social and Psychological Bases of Ideology and System Justification." In *Social and Psychological Bases of Ideology and System Justification*, edited by John T. Jost, Aaron C. Kay, and Hulda Thorisdottir, Chapter 1. New York: Oxford University Press, 2009.
Thurston, Alexander. "Algeria's Gia: The First Major Armed Group to Fully Subordinate Jihadism to Salafism." *Islamic Law and Society* 24 (2017): 412–436.
Thurston, Robert W. "Fear and Belief in the Ussr's 'Great Terror': Response to Arrest, 1935–1939." *Slavic Review* 45, no. 2 (1986): 213–234.
Tileagă, Christian. "Ideologies of Moral Exclusion: A Critical Discursive Reframing of Depersonalization, Delegitimization and Dehumanization." *British Journal of Social Psychology* 46, no. 4 (2007): 717–737.
Tilly, Charles, and Robert E. Goodin. "It Depends." In *The Oxford Handbook of Contextual Political Analysis*, edited by Robert E. Goodin and Charles Tilly, Chapter 1. Oxford: Oxford University Press, 2006.
Timmermann, Wibke Kristin. "The Relationship Between Hate Propaganda and Incitement to Genocide: A New Trend in International Law Towards Criminalization of Hate Propaganda?" *Leiden Journal of International Law* 18, no. 2 (2005): 257–282.
Tomasello, Michael. *The Cultural Origins of Human Cognition*. Cambridge, MA: Harvard University Press, 1999.
Tomuschat, Christian. "Foreword." In *Memory of Silence: The Guatemalan Truth Commission Report*, edited by Daniel Rothenberg, xv–xviii. New York: Palgrave Macmillan, 2012.
Torrey, Norman Lewis. *Les Philosophes. The Philosophers of the Enlightenment and Modern Democracy*. Oakville, Ontario: Capricorn Books, 1961.
Trevor-Roper, Hugh. *The Bormann Letters*. London: Weidenfeld & Nicolson, 1954.
Tschantret, Joshua. "Revolutionary Homophobia: Explaining State Repression Against Sexual Minorities." British Journal of Political Science, Early Release Online Version (2019): 1–22.
Tsintsadze-Maass, Eteri, and Richard W. Maass. "Groupthink and Terrorist Radicalization." *Terrorism and Political Violence* 26, no. 5 (2014): 735–758.
Tuckwood, Christopher. "From Real Friend to Imagined Foe: The Medieval Roots of Anti-Semitism as a Precondition for the Holocaust." *Genocide Studies and Prevention* 5, no. 1 (2010): 89–105.
Tully, James H. "The Pen Is a Mighty Sword: Quentin Skinner's Analysis of Politics." *British Journal of Political Science* 13, no. 4 (1983): 489–509.
Turse, Nick. *Kill Anything That Moves: The Real American War in Vietnam* New York: Picador, 2013.
Ugarriza, Juan E. "Ideologies and Conflict in the Post-Cold War." *International Journal of Conflict Management* 20, no. 1 (2009): 82–104.
Ugarriza, Juan E., and Matthew J. Craig. "The Relevance of Ideology to Contemporary Armed Conflicts: A Quantitative Analysis of Former Combatants in Colombia." *Journal of Conflict Resolution* 57, no. 3 (2013): 445–477.
Ulfelder, Jay, and Benjamin A. Valentino. *Assessing Risks of State-Sponsored Mass Killing*. Washington, DC: Political Instability Task Force, 2008. https://ssrn.com/abstract=1703426.

United Nations. *Framework of Analysis for Atrocity Crimes: A Tool for Prevention*. New York: United Nations Office on Genocide Prevention and the Responsibility to Protect, 2014.

Ussishkin, Daniel. *Morale: A Modern British Histroy*. Oxford: Oxford University Press, 2017.

Uvin, Peter. "Tragedy in Rwanda: The Political Ecology of Conflict." *Environment* 38, no. 3 (1996): 7–29.

Uvin, Peter. "Prejudice, Crisis, and Genocide in Rwanda." *African Studies Review* 40, no. 2 (1997): 91–115.

Uvin, Peter. *Aiding Violence: The Development Enterprise in Rwanda*. West Hartford, CT: Kamarian Press, 1998.

Uvin, Peter. "Reading the Rwandan Genocide." *International Studies Review* 3, no. 3 (2001): 75–99.

Uzonyi, Gary. "Interstate Rivalry, Genocide, and Politicide." *Journal of Peace Research* 55, no. 4 (2018): 476–490.

Uzonyi, Gary, and Burak Demir. "Excluded Ethnic Groups, Conflict Contagion, and the Onset of Genocide and Politicide During Civil War." International Studies Quarterly, Online First Version (2020): 1–10.

Valentino, Benjamin A. *Final Solutions: Mass Killing and Genocide in the 20th Century*. Ithaca, NY: Cornell University Press, 2004.

Valentino, Benjamin A. "Why We Kill: The Political Science of Political Violence Against Civilians." *Annual Review of Political Science* 17, no. 89–103 (2014).

Valentino, Benjamin A., Paul Huth, and Dylan Balch-Lindsay. "'Draining the Sea': Mass Killing and Guerrilla Warfare." *International Organization* 58, no. 2 (2004): 375–407.

van Dijk, Teun. "Ideological Discourse Analysis." *New Courant* 4 (1995): 135–161.

van Dijk, Teun. *Ideology: A Multidisciplinary Approach*. London: Sage Books, 1998.

van Ree, Erik. "Stalin's Organic Theory of the Party." *The Russian Review* 52 (1993): 43–57.

Vatlin, Alexander. *Agents of Terror: Ordinary Men and Extraordinary Violence in Stalin's Secret Police*. Edited by Seth Bernstein. Madison: University of Wisconsin Press, 2016.

Vela Castañeda, Manolo E. "Perpetrators: Specialization, Willingness, Group Pressure and Incentives. Lessons from the Guatemalan Acts of Genocide." *Journal of Genocide Research* 18, no. 2–3 (2016): 225–244.

Verdeja, Ernesto. "The Political Science of Genocide: Outlines of an Emerging Research Agenda." *Perspectives on Politics* 10, no. 2 (2012): 307–321.

Verpoorten, Marijke. "How Many Died in Rwanda?" *Journal of Genocide Research* 22, no. 1 (2020): 94–103.

Verwimp, Philip. "The 1990–92 Massacres in Rwanda: A Case of Spatial and Social Engineering?" *Journal of Agrarian Change* 11, no. 3 (2011): 396–419.

Vetlesen, Arne Johan. *Evil and Human Agency: Understanding Collective Evildoing*. Cambridge: Cambridge University Press, 2005.

Viola, Lynne. "The Second Coming: Class Enemies in the Soviet Countryside, 1927–1935." In *Stalinist Terrror: New Perspectives*, edited by John Arch Getty and Roberta T. Manning, Chapter 3. Cambridge: Cambridge University Press, 1993.

Viola, Lynne. *The Unknown Gulag: The Lost World of Stalin's Special Settlements*. Oxford: Oxford University Press, 2007.

Viola, Lynne. "The Question of the Perpetrator in Soviet History." *Slavic Review* 72, no. 1 (2013): 1–23.

Viola, Lynne. *Stalinist Perpetrators on Trial: Scenes from the Great Terror in Soviet Ukraine*. Oxford: Oxford University Press, 2017.

Viola, Lynne. "New Sources on Soviet Perpetrators of Mass Repression: A Research Note." *Canadian Slavonic Papers/Revue Canadienne des Slavistes* 60, no. 3-4 (2018): 592-604.

Viola, Lynne, V.P. Danilov, N.A. Ivnitskii, and Denis Kozlov, eds. *War Against the Peasantry, 1927-1930: The Tragedy of the Soviet Countryside.* New Haven: Yale University Press, 2005.

Vo, Nghia M. *The Bamboo Gulag: Political Imprisonment in Communist Vietnam.* Jefferson, NC: McFarland & Company, 2003.

Volkan, Vamik. *Killing in the Name of Identity: A Study of Bloody Conflicts.* Charlottesville, VA: Pitchstone Publishing, 2006.

Volkan, Vamik. "Large-Group Identity, International Relations and Psychoanalysis." *International Forum of Psychoanalysis* 18 (2009): 206-213.

Vollhardt, Johanna Ray. "The Role of Social Psychology in Preventing Group-Selective Mass Atrocities." In *Reconstructing Atrocity Prevention*, edited by Sheri P. Rosenberg, Tibi Galis, and Alex Zucker, 95-124. Cambridge: Cambridge University Press, 2015.

Walder, Andrew G. "Collective Behavior Revisited: Ideology and Politics in the Chinese Cultural Revolution." *Rationality and Society* 6, no. 3 (1994): 400-421.

Waldner, David. "Process Tracing and Causal Mechanisms." In *The Oxford Handbook of Philosophy of Social Science*, edited by Harold Kincaid, 65-81. Oxfrod: Oxford University Press, 2012.

Waldron, Jeremy. *The Harm in Hate Speech.* Cambridge, MA: Harvard University Press, 2012.

Walker, J. Samuel. "The Decision to Use the Bomb: A Historiographical Update." *Diplomatic History* 14, no. 1 (1990): 97-114.

Walker, Stephen G., and Mark Schafer. "Belief Systems as Causal Mechanisms in World Politics: An Overview of Operational Code Analysis." In *Beliefs and Leadership in World Politics: Methods and Applications of Operational Code Analysis*, edited by Mark Schafer and Stephen G. Walker, 3-22. New York: Plagrave Macmillan, 2006.

Walker, R.B.J. *Inside/Outside: International Relations as Political Theory.* Cambridge: Cambridge University Press, 1992.

Waller, James. *Becoming Evil: How Ordinary People Commit Genocide and Mass Killing.* Oxford: Oxford University Press, 2007.

Wallis, Andrew. *Silent Accomplice: The Untold Story of France's Role in the Rwandan Genocide.* London: I.B. Tauris, 2006/2014.

Wallis, Andrew. *Stepp'd in Blood: Akazu and the Architects of the Rwandan Genocide Against the Tutsi.* Winchester: Zero Books, 2019.

Walt, Stephen. *Revolution and War.* Ithaca, NY: Cornell University Press, 1996.

Waltz, Kenneth. *Theory of International Politics.* Boston, MA: McGraw HIll, 1979.

Waltz, Kenneth. "Reflections on *Theory of International Politics*: A Response to My Critics." In *Neorealism and Its Critics*, edited by Robert O. Keohane, 322-346. New York: Columbia University Press, 1986.

Waltz, Kenneth N. "Realist Thought and Neorealist Theory." *Journal of International Affairs* 44, no. 1 (1990): 21-37.

Walzer, Michael. *Just and Unjust Wars: A Moral Argument with Historical Illustrations.* 3rd ed. New York: Basic Books, 2000.

Wayman, Frank W., and Atsushi Tago. "Explaining the Onset of Mass Killing, 1949-87." *Journal of Peace Research* 47, no. 1 (2010): 3-13.

Wedeen, Lisa. *Ambiguities of Domination: Politics, Rhetoric and Symbols in Contemporary Syria.* Chicago: Chicago University Press, 1999.

Wedeen, Lisa. "Conceptualizing Culture: Possibilities for Political Science." *American Political Science Review* 96, no. 4 (2002): 713–728.
Wedeen, Lisa. *Authoritarian Apprehensions: Ideology, Judgment and Mourning in Syria*. Chicago: University of Chicago Press, 2019.
Weinstein, Jeremy M. *Inside Rebellion: The Politics of Insurgent Violence*. Cambridge: Cambridge University Press, 2007.
Weiss, Jessica Chen. *Powerful Patriots: Nationalist Protest in China's Foreign Relations*. Oxford: Oxford University Press, 2014.
Weiss, John. *Ideology of Death: Why the Holocaust Happened in Germany*. Chicago: Elephant Paperbacks, 1997.
Weitz, Eric D. *A Century of Genocide: Utopias of Race and Nation*. Princeton: Princeton University Press, 2003.
Welsh, Jennifer M. "'I' Is for Ideology: Conservatism in International Affairs." *Global Society* 17, no. 2 (2003): 165–185.
Welsh, Jennifer M. "Implementing the 'Responsibility to Protect': Where Expectations Meet Reality." *Ethics and International Affairs* 24, no. 4 (2010): 415–430.
Welsh, Jennifer M. "The Responsibility to Protect After Libya & Syria." *Daedalus* 145, no. 4 (2016): 75–87.
Wendt, Alexander. *Social Theory of International Politics*. Cambridge: Cambridge University Press, 1999.
Werth, Nicholas. "The Mechanism of a Mass Crime: The Great Terror in the Soviet Union, 1937–1938." In *The Specter of Genocide: Mass Murder in Historical Perspective*, edited by Robert Gellately and Ben Kiernan, Chapter 10. Cambridge: Cambridge University Press, 2003.
Westad, Odd Arne. *The Global Cold War: Third World Interventions and the Making of Our Times*. Cambridge: Cambridge University Press, 2005.
White, Kenneth R. "Scourge of Racism: Genocide in Rwanda." *Journal of Black Studies* 39, no. 3 (2009): 471–481.
Whitewood, Peter. *The Red Army and the Great Terror: Stalin's Purge of the Soviet Military*. Lawrence: University Press of Kansas, 2015.
Whitewood, Peter. "Victims and Perpetrators Under Stalin." *HISTORY: Review of New Books* 46 (2018): 59–62.
Whiting, Sophie A. "'The Discourse of Defence': 'Dissident' Irish Republican Newspapers and the 'Propaganda War.'" *Terrorism and Political Violence* 24, no. 3 (2012): 483–503.
Wijze, Stephen de. "Political Evil—Warping the Moral Landscape." In *Moral Evil and Practical Ethics*, edited by Sholmit Harrosh and Roger Crisp, 165–198. Abingdon: Routledge, 2019.
Williams, Timothy. "More Lessons Learned from the Holocaust—Towards a Complexity-Embracing Approach to Why Genocide Occurs." *Genocide Studies and Prevention* 9, no. 3 (2016): 137–153.
Williams, Timothy. *The Complexity of Evil: Perpetration and Genocide*. New Brunswick: Rutgers University Press, 2021.
Williams, Timothy, and Dominik Pfeiffer. "Unpacking the Mind of Evil: A Sociological Perspective on the Role of Intent and Motivations in Genocide." *Genocide Studies and Prevention* 11, no. 2 (2017): 72–87.
Wilshire, Bruce. *Get 'Em All, Kill 'Em*. Lanham, MD: Lexington Books, 2006.
Wilson, Chris. "Provocation or Excuse?: Process-Tracing the Impact of Elite Propaganda in a Violent Conflict in Indonesia." *Nationalism and Ethnic Politics* 17, no. 4 (2011): 339–360.

Wilson, Kevin. *Men of Air: The Doomed Youth of Bomber Command, 1944*. London: Weidenfeld & Nicolson, 2007.
Wilson, Richard Ashby. "Propaganda and History in International Criminal Tribunals." *Journal of International Criminal Justice* 14 (2016): 519–541.
Wilson, Richard Ashby. *Incitement on Trial: Prosecuting International Speech Crimes*. New York: Cambridge University Press, 2017.
Wolfers, Arnold. "'National Security' as an Ambiguous Symbol." *Political Science Quarterly* 67, no. 4 (1952): 481–502.
Wolfsfeld, Gadi. *Media and the Path to Peace*. Cambridge: Cambridge University Press, 2004.
Woolf, Linda M., and Michael R. Hulsizer. "Psychosocial Roots of Genocide: Risk, Prevention, and Intervention." *Journal of Genocide Research* 7, no. 1 (2005): 101–128.
Wright, Richard. "The Ness Account of Natural Causation: A Response to Criticisms." In *Critical Essays on "Causation and Responsibility"*, edited by Benedikt Kahmen and Markus Stepanians, Chapter 14. Berlin: De Gruyter, 2013.
Wunderlich, Carmen. "Theoretical Approaches in Norm Dynamics." In *Norm Dynamics in Multilateral Arms Control: Interests, Conflicts and Justice*, edited by Harald Muller and Carmen Wunderlich, 20–48. Athens, GA: University of Georgia Press, 2013.
Xiuyuan, Lü. "A Step Toward Understanding Popular Violence in China's Cultural Revolution." *Pacific Affairs* 67, no. 3 (1994): 533–563.
Yanagizawa-Drott, David. "Propaganda and Conflict: Evidence from the Rwandan Genocide." *The Quarterly Journal of Economics* 129, no. 4 (2014): 1947–1994.
Yashar, Deborah J. *Demanding Democracy: Reform and Reaction in Costa Rica and Guatemala, 1970s–1950s*. Stanford, CA: Stanford University Press, 1997.
Ylikoski, Petri. "Micro, Macro, and Mechanisms." In *The Oxford Handbook of Philosophy of Social Science*, edited by Harold Kincaid, 21–45. Oxford: Oxford University Press, 2012.
Ylikoski, Petri. "Causal and Constitutive Explanation Compared." *Erkenntnis* 78, no. 2 (2013): 277–297.
Ylikoski, Petri, and Jaakko Kuorikoski. "Dissecting Explanatory Power." *Philosophical Studies* 148 (2010): 201–219.
Yom, Sean. "From Methodology to Practice: Inductive Iteration in Comparative Research." *Comparative Political Studies* 48, no. 5 (2015): 614–644.
Young, Marilyn B. "Bombing Civilians from the Twentieth to Twenty-First Centuries." In *Bombing Civilians: A Twentieth-Century History*, edited by Yuki Tanaka and Marilyn B. Young, 154–174. New York: The New Press, 2009.
Zald, Mayer N. "Ideological Structured Action: An Enlarged Agenda for Social Movement Research." *Mobilization: An International Journal* 5, no. 1 (2000): 1–16.
Zaller, John R. *The Nature and Origins of Mass Opinion*. Cambridge: Cambridge University Press, 1992.
Zimbardo, Philip. *The Lucifer Effect: How Good People Turn Evil*. London: Rider Books, 2007.
Žižek, Slavoj. "The Spectre of Ideology." In *Mapping Ideology*, edited by Slavoj Žižek, 1–33. London: Verso, 1994.
Zunes, Stephen. "The Role of Non-Violent Action in the Downfall of Apartheid." *The Journal of Modern African Studies* 37, no. 1 (1999): 137–169.

Index

Adams, Richard 228–30
Adena, Maja 73
Afflitto, Frank 239
African National Congress 58, 130
agency 12, 20, 57, 78, 85–6, 124–7, 149, 158, 166–7, 198, 207, 257, 288, 303, 312, 330
aggression 1–3, 99, 118
Akazu network (Rwanda) 263, 281, 286, 298
Akers, Laura 12, 116
Algeria 56, 220
Allen, Michael Thad 50
Alvarez, Alex 12, 33
Anderson, Kjell 18, 62, 98, 124, 290, 292, 300–1
Anderson, Orvil 219
Anderton, Charles 15
anti-Semitism 5, 38, 71, 78, 81, 85–6, 87
Antonov-Ovseenko, Anton 172
Árbenz Guzmán, Juan Jacobo 222, 231
Arellano, Archbishop Rosell y 231
Arendt, Hannah 5, 59, 89 n.26, 126, 311
Arévalo, Juan José 222
Argentina 56, 57, 79, 108, 259
armed groups, *see* civil war
Armed Islamic Group, *see* Algeria
Armenian Genocide 9, 26, 76–7, 79, 87, 88, 113, 131
Arnold, Henry 197, 199, 200, 202–3, 205, 216
Asch, Solomon 47, 59
atrocity crimes, definition of 1
atrocity speech law 13–14
authoritarianism 5, 23, 89–90, 91, 94, 97, 101 n.39, 159, 228, 232, 250, 259–61, 263, 269, 273–4, 297, 305, 316

Bagosora, Theoneste 273, 275, 281, 283, 286
Balcells, Laia 205
Balch-Lindsay, Dylan 15 n.69, 55
Baldwin, Stanley 187
Bandura, Albert 22, 114–15
Bangladesh 22, 56
Bartov, Omar 118
Bauman, Zygmunt 60–1, 98
Belgium 32, 262, 267–9, 286, 324
Bellamy, Alex 12, 15, 23, 129, 218, 324
Benesch, Susan 14
Berlin, Isaiah 5

Bey, Nazim 131
bin Laden, Osama 31, 58
Bizimana, Augustin 281
Bloxham, Donald 12, 51, 76, 79
Blumer, Hebert 326
Bolshevik Party 89, 90, 134, 136, 139–43, 145–9, 156–7, 159, 174
Bormann, Martin 126
Botswana 129
Brett, Roddy 228, 230, 238, 240, 248, 253
Browning, Christopher 3, 60, 62, 71
Brubaker, Rogers 324–5
brutalization 60–1, 80–1, 118–19, 227, 242–3, 246–50, 288
Bukharin, Nikolai 141
Bulgaria 85
Bulutgil, Zeynep 12, 13, 76
bureaucracy, *see* organizations
Burleigh, Michael 118, 122
Burundi 267–8, 270, 275, 277, 280, 283, 299, 304
Byrd, Scott 51

California 20, 85
Cambodia 3, 47–8, 56, 57, 60, 77, 81–2, 89, 94, 97, 98, 106, 112–13, 119, 128, 219, 311, 314
Carter, Jimmy 257, 259
case studies 24–5, 26–7, 314, 322–4
Catholicism 79, 85, 226–7, 230, 234, 238–9, 261, 268, 298
causal explanation 11, 14, 18–19 n.92, 24–5, 27 n.133, 37, 88 n.182, 94–5, 103–4, 311, 321–3, 326–30
Central Intelligence Agency of the United States 222, 226
Chalk, Frank 32
Chambon-sur-Lignon 62, 85–6
Chandler, David 60
Checkel, Jeffrey 92
Cheka 141
Chennault, Claire Lee 197
Chicherin, Georgii 145
China 1, 5, 19 n.94, 31, 70 n.85, 82, 97, 197
Churchill, Winston 182, 191, 195, 196, 201, 202, 205
Ciardi, John 209

civil war (general) 15, 16, 17, 33, 55, 58–9, 76, 88 n.182, 90, 314–15, 323
civilian immunity, *see* norms, of civilian immunity
civilians, definition of 15 n.69
Coalition pour la Défense de la République (CDR) 276–8, 291, 296, 298
coercion 2–3, 20, 66, 71, 74, 76, 77, 80, 82, 90, 96, 124, 157–9, 160, 165, 167, 172–3, 222, 224, 243, 246, 250, 258, 263, 284, 288–90, 294, 327, 329
  difficulty of micromanaging via 20, 74 n.110, 80–2, 158, 161, 167, 206, 289
  as logic of violence 76, 124, 240
Cold War 36, 37, 45, 230–1, 233, 234, 257, 262
collective punishment, *see* guilt attribution
collectivization of Soviet agriculture 143, 145–9, 155, 162, 171, 177
Collins, Randall 37, 60
colonialism 20, 32, 90, 91, 121, 126, 220, 240, 267–9, 270–1, 286
communism 30, 34, 45, 89, 96, 97, 98, 119, 121, 124, 128, 309, 328
  and anti-communism 37, 49, 83, 85, 90, 107, 122–3, 230–5, 238–9, 244, 246, 248–9, 253–4, 257–9, 260
  in the Soviet Union 46–7, 48, 89, 134–78
comparative historical analysis 24–7, 104, 321–4
complexity 14, 18, 27, 65, 69–70, 77–8, 91, 92–3, 97–8, 103, 307, 310, 313, 321, 329–30
conformity 4, 6, 11, 40–2, 45, 47–8, 49–51, 59–63, 82, 96–7, 108, 250–1, 266, 286, 288, 289, 309, 328–9
Congo, Democratic Republic of 1, 32, 110
conservatism 21, 30–1, 71, 79, 91, 97, 180, 186–7, 190–1, 218, 219–20, 227, 228, 230–3, 235–6, 238–9, 256, 269, 309, 323
constructivism 8, 63, 312, 325–6
Côte d'Ivoire 76, 129
Crawford, Neta 31
crisis 4, 6, 8–9, 18–19, 25, 39, 53, 63–70, 72, 74, 76, 90, 92–3, 96, 98, 102–3, 139–42, 146–7, 159, 178, 181, 213–14, 233–5, 259–61, 262–4, 266, 272, 274–80, 281–2, 286, 296–8, 304–6, 308–9, 312–14, 319
critical theory 8 n.32, 12, 32 n.23, 104
cultural revolution, *see* China
culture, relationship to ideology 5 n.23, 36 n.47
Cunningham, Harry F. 209

Darfur, *see* Sudan
Davis Biddle, Tami 185, 186, 187, 190, 202
de Wijze, Stephen 102
deagentification, *see* destruction of alternatives

dehumanization 12, 39, 40, 62, 82–3, 98, 100 n.36, 106, 111, 113–15, 168–9, 203, 210, 216, 248, 253, 295, 303
deidentification 8, 86, 111–15, 200–1, 203, 215, 232, 246–8, 295
dekulakization 140, 147–9, 155, 178
Des Forges, Allison 263
destruction of alternatives 8, 124–7, 131, 137, 139, 140, 150, 154–5, 171, 184, 189, 195, 197–8, 201, 214, 300, 304, 316
Di Stefano, Paul 294
Dill, Janina 99
Dirty War, *see* Argentina
discourse 10, 12, 17, 29, 40–2, 45–8, 63–7, 70–1, 74, 89, 94–5, 98, 101, 103 n.48, 104, 107, 120, 137, 146, 155, 168, 174–5, 213–16, 232, 250, 256, 260, 266, 296, 316
discourse analysis 25–6, 29, 321, 324–7
Douhet, Gentile 184–5, 188–9
Dower, John 211
Downes, Alexander 55, 124, 180, 192, 201
Downey, Sheridan 197
Dresden, bombing of 200, 208, 217
Dumitru, Diana 86
Dzerzhinskii, Felix 141–3

Eaker, Ira 197, 200, 219
East Pakistan, *see* Bangladesh
East Timor 56
economic factors 6, 13, 19, 28, 29, 31, 32–3, 55–6, 67, 75, 88–92, 118, 143–4, 174, 222, 226 n.29, 230, 240, 252, 262, 265–6, 270, 281, 311–12
Edgerton, David 191
Egypt 20, 84
Ehrenreich, Barbara 81
Eichmann, Adolf 59, 122
Eisenhower, Dwight D. 205
Ejército Guerrillero de los Pobres (EGP) 222, 236, 244, 251
El Salvador 58
elites, *see* political elites
emotions 6, 21–2, 33–4, 42–3, 60–1, 81, 106–7, 112, 115, 116, 210
epistemic aspects of ideology 21–3, 178, 204
epistemic dependence 21, 73, 110, 175, 206, 298, 313, 319
Esteban, Joan 55–6
ethics of war 319–20
ethnonationalism, *see* nationalism
Ewell, Julian Johnson 197
experimental research 24, 60, 99, 322–3, 329

extremism 4–5, 7, 13, 55, 74–5, 89, 95–6, 103, 144, 280, 284, 285, 286, 311, 313, 314, 316, 318–20
Ezhov, Nikolai 151–4, 155, 157, 159, 160, 164–5

Fairbanks, Charles 258–9
Falla, Ricardo 239, 261
fanaticism, *see* ideology, true believer model of
fascism 34–5 n.40, 91 n.200
  in Germany, *see* Nazism
Fein, Helen 112
feminism 8 n.32, 12, 17
Figes, Orlando 139 n.36, 142, 157, 171, 175
Fisher, John 199
Fitzpatrick, Sheila 155, 169
Fjelde, Hanna 55
Fox, Nicole 294
framing 21, 36, 39–40, 44, 109, 117, 123–4, 161, 200, 210, 230, 311, 313–14, 329
Freeden, Michael 32
Fujii, Lee Ann 39–40, 60, 263, 266 n. 31, 291
Fuller, J.F.C. 186
futurization 8, 104, 120–4, 127, 157, 184, 189, 196–7, 236–7, 252

Garcia, Kjell Laugerud 251–2, 259
Garrard-Burnett, Virginia 225, 238, 239–40, 243, 250, 252, 255, 256
Gasana, James 279
Gasasira, James Gasana 294
Gault, William Barry 113
gender 105–6 n.54, 117–19, 131–2, 241
genocide
  character of 12, 16, 19 n.96, 32, 101–2, 238–9, 240–1, 261
  dangers of exclusive focus on 16
  logic of violence of 64, 76–7, 101, 240–1
genocide studies 96, 108, 224, 259–61, 318–19
Gerlach, Christian 51, 87
Germany
  area bombing of 57, 58, 64, 179–80, 193–7, 199–205, 207–8, 211–18
  colonialism of 32, 85, 112, 127, 267
  Nazi, *see* Nazism
  roots of anti-Semitism in 87, 88
  use of air power by 183, 191
Getty, John Arch 136, 146, 158–9
Goldhagen, Daniel 5, 38, 87
Goldman, Wendy 150, 167–8, 172
Gorbachev, Mikhail 47, 58, 89
Gordon, Gregory S. 13–14
Göring, Hermann 195
Gorokhov, Mikhail 165
Gourevitch, Philip 282

GPU (State Political Directorate), *see* NKVD (People's Commissariat for Internal Affairs)
Gramajo Morales, Héctor Alejandro 236
Grandin, Greg 227, 234
Gries, Peter Hays 322, 323
Guatemala
  Archdiocese of 221, 244
  authoritarian governments of 222–3, 232–42
  civil self-defence patrols (PACs) in 223, 235, 243
  Commission for Historical Clarification (CEH) of 221, 225, 233, 254
  democratic period of 222, 228–32
  origins of civil war in 222–3
  question of genocide in 240–1
  US involvement in 256–9
Guichaoua, André 264–5, 277, 280, 284, 286, 287, 289
guilt attribution 8, 21, 104, 108–11, 154, 155–6, 160–1, 163, 164–5, 172, 201–3, 207–9, 211, 215–16, 218, 235–6, 244–6, 248, 251–2, 261, 283, 290, 293, 329
Gutierrez Sanín, Francisco 33, 330

Habimana, Cyasa 285
Habyarimana, Agathe Kanziga 263, 281
Habyarimana, Juvenal 262–3, 271–6, 279–81, 283, 285, 286–7, 292, 295, 296, 298–9
Hagan, John 39
Hagenloh, Paul 153, 158
Halfin, Igal 136, 172
Hall, Norris 189
Hallock, Ted 208
Hamilton, Lee 60
Hansell, Haywood S. 203
Hardin, Russell 49
hardline ideological security doctrines 7–9, 19, 63–71, 75–9, 81–6, 89–92, 100–27, 312–13
Harff, Barbara 32
Harmer, Tanya 257
Harmon, Christopher 180, 192 n.94, 193 n.99
Harris, Arthur 179–80, 196–7, 199, 200, 203, 204, 215, 217
Harris, James 143, 145, 147, 150, 174, 178
Hastings, Max 185, 206, 212
hate speech 13–14, 72
Hay, William Delisle 182
Herero and Namaqua genocide 32, 85, 112, 127
Hiebert, Maureen 34, 128, 324
Hinton, Alexander 119
Hiroshima 193–4, 198–9, 200, 201, 202–3, 211
Hitler, Adolf 35, 43–4, 71, 78, 91, 117, 122, 195, 201
Hoess, Rudolf 118, 122

Holocaust, *see* Nazism
Holodomor 134 n.3
Huddy, Leonie 111–12
Huie, William Bradford 216
Hull, Isabel 109
Hulsizer, Michael 87
Hultman, Linda 55
human rights 1, 127–9, 221, 233, 250, 256, 257–9, 263, 315
Huth, Paul 15 n.69, 55
Hutu
  identity category of 266–8
  revolution 268–9

Iagoda, Genrikh 148, 150, 152
ideas, study of 24–6, 324–30
identity
  ethnic 13, 36–7, 39–40, 42, 76, 77, 87, 111–12, 129, 149, 154, 253, 262, 267–8, 271–3, 324–5
  relationship to ideology of 13, 17, 35–7, 43, 44, 66, 111–15
ideological activism 13–14, 17, 20, 23, 38, 72–5, 101, 168–70, 173–4, 214–17, 297–305, 316–18
ideological heterogeneity, *see* perpetrators of mass killing, diversity of
ideological infrastructure 11, 41–2, 49–52, 70–5, 92, 132, 141, 150, 180–1, 232–5, 266, 268–74, 280–1, 296–8, 305–6, 309–10, 318
ideological instrumentalization, *see* ideological structure
ideological internalization 4–5, 9–11, 29, 34, 35–6, 37–8, 40–5, 49–51, 61–2, 71–4, 83–4, 92–3, 104, 310, 324, 327
ideological radicalization, *see* radicalization
ideological structure 10–11, 29, 35–6, 38–9, 40–1, 45–51, 61–2, 71–2, 74, 75, 82, 90, 92–3, 104, 118, 129, 324, 327
ideologically radicalized security politics 7–8, 12, 23, 100–3, 146, 308
ideology
  causal power of 9–11, 37–51, 309–10
  causes of 87–92
  definitions of 4, 16–17, 30–7
  extraordinary or ordinary 4–6, 16–17, 30, 32–3, 43, 51–2, 95, 103, 313, 318–19
  limits of, *see* non-ideological factors
  personal *vs.* shared 35
  traditional perspectives on, *see* traditional-ideological accounts of mass killing
  true believer model of 9–11, 28–32, 37–8, 83, 86, 171, 309–10

Impuzamugambi 279, 281, 289
Indonesia 20 n.106, 79, 83, 85, 90, 97, 107, 122, 123
Inglis, F.F. 196
insanity 1–2, 9, 56, 178, 297, 318–19
Interahamwe 279, 281, 283, 285, 289–90, 292, 298, 301
internalization, *see* ideological internalization
interpretivism 24–5, 27, 40, 45 n.102, 89–90, 320, 325–7, 328, 330
Iran 20, 99, 123
Iraq 1, 3, 48 n.118, 220
Irish republicanism 31
Islam 22–3, 36, 44, 56, 76, 85, 87, 110, 113
Islamic State of Iraq and Syria 3, 23, 44, 316

Jackson, Karl 113
Japan 192, 217–18
  area bombing of 77, 179–80, 192–4, 197–203, 209–10, 211, 213, 215–17
  atrocities by 114, 202–3
Jervis, Robert 330
jihadism 3, 23, 31, 36, 44, 316
Johnson, Carter 86
Jonassohn, Kurt 32
justificatory mechanisms 8, 95, 103–27, 132, 200, 283, 316, 329
justificatory narrative 7, 18–20, 21, 26, 44, 53–4, 63–6, 70–5, 76, 81–3, 86, 88, 94, 101, 103–28, 131–3, 308, 311, 314

Kaganovich, Lazar 148, 151, 153, 155, 157, 176
Kajuga, Robert 281, 283
Kallis, Aristotle 78
Kalyvas, Stathis 55, 76, 223–4
Kambanda, Jean 281, 284, 286
Kamenev, Lev 150–2, 170
Kanyarengwe, Alexis 273
Kaplan, Robert 87
Karadžić, Radovan 48
Karamira, Froduald 277
Kaufman, Stuart 31
Kayibanda, Gregoire 268–71, 272, 276
Kellow, Christine 297
Kelman, Herbert 60, 98–9
Khmer Rouge, *see* Cambodia
Kiernan, Ben 12, 113
Kim, Dongsuk 55
Kimonyo, Jean-Paul 264 n.18, 276
King, Charles 34
Kirov, Sergei 149–52, 177
Klusemann, Stefan 60, 61
Kołakowski, Leszek 138
Kopelev, Lev 171

Korea 97, 154, 219
Kosovo, *see* Yugoslavia, Former
Kotkin, Stephen 136 n.19, 162, 174, 175
Kravchenko, Viktor 173
Kruijt, Dirk 236
Krylenko, Nikolai 145
Kubota, Yuchi 224–5
Kühne, Thomas 18, 80
kulak, *see* dekulakization

ladino 222, 224, 228–9, 239, 252
Laitin, David 324–5
Lankford, Adam 98
legitimation 4, 7, 9, 11, 19–20, 32, 38, 39, 45, 48, 61, 63, 71–2, 74, 80, 82, 83, 84, 107–8, 109, 117, 123–4, 128, 163–4, 168, 170, 176, 182, 206–7, 210–17, 248–9, 251–6, 286, 290, 292, 296, 303, 317, 318, 329
Legro, Jeffrey 90
LeMay, Curtis Emerson 197–8, 199, 200, 202, 205, 219
Lenin, Vladimir Ilyich 31, 88, 89, 90, 91, 134 n.2, 136, 138–43, 145, 146, 154, 178
Lewandowsky, Stephan 74
Lewis, Bruce 209–10
Li, Darryl 300, 303
liberalism 17, 21, 23, 30, 31, 33–5, 37, 47–8, 51, 58, 66, 79, 85, 89, 97, 125, 156, 163, 180–1, 190–1, 218–20, 252, 276, 313
Lickel, Brian 109
Liddell Hart, Basil 186
limitationist ideological security doctrines 65–6, 67–9, 83, 95, 101, 117, 127–31, 217–18, 257, 294–5, 305, 315, 317
Lindbergh, Charles 190, 205–6
Lindemann, F.A. 196
Lithuania 81
Littman, Rebecca 2
Litvinov, Maksim 145
Lizinde, Théoneste 273
Logeist, Guy 269
Longman, Timothy 266
Lord's Resistance Army 110
Lösener, Bernhard 78
Lovett, Robert 199
Lucas Garcia, Fernando Romeo 223, 225, 234–9, 251–2, 254–5, 256
Lucas Garcia, Manuel Benedicto 236
Luftwaffe 187, 191, 194–5

Mali 76, 129
Mamdani, Mahmood 263, 268 n.45, 289–90
Mandela, Nelson, *see* African National Congress
Mann, Michael 91 n.200, 119–20, 322 n.10

Mao, Zedong 1, 5, 31, 82, 97
Marshall, George C. 199, 202
Marx, Karl 138
Marxism, *see* communism
mass killing
 definition of 1, 15–16
 prediction of 27, 36–7, 91–2, 105, 313, 315–16
 prevention of 13–14, 316–18
 puzzle of 1–4
mass publics, *see* public constituencies
material interests, *see* economic factors
Maya
 identity category of 221 n.3
 mass killing of 3, 123, 221, 224, 236–51, 259–61
 as 'sanctioned' state construct 227
Mbonampeka, Stanislas 283–4
McDoom, Omar 12, 60, 62, 115–16, 269, 299, 301, 313
McLoughlin, Stephen 129, 130
McMahan, Jeff 320
Melchers, Wilhelm 78
Messer, Robert 194 n.109, 198
Mezhinskii, Viacheslav 143–4
Michie, Allan 215
Milgram, Stanley 47, 59, 61
military necessity 7, 121–2, 127, 205, 319
Milošević, Slobodan 38, 48, 84
Mironko, Charles 295, 303
Mitchell, Billy 188–9, 197
Mitchell, Neil 23
Mladić, Ratko 48
Molotov, Vyacheslav 143, 176
Monroe, Kristen Renwick 106
moral disengagement 4–5, 95, 98–100, 102, 112–15, 119, 131, 133, 176–7
Morelli, Massimo 55–6
Moscow Show Trials 145, 151, 168–9, 174
Moses, Dirk 51
motivated reasoning 35, 74, 89–90, 298, 313, 319
motivation 4, 5, 8–9, 10–11, 13, 14, 17, 18, 19, 28–9, 35–6, 37–8, 39, 41, 44–5, 48, 52, 54–6, 63, 70–2, 74, 80–3, 92–3, 109–10, 115–16, 117–18, 128, 132, 137, 158–9, 160, 166, 207, 224, 243, 247, 249–51, 264, 266, 281–2, 288, 290, 292, 294, 296–7, 307
Mouvement Révolutionnaire Nationale pour le Développement (MRND) 271, 273–4, 275, 277–80, 292, 295, 296, 298
Mubarak, Hosni 84
Mugesera, Léon 278–9
Murray, Elisabeth Hope 12, 101
Mussolini, Benito 119

Myanmar 1, 22–3, 316

Nagasaki 193–4, 198, 200, 203, 211
Nahimana, Ferdinand 273, 298
Nahoum-Grappe, Véronique 109
Namaqua genocide, *see* Herero and Namaqua genocide
Nanking, Rape of 114
narratives, *see* justificatory narrative
National Security Doctrine 233
nationalism 12, 34, 38–9, 42, 48–9, 62–3, 71, 79, 81, 82, 85, 95–6, 112, 227, 234, 235–6, 241, 255, 260, 263, 266, 268–71, 272–80, 282–3, 285–6, 297–8, 302–3, 304–6, 309, 312–13, 316
Naumov, Oleg 136, 146
Nazism 1, 5, 8 n.33, 14, 18, 30, 34–6, 37–8, 41–2, 43–4, 46, 49–50, 59, 60–1, 62, 71, 73, 77–8, 80, 81, 84–5, 86, 87–8, 90, 91, 94, 97–8, 103, 106, 107, 113, 115, 118, 119, 121, 122, 123, 126, 194–5, 201, 309, 311, 312–13, 320
Ndadaye, Melchior 275, 277, 283, 304
negative cases 127–31, 315, 323–4
Neilsen, Rhiannon 101, 106
Nelson, Diane 227–8
neo-ideological account of mass killing
  defining features of 6–11, 63–70, 100–3
  precedents in existing scholarship of 12, 310–11
Neumann, Peter 44
Ngeze, Hassan 278, 298
Nixon, Richard 123, 241
NKVD (People's Commissariat for Internal Affairs) 70 n.85, 141, 151–3, 157–8, 160–7, 170, 172
non-ideological factors 35, 311–12
normative extremity 18, 58, 59, 63, 69, 99, 102, 309, 314
norms 5, 17, 19, 20 n.106, 36, 40, 46–7, 58, 59–61, 67, 71, 81–2, 90, 99, 100, 107, 115, 116, 118, 137, 163, 178, 205, 243, 249–50, 286, 288, 303, 326–7, 328
  of civilian immunity 58, 66 n.72, 128–9, 130 n.208, 315
  as source of ideological influence 10–11, 17, 29, 36, 40, 42, 45–6, 48, 61–3, 74, 83–4, 93, 104, 251, 266, 309, 320
Norton, Roy 182
Nyandwi, Justin 285
Nyseth Brehm, Hollie 294, 301
Nzirorera, Joseph 281

Oberschall, Anthony 39
Odell, S.W. 182

Oglesby, Elizabeth 227–8
one-sided violence 16
Opotow, Susan 112
Ordzhonikidze, Sergo 145
Organización del Pueblo en Armas (ORPA) 222–3
organizations 2, 4, 6–7, 9–11, 17, 29, 32, 41, 45–8, 50, 59–61, 62, 64, 78–9, 82–3, 84, 92, 125, 129–30, 132, 144, 158, 160–4, 166, 167–8, 205, 233, 250, 316–18, 328–9
Ortiz, Frank V. 253
Ottoman Empire 9, 26, 76, 77, 79, 88, 113, 121–2, 131
Overy, Richard 50, 186, 187, 193, 194, 204, 205, 213

pacifism 66, 99, 129 n.205, 138, 181–2, 213
Paluck, Elizabeth Levy 2
Pasternak, Boris 171
peer-pressure, *see* conformity
Peirse, Sir Richard 196, 200
permissive ideological security doctrines 66–70, 127–8
perpetrators of mass killing
  coalitions of 11, 18–20, 35, 70–4, 84–5, 132, 234, 235, 255, 257, 277
  diversity of 11, 18, 29, 38, 41, 44–5, 49–51, 70–1, 93, 166, 208, 235, 251
  elite level, *see* political elites
  ordinariness of 2, 51, 59–60, 62, 101, 288, 313, 318–19
Perrera, Guido 201
Petrova, Maria 49
Philippines 112, 202
Pion-Berlin, David 108
Poland 78, 147, 162
political crisis, *see* crisis
political elites 7, 10, 18, 19–20, 30, 39 n.65, 45–6, 55, 67, 70, 71, 74, 75–9, 80, 83–4, 89, 127, 132–3, 145–59, 191–205, 236–42, 280–7, 309–10, 316–17, 319
  apex and intermediary 75
political psychology 14, 29, 43, 44, 89–90
Popper, Karl 5
Popular Army of Liberation, *see* El Salvador
Portal, Sir Charles 187, 196, 201
post-hoc rationalization 5, 19, 23, 28–9, 63–4, 72, 75, 151, 162, 165, 239, 246, 264, 310, 317, 327, 328–9
pragmatism, *see* strategy
prejudice, *see* motivated reasoning
Press, Daryl 99
Priestland, David 135, 138, 168
propaganda, *see* ideological activism

Prunier, Gerard 263, 267, 273, 282, 284, 288, 299–300
public constituencies 20, 33 n.30, 54, 70, 84–6, 167–76, 210–17, 251–6, 297–305, 309, 316–17

Querido, Chyanda 56

racism 12, 26–7, 78, 85, 96, 101, 185, 213, 222, 224, 227–30, 232, 234, 236, 240, 246–9, 253, 254, 255, 259–61, 263, 267, 276, 278, 286, 298–9
radicalization 7, 13 n.62, 14, 19, 22, 35, 60, 64–5, 68–9, 72–5, 87–92, 100–3, 115–16, 118, 132, 140, 141, 143–5, 150–1, 198, 228–35, 260–1, 274–80, 286, 297–301, 312–14, 316, 318, 328–9
   definition of 88
   of security politics, see ideologically radicalized security politics
Radio Rwanda 270, 278, 298, 304
Radio Television Libre des Milles Collines (RTLM) 273, 278, 297–304
rank-and-file perpetrators 18–20, 42, 54, 70, 73, 79–84, 100, 102, 160–7, 205–10, 242–51, 287–97, 311
rationalist accounts of mass killing 5–6, 7–8, 12, 38, 54–9, 77, 88, 102, 105, 120, 124, 134–6, 140, 141, 180–1, 192–4, 224–7, 264–6, 308–9, 310–11, 323, 328
rationalization, see post-hoc rationalization
Reagan, Ronald 259
Red Army (Soviet) 3, 56, 135, 164
Red Terror, see Russian Civil War, mass killing in
resistance and rescue 48, 56, 62, 63, 70, 79, 80, 81–2, 84–6, 127–31, 132, 188, 205, 250, 254, 256, 261, 281, 285, 294–5
restraints on mass killing 2, 58, 69, 83–4, 84–5, 130 n.209, 131–2, 159, 191, 210, 217–18, 220, 257, 313, 315
revolutionary politics 5, 7, 25, 65, 91, 95–103, 121, 136–7, 138–41, 142, 176–7, 269, 271, 284, 308–9, 311, 316, 323
Ríos Montt, Efraín 223, 225, 233–42, 252–6, 259
Robinson, Geoffrey 79
Rohingya 22–3
Rohner, Dominic 55–6
Roma 78, 94
Romania 85–6
Roniger, Luis 257
Roosevelt, Franklin Delano 182, 197, 202
Roth, Paul 60–1
Rummel, Rudolph 218
Runciman, David 319

Russia
   Soviet, see Union of Soviet Socialist Republics
   Tsarist 139, 141
Russian Civil War
   consequences of 90, 91, 139
   mass killing in 139–43, 145
   origins of 89, 90, 139
Rwanda
   colonial period of 277–9
   competing explanations of mass killing in 39, 60, 61–2, 76, 79, 263–6
   ethnonationalism in 39, 83, 268–74, 275–80, 281–3
   genocide of the Tutsi in 1, 20 n.106, 26–7, 56, 60, 61–2, 64, 70 n.85, 76, 77, 79, 83, 107, 262–3, 280–306
   independence of 269
   pre-colonial history of 266–7
Rwandan Patriotic Front (RPF) 262–3, 264–5, 275, 276, 277, 280, 283, 285, 292, 301, 302, 303
Ryan, James 119, 139, 142
Rymond-Richmond, Wenona 39

Saucier, Gerard 12, 116
Savage, Rowan 39–40, 85, 99, 115
scapegoating, see guilt attribution
Schaffer, Ronald 184, 202–3, 216
Scharpf, Adam 79
Schirmer, Jennifer 227, 241, 242
Schissler, Matthew 316
scientific realism 321, 326
Scott, Robert, Jr. 209
Searle, Thomas 180
self-interest 4, 11, 15 n.74, 23, 28, 30–1, 38, 52, 58, 63–4, 75, 82, 108, 116, 137, 158–9, 160, 165, 242–3, 249–51, 260, 263–6, 281, 305, 310, 327, 328
Semelin, Jacques 12, 85
Senegal 76, 129
Serbia, see Yugoslavia, Former
Shaw, Martin 12, 101, 105
Shearer, David 145, 165
Sherman, William 127
Sherry, Michael 190, 205, 207, 210, 213, 214–15, 217
Sibomana, André 270
Simonov, Konstantin 175
Sinti 78, 95
situationist accounts of mass killing 6, 7, 12, 24, 59–63, 88–9, 98–9, 137, 181, 209, 243, 290, 309–10, 315, 323, 328–9
Slim, Hugo 12, 111, 317
Snow, David 51

social conformity, *see* conformity
social psychology 47–8, 59–62
Sofsky, Wolfgang 9 n.37, 61
South Africa 130
Soviet Union, *see* Union of Soviet Socialist Republics
speech, *see* discourse
Speer, Albert 192
Spitulnik, Debra 300
Stalin, Joseph
  personality and ideology of 97, 134–7, 138, 141–5, 148–52, 155–7, 159, 177–8
  rise to power of 143–5
Stalinism 5, 134–9, 143–59, 169–78, 313
Staniland, Paul 43, 105
Staub, Ervin 98
Steeves, Leslie 297
Steizinger, Johannes 35
strategic indeterminacy 18, 57–9, 63, 69, 105, 120–1, 134–6, 181, 192–4, 224–7, 264–6, 308–9, 314
strategy
  definition of 8–9 n.36
  interdependence with ideology 7–9, 20–1, 29–32, 37, 52, 54–6, 63–70, 75, 77–8, 100–3, 105–8, 120–7, 129–30, 137, 139, 178, 180–1, 191–7, 224–8, 237–42, 246, 252–3, 260, 266, 286, 305–6, 308–11, 314–15
  as kind of incentive for mass killing 6, 77–8, 92, 140, 146, 180, 218–19, 223–4, 264–6, 308–10
Straus, Scott 12, 68, 75–6, 79, 94, 129, 263, 264, 272, 282, 284, 285, 291, 295, 297, 299, 300, 301–2, 313, 324
Su, Yang 81
Sudan 1, 39, 76, 113 n.109
Sullivan, Christopher M. 224
Syria 1, 3, 316

Tanzania 129
Taylor, Kathleen 101 n.39, 106
terrorism 15–16, 17, 29–30, 58–9, 314
Theis, Edouard 61
Thi, Phyu Phyu 316
threat construction 7, 8, 66–7, 102, 105–8, 121, 122–3, 141, 146–9, 177, 184, 189, 201, 215–17, 239, 253–4, 275–9, 282–3, 293, 309–11, 329
Thurston, Robert 172
torture 60, 81, 150, 160, 163, 211, 246, 248, 257
total war 16, 64, 130, 198, 199, 200
totalitarianism 2, 3, 5, 23, 96, 136, 176–7, 273–4, 282, 311, 323

traditional-ideological accounts of mass killing 4–5, 6–7, 9–11, 12, 20–1, 23, 28–9, 87, 93, 95–100, 102, 136–7, 138–9, 145–6, 147–8, 160, 264, 272, 297–8, 305, 310–11, 323, 328
Trenchard, Viscount Hugh 185–6, 187–8, 189
Trevor-Roper, Hugh 84
Trocmé, André 62
Trotsky, Leon 140, 141, 151–2
true believer model, *see* ideology, true believer model of
Truman, Harry S. 194, 198, 199, 201
Tukhachevsky, Mikhail 175
Tutsi
  Genocide of, *see* Rwanda, genocide of the Tutsi in
  identity category of 266–8

ultrahardline ideological security doctrines 66–8, 71, 101–2, 137, 142, 143, 177, 178, 217–18, 235, 276, 311
Ukraine 148, 161–2
Ulfelder, Jay 15
Union of Soviet Socialist Republics 48, 50, 86, 88, 107, 112, 114, 124, 128, 145–78, 219, 309, 313
  collapse of 46–7, 89
  formation of 139–43
  patterns of violence in 134–5
United Kingdom 64, 114, 143, 179–81, 191–220
  evolution of bombing doctrine in 185–8
  in World War I 183–4
United Nations Organization 13, 263
United States of America 31, 32, 44, 48, 58, 65, 68–9, 81, 83, 85, 89, 112, 113, 123, 127, 128, 130, 180–1, 191–220
  evolution of bombing doctrine in 188–91
  involvement in Latin America 20 n.106, 222, 226, 233, 234, 236, 239, 251–2, 256–9
Uritskii, Moise 141
utopianism 5, 7, 12, 21, 25, 29, 30–1, 65, 95–8, 121–2, 136, 138, 142, 145, 147–8, 157, 176–7, 282, 311, 319
Uvin, Peter 278, 286
Uwilingiyimana, Agathe 279, 281

Valentino, Benjamin 8–9, 12, 15, 55, 96, 99, 124, 223–4
valorization 7, 8, 25, 49, 58, 102, 105–6 n.54, 116–20, 125, 129, 131, 155–7, 163, 169, 196, 198–200, 214–15, 227, 236, 250, 252, 284, 288–9, 293, 303, 310, 319
Vansittart, Baron Robert 203
Vatlin, Alexander 158, 160–1, 163–4, 166–7, 171

Vela Casteñeda, Manolo 234, 243, 244, 247–8, 249, 254
Verne, Jules 182
Verwimp, Philip 282
Vietnam 77, 97, 128
  US war in 81, 83, 97, 113, 130, 210, 219–20
Viola, Lynne 143, 147, 158, 160, 164, 166, 167, 171
virtuetalk, *see* valorization
von Trotha, Lothar 127
Voroshilov, Klimint 152
Vyshinsky, Andrey 159, 168–9

Wagner, Michael 107
Walker, J. Samuel 194
Walker, Kenneth 190
Waller, James 2, 20, 51
Wallis, Andrew 272–3, 285, 286, 290
Walton, Matthew 316
Waltz, Kenneth 329
Wedeen, Lisa 11, 36 n.47
Weicker, Lowell 199, 203

Weiss, John 87, 201
Weitz, Eric 12, 82, 96, 311
Wells, H.G. 182
Whitbeck, Harris 237
Whitewood, Peter 136
Williams, Timothy 80
Wilson, Kevin 206
Wilson, Richard Ashby 14
Wood, Elisabeth Jean 33, 330
Woolf, Linda 87
Wright, Orville 183

Yanagizawa-Drott, David 49, 301
Yashar, Deborah J. 232
Yom, Sean 322, 323
Yugoslavia, Former 38–9, 42, 48–9, 56, 82, 84, 85, 87, 110, 112, 113, 325

Zambia 129
Zigiranyirazo, Protais 273, 281
Zimbardo, Philip 47, 59, 61
Zinoviev, Grigory 150–2, 170
Zuckerman, Solly 204